AF368533

Application of Geophysical Data Interpretation in Geological and Mineral Potential Mapping

Application of Geophysical Data Interpretation in Geological and Mineral Potential Mapping

Editors

Luan Thanh Pham
Saulo Pomponet Oliveira
Le Van Anh Cuong

Basel • Beijing • Wuhan • Barcelona • Belgrade • Novi Sad • Cluj • Manchester

Editors

Luan Thanh Pham	Saulo Pomponet Oliveira	Le Van Anh Cuong
Department of Geophysics	Department of Mathematics	Department of Geophysics
Vietnam National University	Federal University of Paraná	VNUHCM-University of Science
Hanoi	Curitiba	Ho Chi Minh
Vietnam	Brazil	Vietnam

Editorial Office
MDPI
St. Alban-Anlage 66
4052 Basel, Switzerland

This is a reprint of articles from the Special Issue published online in the open access journal *Minerals* (ISSN 2075-163X) (available at: www.mdpi.com/journal/minerals/special_issues/293OL7SQT9).

For citation purposes, cite each article independently as indicated on the article page online and as indicated below:

Lastname, A.A.; Lastname, B.B. Article Title. *Journal Name* **Year**, *Volume Number*, Page Range.

ISBN 978-3-7258-0090-2 (Hbk)
ISBN 978-3-7258-0089-6 (PDF)
doi.org/10.3390/books978-3-7258-0089-6

Contents

About the Editors

Luan Thanh Pham

Luan Thanh Pham graduated from VNU University of Education in 2013 with a Bachelor's degree in Physics Teacher Education. He joined the Faculty of Physics at VNU University of Science as a Lecturer in 2014. He received a Master's and PhD in Geophysics from VNU University of Science in 2016 and 2021. During 2021–2023, he was a postdoc at the Department of Geophysics at the Faculty of Physics of VNU University of Science. His research interests include the inversion and enhancement methods of gravity and magnetic data and their applications in mapping geologic structures.

Saulo Pomponet Oliveira

Saulo Pomponet Oliveira holds a B.Sc. and M.Sc. in Applied Mathematics (State University of Campinas, Brazil, 1996) and a Ph.D. in Applied Mathematics (University of Colorado at Denver, USA, 2002). After graduation, he was a post-doctoral visitor at the Geophysics Department of Federal University of Bahia, Brazil (2003–2006) and the National Institute of Oceanography and Applied Geophysics, Italy (2006–2008), and has worked at the Federal University of Parana, Brazil, since 2008, currently appointed as Associate Professor. His main research interest is mathematical and computational methods for processing and inversion of potential field and radiometric data.

Le Van Anh Cuong

Le Van Anh Cuong holds a PhD in Exploration Geophysics from Curtin University, Australia (2018). He is a lecturer in the Department of Geophysics, Faculty of Physics and Engineering Physics at VNUHCM-University of Science since 2007. His research interests consist of (i) Earth Sciences as geophysical methods (i.e., seismic, electric, and electromagnetic) and applications of geophysics in civil and environmental engineering, geology, and mineral exploration, and (ii) Information and Computing Sciences as applications of supercomputing, data mining, and machine learning.

Preface

Geophysical methods are used as a powerful tool for investigating the geological structures of the Earth's crust, which often control the locations and geometries of natural resources such as oil, gas, and minerals. A variety of geophysical methods have been introduced using different physical characteristics of the Earth materials. The scientific exploration and understanding of the geophysical methods and their applications have become imperative for industrialists and academics.

This reprint, *Application of Geophysical Data Interpretation in Geological and Mineral Potential Mapping*, published by MDPI, is a compilation of scientific papers on new interpretation results in geological and mineral potential mapping using geophysical methods such as seismic, electrical resistivity, gravity and magnetic methods. This reprint also provides some new methodologies for a variety of geophysical methods to aid geological methods with an emphasis on mineral exploration.

In conclusion, *Application of Geophysical Data Interpretation in Geological and Mineral Potential Mapping* is the result of a cooperative effort to compile and disseminate knowledge in the geophysical field. All of the scientific papers in this book are original contributions that together provide a thorough resource on applications of geophysical methods. The guest editors extend their sincere gratitude to the contributing authors who believed in this reprint project from the beginning.

Luan Thanh Pham, Saulo Pomponet Oliveira, and Le Van Anh Cuong
Editors

Editorial

Editorial for the Special Issue "Application of Geophysical Data Interpretation in Geological and Mineral Potential Mapping"

Luan Thanh Pham [1,*], Saulo Pomponet Oliveira [2,*] and Cuong Van Anh Le [3]

[1] Department of Geophysics, Faculty of Physics, University of Science, Vietnam National University, Hanoi 100000, Vietnam

[2] Graduate Program in Geology, Department of Mathematics, Federal University of Paraná, Curitiba 81531-980, PR, Brazil

[3] Department of Geophysics, Faculty of Physics and Engineering Physics, University of Science, Vietnam National University Ho Chi Minh City, Ho Chi Minh City 700000, Vietnam; lvacuong@hcmus.edu.vn

* Correspondence: luanpt@hus.edu.vn (L.T.P.); saulopo@ufpr.br (S.P.O.)

Citation: Pham, L.T.; Oliveira, S.P.; Le, C.V.A. Editorial for the Special Issue "Application of Geophysical Data Interpretation in Geological and Mineral Potential Mapping". *Minerals* **2024**, *14*, 63. https://doi.org/ 10.3390/min14010063

Received: 27 December 2023
Accepted: 3 January 2024
Published: 4 January 2024

Geological structures often control the distribution of natural resources such as oil, gas, and minerals [1–3]. Interpreting geophysical data can help in mapping these structures [4–8]. To better understand the application of geophysical data interpretation in mapping geological structures and mineral deposits, we collected 10 contributions for this Special Issue. These contributions will be briefly presented here.

In the first contribution by Xu and co-authors, "Research on the Tectonic Characteristics and Hydrocarbon Prospects in the Northern Area of the South Yellow Sea Based on Gravity and Magnetic Data", potential field and seismic data are employed to locate fault lineaments and estimate the basement depth in the northern basin and the middle uplift of the South Yellow Sea, China.

In "Low-Dimensional Multi-Trace Impedance Inversion in Sparse Space with Elastic Half Norm Constraint", Lan et al. aim to improve the computational efficiency in multi-trace impedance inversion. They also proposed an inversion constraint based on an elastic half norm, and tested the proposed approach on synthetic and field seismic 2D profiles.

Park and co-authors employ paleomagnetic data and analyze the anisotropy of magnetic susceptibility in their work "Preferred Orientations of Magnetic Minerals Inferred from Magnetic Fabrics of Hantangang Quaternary Basalts", contributing to the knowledge about the eruptive origin of the Hantangang River Volcanic Field, Korea.

The methodological background of the article entitled "Particle Swarm Optimization (PSO) of High-Quality Magnetic Data of the Obudu Basement Complex, Nigeria" by Ekwok and collaborators is a stochastic optimization approach to the inversion of magnetic profiles. The inversion considers a universal model that comprises a variety of simple geometric sources (spheres, cylinders, thin sheets, and geological contacts). The inversion technique was applied to four magnetic profiles from the Precambrian Obudu basement complex, Nigeria, revealing depositional zones for igneous-related minerals and migratory pathways for hydrothermal fluids.

Another methodological strand, data-driven machine learning algorithms, is the tool used by Behnia et al. in their work, "Mineral Prospectivity Mapping for Orogenic Gold Mineralization in the Rainy River Area, Wabigoon Subprovince". Specifically, the Random Forest algorithm is employed to produce prospectivity maps from geological, gravity, and magnetic data from the Rainy River Area, Canada.

In "Mapping of the Structural Lineaments and Sedimentary Basement Relief Using Gravity Data to Guide Mineral Exploration in the Denizli Basin", Altinoğlu applies depth-estimation and edge detection techniques to gravity data with the purpose of designing a basin depth model and delineate geological structures of the Denizli Basin, Turkey. These results are correlated especially with geothermal occurrences.

Alvandi et al. also consider edge detection methods in their work, "Enhancement of Potential Field Source Boundaries Using the Hyperbolic Domain (Gudermannian Function)", where new techniques are proposed and compared with classical ones in synthetic and real-potential-field data, in particular, gravity data from the Jalal Abad iron mine, Iran.

Induced polarization was the geophysical method of choice by do Amaral and collaborators in "Electrical Prospecting of Gold Mineralization in Exhalites of the Digo-Digo VMS Occurrence, Central Brazil". The geological–geophysical model obtained from the processing and inversion of the acquired data, in addition to the correlation of electrical and surface geological data, has made it possible to identify four anomalous areas related to potential mineralized zones.

Egorov and co-authors, in "Impact of the Regional Pai-Khoi-Altai Strike-Slip Zone on the Localization of Hydrocarbon Fields in Pre-Jurassic Units of West Siberia", carry out the interpretation of gravity, magnetic, and seismic data from the Pai-Khoi-Altai strike-slip zone, Russia, for the automated forecasting of prospective deep-seated hydrocarbon deposits in the study area.

In "Gravity Data Enhancement Using the Exponential Transform of the Tilt Angle of the Horizontal Gradient", Pham et al. introduced an improved enhancement technique that uses the exponential transform of the tilt angle of the horizontal gradient to improve the edge detection results. The robustness of the presented method is tested on synthetic models before applying to real gravity datasets to determine the geological features of the Tuan Giao area (Vietnam) and the boundaries of the Voisey's Bay Ni–Cu–Co deposit (Canada).

In conclusion, the papers collected in this Special Issue provided some applications and new methodologies for a variety of geophysical methods to aid geological methods with an emphasis on mineral exploration. We hope that these papers will further stimulate the integration of geological and geophysical information, as well as the development of processing and inversion techniques suited for mineral exploration.

Acknowledgments: The Guest Editors thank the authors, reviewers, as well as the Editorial Board and staff from *Minerals* for their contributions to this Special Issue.

Conflicts of Interest: The authors declare no conflicts of interest.

List of Contributions:

1. Xu, W.; Yao, C.; Yuan, B.; An, S.; Yin, X.; Yuan, X. Research on the Tectonic Characteristics and Hydrocarbon Prospects in the Northern Area of the South Yellow Sea Based on Gravity and Magnetic Data. *Minerals* **2023**, *13*, 893. https://doi.org/10.3390/min13070893.
2. Lan, N.; Zhang, F.; Xiao, K.; Zhang, H.; Lin, Y. Low-Dimensional Multi-Trace Impedance Inversion in Sparse Space with Elastic Half Norm Constraint. *Minerals* **2023**, *13*, 972. https://doi.org/10.3390/min13070972.
3. Park, J.; Shin, J.; Shin, S.; Park, Y. Preferred Orientations of Magnetic Minerals Inferred from Magnetic Fabrics of Hantangang Quaternary Basalts. *Minerals* **2023**, *13*, 1011. https://doi.org/10.3390/min13081011.
4. Ekwok, S.; Eldosouky, A.; Essa, K.; George, A.; Abdelrahman, K.; Fnais, M.; Andráš, P.; Akaerue, E.; Akpan, A. Particle Swarm Optimization (PSO) of High-Quality Magnetic Data of the Obudu Basement Complex, Nigeria. *Minerals* **2023**, *13*, 1209. https://doi.org/10.3390/min13091209.
5. Behnia, P.; Harris, J.; Sherlock, R.; Naghizadeh, M.; Vayavur, R. Mineral Prospectivity Mapping for Orogenic Gold Mineralization in the Rainy River Area, Wabigoon Subprovince. *Minerals* **2023**, *13*, 1267. https://doi.org/10.3390/min13101267.
6. Altinoğlu, F. Mapping of the Structural Lineaments and Sedimentary Basement Relief Using Gravity Data to Guide Mineral Exploration in the Denizli Basin. *Minerals* **2023**, *13*, 1276. https://doi.org/10.3390/min13101276.
7. Alvandi, A.; Su, K.; Ai, H.; Ardestani, V.; Lyu, C. Enhancement of Potential Field Source Boundaries Using the Hyperbolic Domain (Gudermannian Function). *Minerals* **2023**, *13*, 1312. https://doi.org/10.3390/min13101312.

8. do Amaral, P.; Borges, W.; Toledo, C.; Silva, A.; de Godoy, H.; Leão Santos, M. Electrical Prospecting of Gold Mineralization in Exhalites of the Digo-Digo VMS Occurrence, Central Brazil. *Minerals* **2023**, *13*, 1483. https://doi.org/10.3390/min13121483.
9. Egorov, A.; Antonchik, V.; Senchina, N.; Movchan, I.; Oreshkova, M. Impact of the Regional Pai-Khoi-Altai Strike-Slip Zone on the Localization of Hydrocarbon Fields in Pre-Jurassic Units of West Siberia. *Minerals* **2023**, *13*, 1511. https://doi.org/10.3390/min13121511.
10. Pham, L.T.; Oliveira, S.P.; Le, C.V.A; Bui, N.T.; Vu, A.H.; Nguyen, D.A. Gravity Data Enhancement Using the Exponential Transform of the Tilt Angle of the Horizontal Gradient. *Minerals* **2023**, *13*, 1539. https://doi.org/10.3390/min13121539.

References

1. Codeço, M.S.; Weis, P.; Andersen, C. Numerical modeling of structurally controlled ore formation in magmatic-hydrothermal systems. *Geochem. Geophys. Geosyst.* **2022**, *23*, e2021GC010302. [CrossRef]
2. Xie, W.; Chen, S.; Gan, H.; Wang, H.; Wang, M.; Vandeginste, V. Preservation conditions and potential evaluation of the Longmaxi shale gas reservoir in the Changning area, southern Sichuan Basin. *Geosci. Lett.* **2023**, *10*, 36. [CrossRef]
3. Oksum, E. Grav3CH_inv: A GUI-based MATLAB code for estimating the 3-D basement depth structure of sedimentary basins with vertical and horizontal density variation. *Comput. Geosci.* **2021**, *155*, 104856. [CrossRef]
4. Montsion, R.M.; Perrouty, S.; Lindsay, M.D.; Jessell, M.W.; Frieman, B.M. Mapping structural complexity using geophysics: A new geostatistical approach applied to greenstone belts of the southern Superior Province, Canada. *Tectonophysics* **2021**, *812*, 228889. [CrossRef]
5. Balkaya, Ç.; Ekinci, Y.L.; Çakmak, O.; Blömer, M.; Arnkens, J.; Kaya, M.A. A challenging archaeo-geophysical exploration through GPR and ERT surveys on the Keber Tepe, City Hill of Doliche, Commagene (Gaziantep, SE Turkey). *J. Appl. Geophys.* **2021**, *186*, 104272. [CrossRef]
6. Ekinci, Y.L.; Yigitbas, E. Interpretation of gravity anomalies to delineate some structural features of Biga and Gelibolu peninsulas, and their surroundings (north-west Turkey). *Geodin. Acta* **2015**, *27*, 300–319. [CrossRef]
7. Ai, H.; Ekinci, Y.L.; Balkaya, Ç.; Essa, K.S. Inversion of geomagnetic anomalies caused by ore masses using Hunger Games Search algorithm. *Earth Space Sci.* **2023**, *10*, e2023EA003002. [CrossRef]
8. Ekinci, Y.L.; Yiğitbaş, E. A geophysical approach to the igneous rocks in the Biga Peninsula (NW Turkey) based on airborne magnetic anomalies: Geological implications. *Geodin. Acta* **2012**, *25*, 267–285. [CrossRef]

Article

Gravity Data Enhancement Using the Exponential Transform of the Tilt Angle of the Horizontal Gradient

Luan Thanh Pham [1], Saulo Pomponet Oliveira [2], Cuong Van Anh Le [3], Nhung Thi Bui [4], An Hoa Vu [4] and Duong Anh Nguyen [4,*]

1   Department of Geophysics, Faculty of Physics, University of Science, Vietnam National University, Hanoi 100000, Vietnam; luanpt@hus.edu.vn

2   Graduate Program in Geology, Department of Mathematics, Federal University of Paraná, Curitiba 81531-980, PR, Brazil; saulopo@ufpr.br

3   Department of Geophysics, Faculty of Physics and Engineering Physics, University of Science, Vietnam National University, Ho Chi Minh City 700000, Vietnam; lvacuong@hcmus.edu.vn

4   Institute of Geophysics, Vietnam Academy of Science and Technology, Hanoi 100000, Vietnam; btnhung.vast@gmail.com (N.T.B.); hoanvt84@gmail.com (A.H.V.)

*   Correspondence: naduong@igp.vast.vn

Abstract: Detecting the boundaries of geologic structures is one of the main tasks in interpreting gravity anomalies. Many methods based on the derivatives of gravity anomalies have been introduced to map the source boundaries. The drawbacks of traditional methods are that the estimated boundaries are divergent or false boundaries appear in the output map. Here, we use the exponential transform of the tilt angle of the horizontal gradient to improve the edge detection results. The robustness of the presented method is illustrated using synthetic data and real examples from the Voisey's Bay Ni-Cu-Co deposit (Canada) and the Tuan Giao (Vietnam). The findings show that the presented technique can produce more precise and clear boundaries.

Keywords: exponential transform; tilt angle; horizontal gradient; edge detection

Citation: Pham, L.T.; Oliveira, S.P.; Le, C.V.A.; Bui, N.T.; Vu, A.H.; Nguyen, D.A. Gravity Data Enhancement Using the Exponential Transform of the Tilt Angle of the Horizontal Gradient. *Minerals* **2023**, *13*, 1539. https://doi.org/10.3390/min13121539

Academic Editors: Michael S. Zhdanov and Behnam Sadeghi

Received: 31 October 2023
Revised: 1 December 2023
Accepted: 6 December 2023
Published: 11 December 2023

1. Introduction

Geophysical methods are known as a powerful tool in mapping geological structures and minerals [1–6]. The gravity method is characterized by low cost and broad coverage compared to other geophysical surveys [7,8]. Interpreting gravity data provides important information about subsurface geological features [9–11]. The enhancement techniques of gravity anomalies can quickly determine the boundaries of the structures, and bring more abundant information for interpreting geologic formations [12–15].

Many techniques have been developed for enhancing gravity data [16–22]. These techniques are based on gradients of the anomalous field [23–26]. Unbalanced and balanced edge detection techniques are the two primary types of edge enhancement techniques [27]. The horizontal gradient [28], analytic signal [29], enhanced horizontal derivative [30], and analytic signals of gravity gradient tensor [31] are the unbalanced filters that are most often used for enhancing gravity data. The unbalanced methods can delineate the boundaries of shallow sources with high amplitudes, but they have limited detection effects on the boundaries of low amplitude anomalies [12,32].

To outline the boundaries of sources located at different depths, some balanced techniques have been developed. Most of these methods have been based on trigonometric functions, such as the tilt angle [33], theta map [34], exponential transform of the theta map [35], and normalized horizontal gradient [36]. A second generation of these methods involved high-order derivatives, for example, the tilt angle of the horizontal gradient [37], horizontal directional theta map [38], the horizontal gradient of the Ntilt [39], directional theta [40], logistic functions [41,42], enhanced horizontal gradient [43], and horizontal gradient of the improved normalized horizontal gradient [44]. The effectiveness of the edge detection techniques in terms of their precision in the determination of edges has been estimated in some recent studies [45–48]. Most of these studies showed that the tilt angle of the horizontal gradient is a powerful tool in mapping geological structures, but its edge map has a low resolution [46–48].

In this paper, we present a method to improve the edge detection results. Our method uses the exponential transform of the tilt angle of the horizontal gradient to bring the edge detection results with a high resolution and avoid producing additional edges in the output map. The application of the presented method is shown on real examples from the Voisey's Bay Ni-Cu-Co deposit (Canada) and the Tuan Giao (Vietnam).

2. Method

The theta map is a popular method in edge detection of potential field data, which normalizes the horizontal gradient by the analytic signal [34]. This method is defined as:

$$TM = cos\frac{\sqrt{\left(\frac{\partial F}{\partial x}\right)^2 + \left(\frac{\partial F}{\partial y}\right)^2}}{\sqrt{\left(\frac{\partial F}{\partial x}\right)^2 + \left(\frac{\partial F}{\partial y}\right)^2 + \left(\frac{\partial F}{\partial z}\right)^2}}. \tag{1}$$

where F is the gravity field.

To delineate the source boundaries more clearly, in 2013, Li suggested using the exponential transform of the theta map that is given by [35]:

$$ETM = exp\left(p \times cos\frac{\sqrt{\left(\frac{\partial F}{\partial x}\right)^2 + \left(\frac{\partial F}{\partial y}\right)^2}}{\sqrt{\left(\frac{\partial F}{\partial x}\right)^2 + \left(\frac{\partial F}{\partial y}\right)^2 + \left(\frac{\partial F}{\partial z}\right)^2}}\right). \tag{2}$$

where p is a constant decided by the interpreter. The use of $p = 4$ or 8 can make the edges more clearly [35]. The maxima of the *ETM* correspond to the source edges. Although the *ETM* can improve the resolution of the *TM*, it does not remove false edges in the edge map of the *TM*.

Another popular method is the tilt angle of the horizontal gradient that is given by [37]:

$$TAHG = atan\frac{\frac{\partial HG}{\partial z}}{\sqrt{\left(\frac{\partial HG}{\partial x}\right)^2 + \left(\frac{\partial HG}{\partial y}\right)^2}}. \tag{3}$$

where the horizontal gradient (*HG*) is given by:

$$HG = \sqrt{\left(\frac{\partial F}{\partial x}\right)^2 + \left(\frac{\partial F}{\partial y}\right)^2}. \tag{4}$$

Although the use of the *TAHG* can avoid bringing false information, its estimated boundaries are divergent. Here, we follow Li [35] to improve the resolution of the tilt angle of the horizontal gradient. The method is defined as:

$$ETAHG = exp\left(p \times atan\frac{\frac{\partial HG}{\partial z}}{\sqrt{\left(\frac{\partial HG}{\partial x}\right)^2 + \left(\frac{\partial HG}{\partial y}\right)^2}}\right). \tag{5}$$

The maxima of the *ETAHG* correspond to the source edges. Similar to the *TAHG*, it does not produce false information in the edge map. However, it can yield the edges with a higher resolution compared to the *TAHG*.

3. Methods Used for Comparison

To estimate the robustness of the presented method, we compared it to popular methods such as the horizontal gradient (*HG*), analytic signal amplitude (*AS*), theta map (*TM*), and some recent methods such as the exponential transform of the *TM* (*ETM*), tilt angle of the horizontal gradient (*TAHG*), horizontal gradient of *NTilt* (*HGNTilt*) and horizontal gradient of *impTDX* (*HGimpTDX*). The *HG*, *TM*, *ETM* and *TAHG* formulas are given in Section 2, while the *AS*, *HGNTilt* and *HGimpTDX* are shortly summarized below.

The *AS* is one of the most commonly used filters, which uses the peaks to extract the edges, and is defined as [29]:

$$AS = \sqrt{\left(\frac{\partial F}{\partial x}\right)^2 + \left(\frac{\partial F}{\partial y}\right)^2 + \left(\frac{\partial F}{\partial z}\right)^2}. \tag{6}$$

The *HGNTilt* uses the horizontal gradient of the *NTilt* to enhance the edges. The method is given by [39]:

$$HGNTilt = \sqrt{\left(\frac{\partial NTilt}{\partial x}\right)^2 + \left(\frac{\partial NTilt}{\partial y}\right)^2}, \tag{7}$$

where *NTilt* is defined as:

$$NTilt = atan\left(k^2 \frac{\frac{\partial^2 F}{\partial z^2}}{\sqrt{\left(\frac{\partial AS_2}{\partial x}\right)^2 + \left(\frac{\partial AS_2}{\partial y}\right)^2}}\right), \tag{8}$$

with AS_2 and k given by:

$$AS_2 = \sqrt{\left(\frac{\partial^3 F}{\partial z \partial z \partial x}\right)^2 + \left(\frac{\partial^3 F}{\partial z \partial z \partial y}\right)^2 + \left(\frac{\partial^3 F}{\partial z \partial z \partial z}\right)^2}, \tag{9}$$

$$k = \frac{M}{\sqrt{dx^2 + dy^2}}, \tag{10}$$

and M is the regional gravity value.

Recently, the *HGimpTDX* method was introduced to improve the resolution of the edges. The technique is based on the hyperbolic tangent function, and is given by [43]:

$$HGimpTDX = \sqrt{\left(\frac{\partial impTDX}{\partial x}\right)^2 + \left(\frac{\partial impTDX}{\partial y}\right)^2}, \tag{11}$$

where *impTDX* is defined as:

$$impTDX = tanh \frac{M\frac{\partial^2 F}{\partial z^2}}{\sqrt{\left(\frac{\partial TDX}{\partial x}\right)^2 + \left(\frac{\partial TDX}{\partial y}\right)^2}}. \tag{12}$$

with TDX given by [36]:

$$TDX = atan\frac{\sqrt{\left(\frac{\partial F}{\partial x}\right)^2 + \left(\frac{\partial F}{\partial y}\right)^2}}{\left|\frac{\partial F}{\partial z}\right|}. \tag{13}$$

4. Results and Discussion

4.1. Synthetic Examples

In this section, we estimate the effectiveness of the *ETAHG* through synthetic gravity examples with and without noise. The synthetic model includes five prisms, as shown in Figure 1a. The parameters of the model are presented in Table 1. Using these parameters, the gravity anomaly of the model is calculated and shown in Figure 1b.

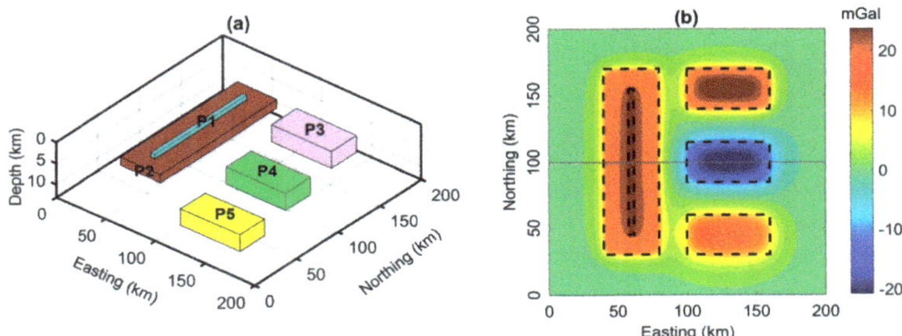

Figure 1. (**a**) The model. (**b**) Gravity anomaly of the model. The gray line denotes a profile.

Table 1. Parameters of the model.

Parameters	P1	P2	P3	P4	P5
Center coordinates (km; km)	60; 100	60; 100	130; 155	130; 100	130; 45
Width (km)	4	40	30	30	30
Length (km)	110	140	60	60	60
Depth of top (km)	2	3	3	6	9
Depth of bottom (km)	3	5	7	10	13
Density contrast (g/cm^3)	0.2	0.3	0.2	−0.2	0.2

In the first example, we applied the selected methods to gravity data in Figure 1b. Figure 2a presents the result of the *HG* method. It can be observed that the *HG* cannot equalize the different anomalies. The *HG* can determine the edges of the sources P2 and P3, but responses from other sources are faint. Figure 2b displays the edges outlined by the *AS*. It is observed that the *AS* is less effective in mapping the edges of the thin or deep sources. The results obtained from the method are fairly faint. Figure 2c presents the edges determined by the *TM* method. The method can equalize anomalies with different amplitudes, but it generates some false edges around the body P4. Figure 2c displays the edges determined by the *ETM* with $p = 1$. It is obvious that the *ETM* result is similar to the *TM* but has a higher resolution. The *ETM* was also computed using $p = 4$ (Figure 2e) and 8 (Figure 2f), as recommended in [35]. Clearly, the use of the *ETM* with $p = 8$ can generate sharper signals over the edges but the edge information of the body P1 is lost. Figure 2g,f display the edges delineated by the *HGNTilt* and *HGimpTDX* methods, respectively. Both methods generate the edges with a very high resolution, but some additional edges appear in the output maps of these methods. Figure 2i presents the edges determined by the *TAHG* method. Although the method can detect all the edges, the edges obtained from this method are divergent. Figure 2j–o display the edges delineated by the *ETAHG* with $p = 1, 2,$

3, 4, 6 and 8, respectively. As can be observed from these figures, the *ETAHG* maps have a higher resolution compared to the *TAHG*. Although the resolution of the *ETAHG* map increases when using larger values of p, the edges of the body P1 are lost or faint.

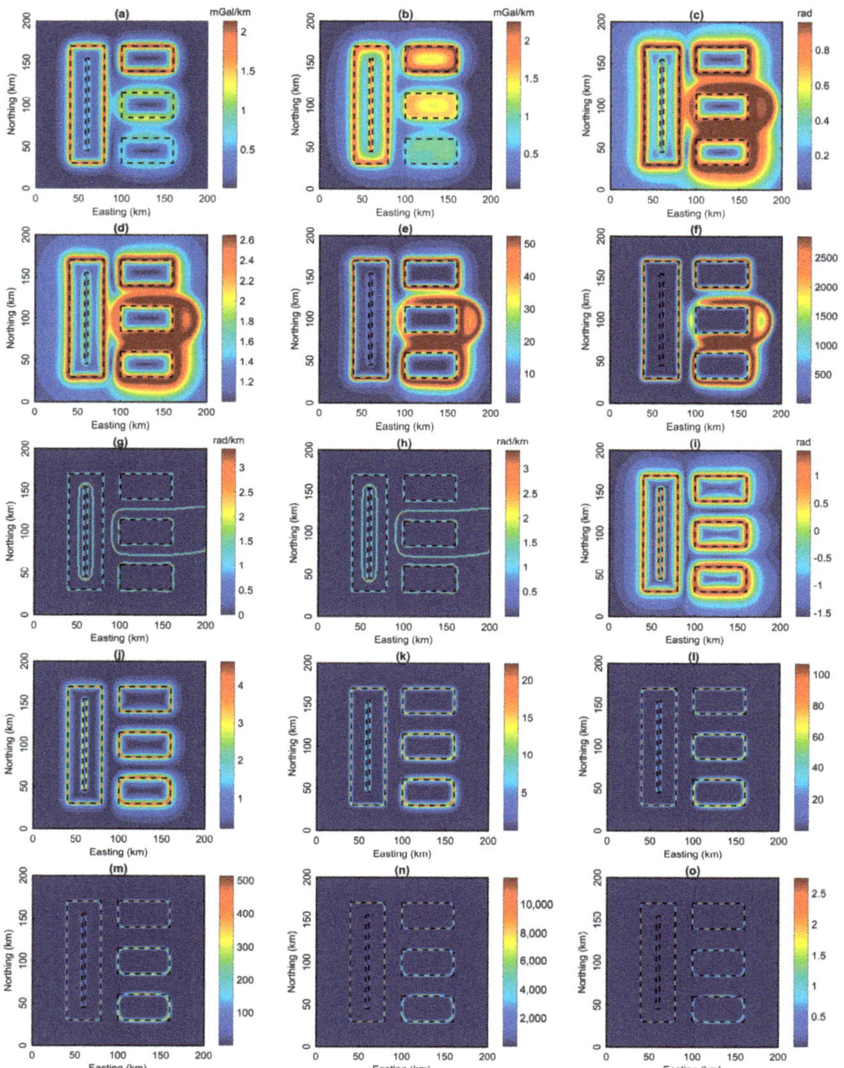

Figure 2. Results of data in Figure 1b. (**a**) *HG*. (**b**) *AS*. (**c**) *TM*. (**d**) *ETM* with $p = 1$. (**e**) *ETM* with $p = 4$. (**f**) *ETM* with $p = 8$. (**g**) *HGNTilt*. (**h**) *HGimpTDX*. (**i**) *TAHG*. (**j**) *ETAHG* with $p = 1$. (**k**) *ETAHG* with $p = 2$. (**l**) *ETAHG* with $p = 3$. (**m**) *ETAHG* with $p = 4$. (**n**) *ETAHG* with $p = 6$. (**o**) *ETAHG* with $p = 8$.

The results of the *TAHG* and *ETAHG* from Figure 2 were also compared in a horizontal profile (Figure 1b). Figure 3a shows the gravity anomaly along this profile. Figure 3b–h display the edges determined by the *TAHG* and *ETAHG* with $p = 1, 2, 3, 4, 6$ and 8, respectively. One can note from these figures that the signals over the edges along the *ETAHG* profiles are sharper than those of the *TAHG*. However, the amplitude of the transformed signal over the source P1 decreases as the p value increases. The *ETAHG* produces weak amplitude responses when the p value is greater than or equal to 2. For this reason, we used $p = 1$ in the subsequent *ETAHG* calculations.

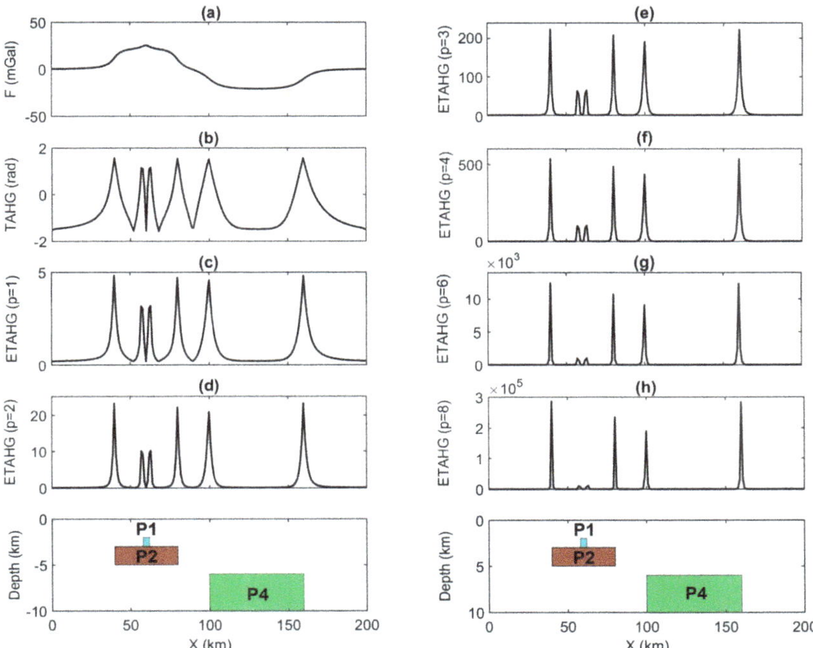

Figure 3. (**a**) Gravity data along the profile in Figure 1b. (**b**) *TAHG*. (**c**) *ETAHG* with $p = 1$. (**d**) *ETAHG* with $p = 2$. (**e**) *ETAHG* with $p = 3$. (**f**) *ETAHG* with $p = 4$. (**g**) *ETAHG* with $p = 6$. (**h**) *ETAHG* with $p = 8$.

To estimate the sensitivity of the *ETAHG* to random noise, we consider the second example where gravity data in Figure 2b was corrupted with 3% Gaussian noise (Figure 4a). Figure 5a–h display the edges delineated by applying the *HG*, *AS*, *TM*, *ETM*, *HGNTilt*, *HGimpTDX*, *TAHG* and *ETAHG* to gravity data in Figure 4a, respectively. As can be observed from these figures, the *HG* and *AS* are less sensitive to noise than others. However, these methods are dominated by the bodies P2 and P3. The *TM*, *ETM*, *HGNTilt*, *HGimpTDX*, *TAHG* and *ETAHG* are less sensitive to the depth of the bodies. As the *HGNTilt* and *HGimpTDX* are based on the third derivatives and/or fourth derivatives, they are more sensitive to noise than the *TM*, *ETM*, *TAHG* and *ETAHG*. In this case, the *ETAHG* still shows the edges more clearly than other methods.

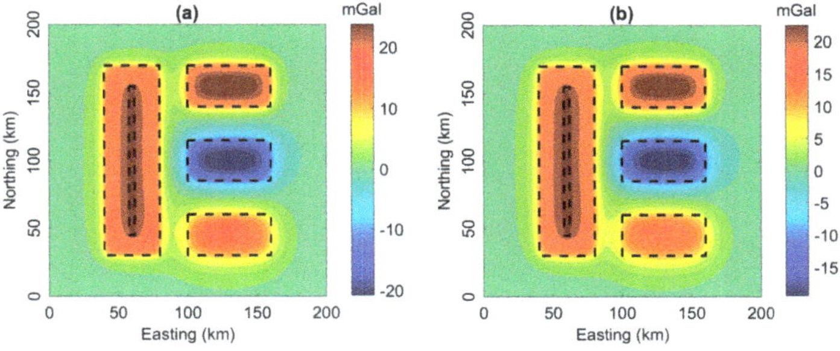

Figure 4. (**a**) Gravity data corrupted with 3% Gaussian noise. (**b**) Upward continued gravity data.

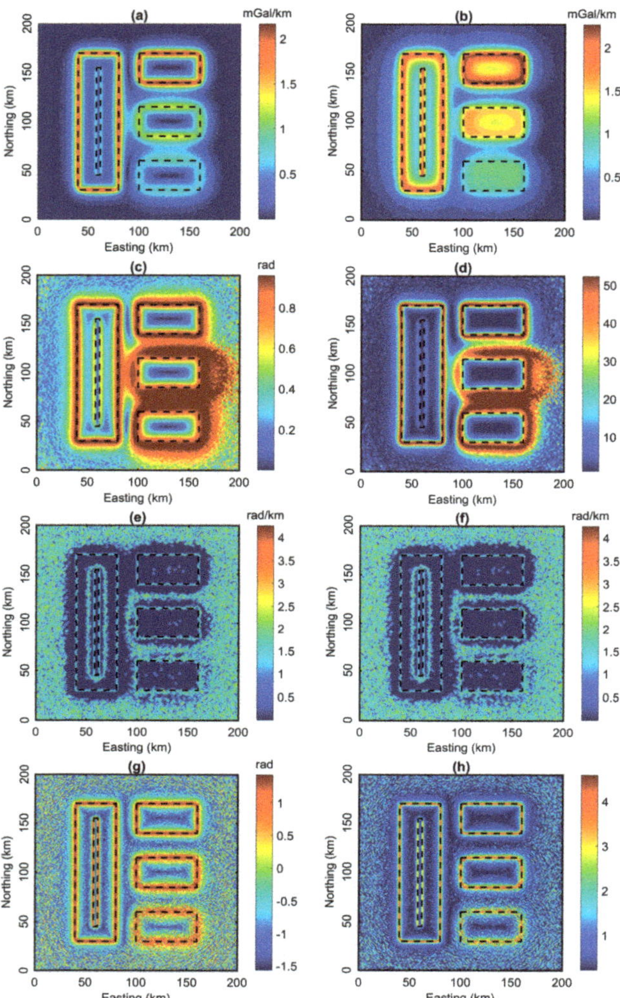

Figure 5. Results of data in Figure 4a. (**a**) *HG*. (**b**) *AS*. (**c**) *TM*. (**d**) *ETM* with $p = 4$. (**e**) *HGNTilt*. (**f**) *HGimpTDX*. (**g**) *TAHG*. (**h**) *ETAHG* with $p = 1$.

Since the enhancement techniques of gravity data are based on derivatives of the field, they amplify the noise. To attenuate the noise effect, the noise-corrupted data were subjected to an upward continuation filter of 1 km before using the techniques (Figure 4b). Figure 6a–h present the edges delineated by applying the *HG*, *AS*, *TM*, *ETM*, *HGNTilt*, *HGimpTDX*, *TAHG* and *ETAHG* to gravity data in Figure 4b, respectively. It is obvious that the *HG* can clearly outline the edges of the bodies P2 and P3, but the responses over the other sources are faint (Figure 6a). The *AS* cannot outline the edges of the dike P1 and deep sources P4 and P5 (Figure 6b). The *ETM* shows sharper edges than the *TM*, but both methods still generate false boundaries around the body P4 (Figure 6c,d). Although the *HGNTilt* and *HGimpTDX* can determine most of the edges with a very high resolution, some additional edges still appear along the north-south edges of the body G1, and around the body P4 output maps of these methods (Figure 6e,f). The *TAHG* and *ETAHG* can highlight all the source boundaries without any false information. However, the *ETAHG* generates higher resolution boundaries of the sources.

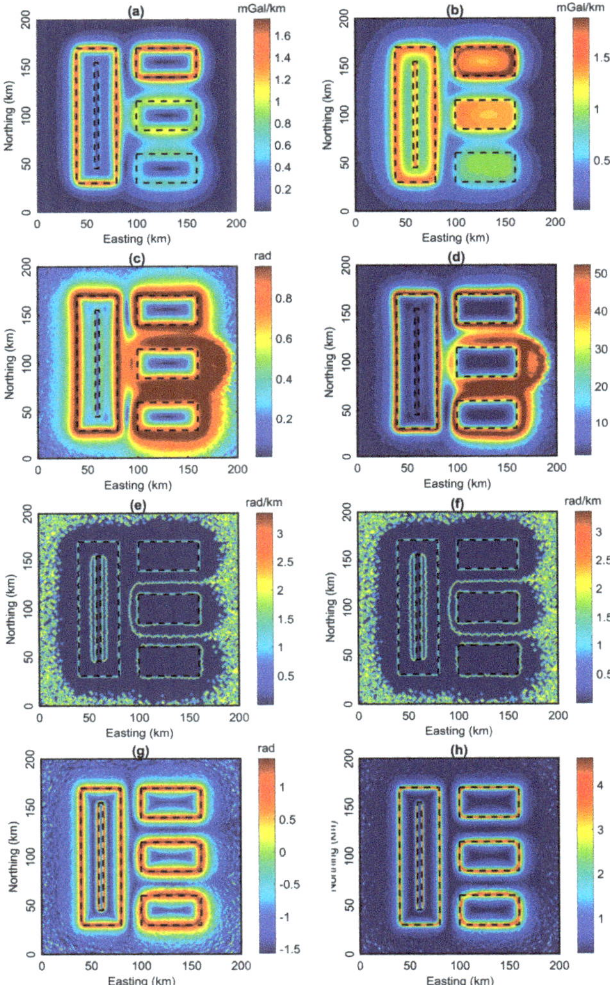

Figure 6. Results of data in Figure 4b. (**a**) *HG*. (**b**) AS. (**c**) *TM*. (**d**) *ETM* with $p = 4$. (**e**) *HGNTilt*. (**f**) *HGimpTDX*. (**g**) *TAHG*. (**h**) *ETAHG* with $p = 1$.

4.2. Real Examples

4.2.1. Voisey's Bay Ni-Cu-Co Deposit

One of the most important mineral discoveries in Canada over the last few decades is the Voisey's Bay Ni–Cu–Co deposit, which is situated on the northeast coast of Labrador (Figure 7) [49]. The primary ore body is the ovoid that is being mined at the moment (Figure 7). With horizontal dimensions of 650 by 350 m and a maximum depth extension of 120 m, it is a massive sulphide lens with an elliptical shape. It is estimated that there are 30 million tons of proven and probable reserves, grading 2.9% nickel, 1.7% copper, and 0.14% cobalt [49]. The deposit is linked to the Voisey's Bay intrusion, which crosses the 1.85 Ga east-dipping collisional boundary between the Archean Nain Province to the east and the Proterozoic Churchill Province to the west (Figure 7) [50,51].

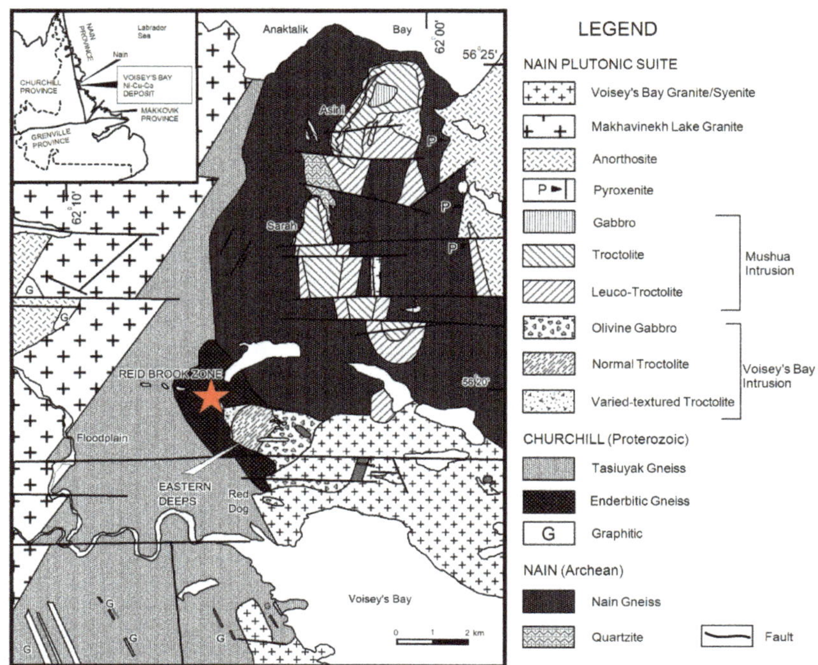

Figure 7. Geology of the Voisey's Bay area showing the location of the Voisey's Bay Ni-Cu-Co deposit (red star) (adapted from [50,51]).

The Bouguer gravity map of the Voisey's Bay is shown in Figure 8a [52]. The Bouguer gravity map of the Voisey's Bay comprises the primary ore body. Figure 8b–i present the edges delineated by applying the *HG, AS, TM, ETM, HGNTilt, HGimpTDX, TAHG* and *ETAHG* to gravity data in Figure 8a, respectively. It can be observed that the peaks of the *HG, HGNTilt, HGimpTDX, TAHG* and *ETAHG* demonstrated the presence of a primary ore body with an approximate ellipsoidal form, as reported by some other studies (Figure 8b,f–i) [49,53]. However, the *HG* and *TAHG* results are divergent. The *HGNTilt* and *HGimpTDX* are very effective in providing high resolution boundaries, but they bring some additional boundaries at the edges of the study area. In this case, the *AS* does not provide a clear image of the main ore body (Figure 8c), while the *TM* and *ETM* bring false maxima in the northeastern region. Comparing the results, one can observe that the *ETAHG* does not yield additional edges, and it can provide the edges more clearly compared to others.

4.2.2. Tuan Giao Area

The Tuan Giao area is located between the South China block and the Sundaland block. It is considered as a part of the transition boundary zone between these two blocks [54]. Two main factors, the (i) collision of the Indo-Australian and Eurasian plates and (ii) subduction of the Pacific plate under Eurasian plate, explain that the Tuan Giao area has high dynamic activities. The Tuan Giao is a mountainous area with a complicated geological structure (Figure 9), dominated by many active faults, such as the Dien Bien Phu fault, Son La fault, Song Da fault and Song Ma fault [55–57]. From the works of many researchers, young materials (i.e., magmatic rocks) intruded into old sedimentary rocks environment through the faults (Figure 9) [55,56]. For this area, Permian–Triassic sediment rocks dominate the most area while the rest as thin stripes disseminate along faults. In this area, there were at least seven earthquakes, with a magnitude of above five occurring from 1914 to 1983. These earthquakes are shown in Figure 5. The two biggest earthquakes occurred at Dien Bien in

1935 (M 6.8) and Tuan Giao in 1983 (M 6.7) (Figure 9). Both earthquakes severely damaged homes and infrastructure and killed or injured dozens of people in landslides [54]. Since then, no earthquakes with a magnitude of 5 have occurred in the study area; thus, this is a high-risk area for earthquakes. Therefore, accurately determining the location of faults in high-risk earthquake areas is necessary to have accurate earthquake hazard assessments and earthquake forecasts in the area.

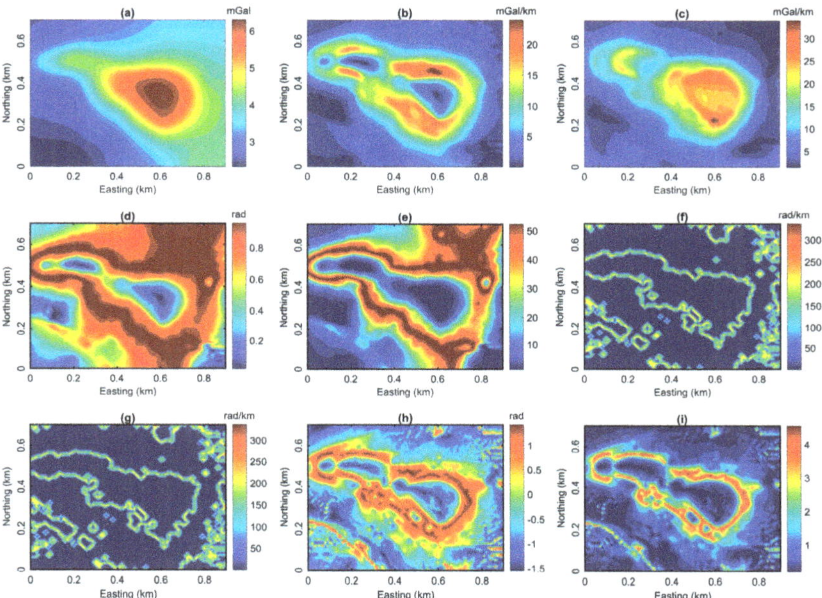

Figure 8. (**a**) Bouguer data of the Voisey's Bay Ni–Cu–Co deposit. (**b**) *HG.* (**c**) *AS.* (**d**) *TM.* (**e**) *ETM* with *p* = 4. (**f**) *HGNTilt.* (**g**) *HGimpTDX.* (**h**) *TAHG.* (**i**) *ETAHG* with *p* = 1.

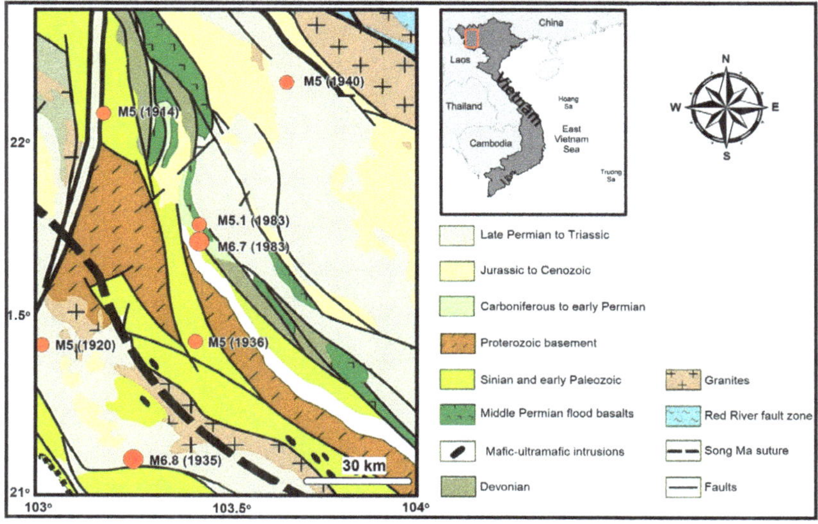

Figure 9. Geology map of the Tuan Giao [55].

Figure 10a displays the Bouguer gravity map of the Tuan Giao [58]. Figure 10b–i display the boundaries delineated by applying the *HG, AS, TM, ETM, HGNTilt, HGimpTDX, TAHG* and *ETAHG* to Bouguer gravity data in Figure 10a, respectively. One can observe that the *HG* and *AS* are dominated by anomalies at the northern part of the area, and these methods do not yield images of the structural boundaries. The obtained image maps from the application of the *TM, ETM, HGNTilt, HGimpTDX* and *TAHG* allow us to extract the structural boundaries, and show the boundaries of the large and small signals clearly (Figure 10d–h). While the edges in the *TM, ETM* and *TAHG* are divergent, the *HGNTilt* and *HGimpTDX* produce very sharp edges. However, the use of the *HGNTilt* and *HGimpTDX* may bring some additional edges, as shown in the synthetic examples. In this case, the edges determined by the *ETAHG* are more precise and clearer (Figure 10i). It can be observed from Figure 10i that the *ETAHG* map shows a dominant NW-SE structural trend of density bodies that correspond favorably to the geological formations of the Tuan Giao. In addition, the maximum locations in the *ETAHG* map exhibit a strong correlation with a large number of NW-NW-SE trending faults in the region.

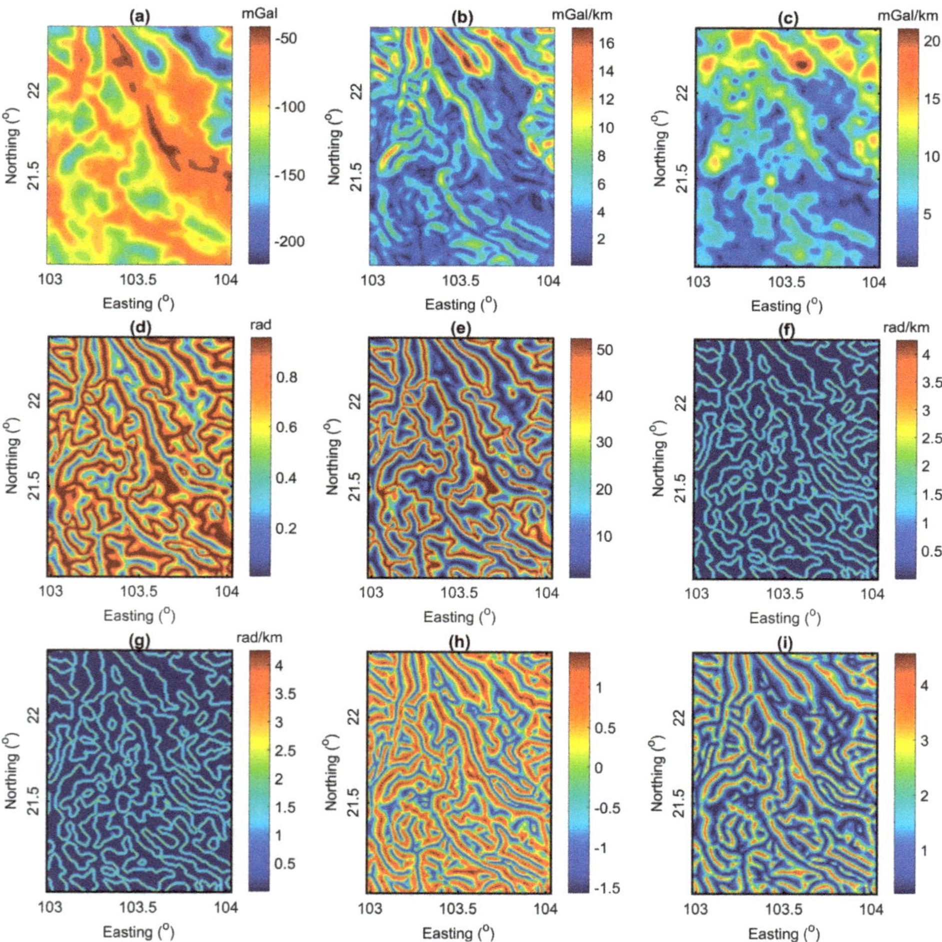

Figure 10. (**a**) Bouguer data of the Tuan Giao. (**b**) *HG*. (**c**) *AS*. (**d**) TM. (**e**) *ETM* with $p = 4$. (**f**) *HGNTilt*. (**g**) *HGimpTDX*. (**h**) *TAHG*. (**i**) *ETAHG* with $p = 1$.

5. Conclusions

In this paper, we have presented an improved method to extract the edges of gravity data. The method uses the exponential transform of the tilt angle of the horizontal gradient to enhance the edges. The theoretical tests show that the presented method can extract the edges of shallow and deep bodies simultaneously. In addition, this method produces results with more precise and clear boundaries compared to other methods. The application of the presented method is illustrated in mapping structures of the Tuan Giao (Vietnam) and boundaries of the Voisey's Bay Ni-Cu-Co deposit (Canada). The findings from the real examples are in agreement with the known structures of the study areas. Assuming that the magnetization direction is known, we can compute RTP magnetic data. Then, the presented method can be used for interpreting magnetic data.

Author Contributions: Conceptualization, L.T.P.; methodology, L.T.P. and S.P.O.; software, L.T.P. and S.P.O.; validation, N.T.B., A.H.V. and D.A.N.; investigation, L.T.P., S.P.O., C.V.A.L., N.T.B., A.H.V. and D.A.N.; data curation, N.T.B., A.H.V. and D.A.N.; writing—original draft preparation, L.T.P., S.P.O., C.V.A.L. and D.A.N.; writing—review and editing, L.T.P., S.P.O., C.V.A.L., N.T.B., A.H.V. and D.A.N.; supervision, C.V.A.L., S.P.O. and D.A.N. All authors have read and agreed to the published version of the manuscript.

Funding: This research was funded by PAS-VAST, QTPL01.01/22-23; and CNPq, 316376/2021-3.

Data Availability Statement: The data presented in this study are available on reasonable request from the corresponding author.

Acknowledgments: Authors thank for the result discussion of Phan Trong Trinh and the support of the senior researcher, NVCC 11.01/22-23. The authors thank two anonymous reviewers for their constructive comments.

Conflicts of Interest: The authors declare no conflict of interest.

References

1. Ai, H.; Ekinci, Y.L.; Balkaya, Ç.; Essa, K.S. Inversion of geomagnetic anomalies caused by ore masses using Hunger Games Search algorithm. *Earth Space Sci.* **2023**, *10*, e2023EA003002. [CrossRef]
2. Sampietro, D.; Capponi, M.; Maurizio, G. 3D Bayesian inversion of potential fields: The Quebec Oka carbonatite complex case study. *Geosciences* **2022**, *12*, 382. [CrossRef]
3. Boszczuk, P.; Cheng, L.Z.; Hammouche, H.; Roy, P.; Lacroix, S.; Cheilletz, A. A 3D gravity data interpretation of the Matagami mining camp, Abitibi Subprovince, Superior Province, Québec, Canada: Application to VMS deposit exploration. *J. Appl. Geophys.* **2011**, *75*, 77–86. [CrossRef]
4. Wijanarko, E.; Arisbaya, I.; Sumintadireja, P.; Grandis, H. Magnetotellurics study of Atambua area, West Timor, Indonesia. *Vietnam J. Earth Sci.* **2022**, *45*, 67–81. [CrossRef] [PubMed]
5. Xayavong, V.; Duc, M.V.; Singsoupho, S.; Anh, D.N.; Prasad, K.N.D.; Minh, T.V.; Do Anh, C. Combination of 2D-Electrical Resistivity Imaging and Seismic Refraction Tomography methods for groundwater potential assessments: A case study of Khammouane province, Laos. *Vietnam J. Earth Sci.* **2023**, *45*, 238–250. [CrossRef] [PubMed]
6. Marques, F.; Matos, J.X.; Sousa, P.; Represas, P.; Araujo, A.; Carvalho, J.; Morais, I.; Pacheco, N.; Albardeiro, L.; Goncalves, P. The role of land gravity data in the Neves-Corvo mine discovery and its use in present-day exploration and new target generation. *First Break* **2019**, *37*, 97–102. [CrossRef]
7. Nabighian, M.N.; Ander, M.E.; Grauch, V.J.S.; Hansen, R.O.; LaFehr, T.R.; Li, Y.; Pearson, W.C.; Peirce, J.W.; Phillips, J.D.; Ruder, M.E. Historical development of the gravity method in exploration. *Geophysics* **2005**, *70*, 63–89. [CrossRef]
8. Hinze, W.; Frese, R.; Saad, A. *Gravity and Magnetic Exploration: Principles, Practices, and Applications*; Cambridge University Press: Cambridge, UK, 2013.
9. Narayan, S.; Kumar, U.; Pal, S.K.; Sahoo, S.D. New insights into the structural and tectonic settings of the Bay of Bengal using high-resolution earth gravity model data. *Acta Geophys.* **2021**, *69*, 2011–2033. [CrossRef]
10. Narayan, S.; Sahoo, S.D.; Pal, S.K.; Kumar, U. Comparative evaluation of five global gravity models over a part of the Bay of Bengal. *Adv. Space Res.* **2022**, *71*, 2416–2436. [CrossRef]
11. Oksum, E. Grav3CH_inv: A GUI-based MATLAB code for estimating the 3-D basement depth structure of sedimentary basins with vertical and horizontal density variation. *Comput. Geosci.* **2021**, *155*, 104856. [CrossRef]
12. Kamto, P.G.; Oksum, E.; Pham, L.T.; Kamguia, J. Contribution of advanced edge detection filters for the structural mapping of the Douala Sedimentary Basin along the Gulf of Guinea. *Vietnam J. Earth Sci.* **2023**, *45*, 287–302.
13. Narayan, S.; Sahoo, S.D.; Pal, S.K.; Kumar, U.; Pathak, V.K.; Majumdar, T.J.; Chouhan, A. Delineation of structural features over a part of the Bay of Bengal using total and balanced horizontal derivative techniques. *Geocarto Int.* **2016**, *32*, 351–366. [CrossRef]

14. Ekinci, Y.L.; Yigitbas, E. Interpretation of gravity anomalies to delineate some structural features of Biga and Gelibolu peninsulas, and their surroundings (north-west Turkey). *Geodin. Acta* **2015**, *27*, 300–319. [CrossRef]

15. Eldosouky, A.M.; El-Qassas, R.A.Y.; Pham, L.T.; Abdelrahman, K.; Alhumimidi, M.S.; El Bahrawy, A.; Mickus, K.; Sehsah, H. Mapping Main Structures and Related Mineralization of the Arabian Shield (Saudi Arabia) Using Sharp Edge Detector of Transformed Gravity Data. *Minerals* **2022**, *12*, 71. [CrossRef]

16. Melouah, O.; Ebong, E.D.; Abdelrahman, K.; Eldosouky, A.M. Lithospheric structural dynamics and geothermal modeling of the Western Arabian Shield. *Sci. Rep.* **2023**, *13*, 11764. [CrossRef] [PubMed]

17. Kumar, U.; Pal, S.K.; Sahoo, S.D.; Narayan, S.; Saurabh, M.S.; Ganguli, S.S. Lineament mapping over Sir Creek offshore and its surroundings using high resolution EGM2008 gravity data: An integrated derivative approach. *J. Geol. Soc. India* **2018**, *91*, 671–678. [CrossRef]

18. Melouah, O.; Eldosouky, A.M.; Ebong, W.D. Crustal architecture, heat transfer modes and geothermal energy potentials of the Algerian Triassic provinces. *Geothermics* **2021**, *96*, 102211. [CrossRef]

19. Prasad, K.N.D.; Pham, L.T.; Singh, A.P. A novel filter "ImpTAHG" for edge detection and a case study from Cambay Rift Basin, India. *Pure Appl. Geophys.* **2022**, *179*, 2351–2364. [CrossRef]

20. Cooper, G.R.J. Novel Second-Order Derivative-Based Filters for Edge and Ridge/Valley Detection in Geophysical Data. *Minerals* **2023**, *13*, 1229. [CrossRef]

21. Cooper, G.R.J. Amplitude-Balanced Edge Detection Filters for Potential Field Data. *Explor. Geophys.* **2023**, *54*, 544–552. [CrossRef]

22. Kafadar, Ö. Applications of the Kuwahara and Gaussian filters on potential field data. *J. Appl. Geophys.* **2022**, *198*, 104583. [CrossRef]

23. Ekinci, Y.L.; Yiğitbaş, E. A geophysical approach to the igneous rocks in the Biga Peninsula (NW Turkey) based on airborne magnetic anomalies: Geological implications. *Geodin. Acta* **2012**, *25*, 267–285. [CrossRef]

24. Sahoo, S.D.; Narayan, S.; Pal, S.K. Fractal analysis of lineaments using CryoSat-2 and Jason-1 satellite-derived gravity data: Evidence of a uniform tectonic activity over the middle part of the Central Indian Ridge. *Phys. Chem. Earth* **2022**, *128*, 103237. [CrossRef]

25. Yuan, Y.; Yu, Q. Edge detection in potential-field gradient tensor data by use of improved horizontal analytic signal methods. *Pure Appl. Geophys.* **2015**, *172*, 461–472. [CrossRef]

26. Eldosouky, A.M.; Mohamed, H. Edge detection of aeromagnetic data as effective tools for structural imaging at Shilman area, South Eastern Desert, Egypt. *Arab. J. Geosci.* **2021**, *14*, 1–10. [CrossRef]

27. Eldosouky, A.M.; Pham, L.T.; Duong, V.H.; Ghomsi, F.E.K.; Henaish, A. Structural interpretation of potential field data using the enhancement techniques: A case study. *Geocarto Int.* **2022**, *37*, 16900–16925. [CrossRef]

28. Cordell, L. Gravimetric expression of graben faulting in Santa Fe Country and the Espanola Basin. In *30th Field Conference, Socorro, NM, USA, 4–6 October 1979: New Mexico Geological Society Guidebook*; New Mexico Geological Society: Socorro, NM, USA, 1979; pp. 59–64.

29. Roest, W.R.; Verhoef, J.; Pilkington, M. Magnetic interpretation using the 3-D analytic signal. *Geophysics* **1992**, *57*, 116–125. [CrossRef]

30. Fedi, M.; Florio, G. Detection of potential fields source boundaries by enhanced horizontal derivative method. *Geophys. Prospect.* **2001**, *49*, 40–58. [CrossRef]

31. Beiki, M. Analytic signals of gravity gradient tensor and their application to estimate source location. *Geophysics* **2010**, *75*, 159–174. [CrossRef]

32. Alvandi, A.; Ardestani, V.E. Edge detection of potential field anomalies using the Gompertz function as a high-resolution edge enhancement filter. *Bull. Geophys. Oceanogr.* **2023**, *64*, 279–300.

33. Miller, H.G.; Singh, V. Potential field tilt a new concept for location of potential field sources. *J. Appl. Geophys.* **1994**, *32*, 213–217. [CrossRef]

34. Wijns, C.; Perez, C.; Kowalczyk, P. Theta map: Edge detection in magnetic data. *Geophysics* **2005**, *70*, 39–43. [CrossRef]

35. Li, L. Improved edge detection tools in the interpretation of potential field data. *Explor. Geophys.* **2013**, *44*, 128–132. [CrossRef]

36. Cooper, G.; Cowan, D. Enhancing potential field data using filters based on the local phase. *Comput. Geosci.* **2006**, *32*, 1585–1591. [CrossRef]

37. Ferreira, F.J.F.; de Souza, J.; Bongiolo, A.B.S.; de Castro, L.G. Enhancement of the total horizontal gradient of magnetic anomalies using the tilt angle. *Geophysics* **2013**, *78*, 33–41. [CrossRef]

38. Yuan, Y.; Gao, J.Y.; Chen, L.N. Advantages of horizontal directional Theta method to detect the edges of full tensor gravity gradient data. *J. Appl. Geophys.* **2016**, *130*, 53–61. [CrossRef]

39. Nasuti, Y.; Nasuti, A.N. Tilt as an improved enhanced tilt derivative filter for edge detection of potential field anomalies. *Geophys. J. Int.* **2018**, *214*, 36–45. [CrossRef]

40. Zareie, V.; Moghadam, R.H. The application of theta method to potential field gradient tensor data for edge detection of complex geological structures. *Pure Appl. Geophys.* **2019**, *176*, 4983–5001. [CrossRef]

41. Pham, L.T.; Oksum, E.; Do, T.D. Edge enhancement of potential field data using the logistic function and the total horizontal gradient. *Acta Geod. Geophys.* **2019**, *54*, 143–155. [CrossRef]

42. Pham, L.T.; Vu, T.V.; Le, T.S.; Trinh, P.T. Enhancement of potential field source boundaries using an improved logistic filter. *Pure Appl. Geophys.* **2020**, *177*, 5237–5249. [CrossRef]

43. Pham, L.T.; Eldosouky, A.M.; Oksum, E.; Saada, S.A. A new high resolution filter for source edge detection of potential field data. *Geocarto Int.* **2022**, *37*, 3051–3068. [CrossRef]
44. Ibraheem, I.M.; Tezkan, B.; Ghazala, H.; Othman, A.A. A New Edge Enhancement Filter for the Interpretation of Magnetic Field Data. *Pure Appl. Geophys.* **2023**, *180*, 2223–2240. [CrossRef]
45. Ekinci, Y.L.; Ertekin, C.; Yigitbas, E. On the effectiveness of directional derivative based filters on gravity anomalies for source edge approximation: Synthetic simulations and a case study from the Aegean graben system (western Anatolia, Turkey). *J. Geophys. Eng.* **2013**, *10*, 035005. [CrossRef]
46. Pham, L.T.; Oksum, E.; Kafadar, O.; Trinh, P.T.; Nguyen, D.V.; Vo, Q.T.; Le, S.T.; Do, T.D. Determination of subsurface lineaments in the Hoang Sa islands using enhanced methods of gravity total horizontal gradient. *Vietnam J. Earth Sci.* **2022**, *44*, 395–409.
47. Nasuti, Y.; Nasuti, A.; Moghadas, D. STDR: A novel approach for enhancing and edge detection of potential field data. *Pure Appl. Geophys.* **2019**, *176*, 827–841. [CrossRef]
48. Liu, J.; Li, S.; Jiang, S.; Wang, X.; Zhang, J. Tools for Edge Detection of Gravity Data: Comparison and Application to Tectonic Boundary Mapping in the Molucca Sea. *Surv. Geophys.* **2023**, *44*, 1781–1810. [CrossRef]
49. Farquharson, C.G.; Ash, M.R.; Miller, H.G. Geologically constrained gravity inversion for the Voisey's Bay ovoid deposit. *Lead. Edge* **2008**, *27*, 64–69. [CrossRef]
50. Li, C.; Naldrett, A. Geology and petrology of the Voisey's Bay intrusion: Reaction of olivine with sulfide and silicate liquids. *Lithos* **1999**, *47*, 1–31. [CrossRef]
51. Ryan, B. *Geological Map of the Nain Plutonic Suite and Surrounding Rocks Nain-Nutak, NTS 14SW. Geological Survey Branch, Newfoundland Department of Mines and Energy, Scale 1:500,000*; Government of Newfoundland and Labrador: St. John's, NL, Canada, 1990.
52. King, A. Review of Geophysical Technology for Ni-Cu-PGE deposits. In Proceedings of the Exploration 07: Fifth Decennial International Conference on Mineral Exploration, Toronto, ON, Canada, 9–12 September 2007; Milkereit, B., Ed.; CVRD Exploration Canada Inc.: Cliff, ON, Canada, 2007; pp. 647–665.
53. Lelièvre, P.; Carter-McAuslan, A.; Farquharson, C.; Hurich, C. Unified geophysical and geological 3D Earth models. *Lead. Edge* **2012**, *31*, 322–328. [CrossRef]
54. Duong, N.A.; Sagiya, T.; Kimata, F.; Tran, D.T.; Vy, Q.H.; Cong, D.C.; Binh, N.X.; Xuyen, N.D. Contemporary horizontal crustal movement estimation for northwestern Vietnam inferred from repeated GPS measurements. *Earth Planets Space* **2013**, *65*, 1399–1410. [CrossRef]
55. Zhou, M.F.; Chen, W.T.; Wang, C.Y.; Prevec, S.A.; Liu, P.P.; Howarth, G.H. Two stages of immiscible liquid separation in the formation of Panzhihua-type Fe–Ti–V oxide deposits, SW China. *Geosci. Front.* **2013**, *4*, 481–502. [CrossRef]
56. Roger, F.; Maluski, H.; Lepvrier, C.; Van, T.V.; Paquette, J.-L. LA-ICPMS zircons U/Pb dating of Permo-Triassic and Cretaceous magmatisms in Northern Vietnam–Geodynamical implications. *J. Asian Earth Sci.* **2012**, *48*, 72–82. [CrossRef]
57. Tran, T.A.; Tran-Trong, H.; Pham-Ngoc, C.; Shellnutt, J.G.; Pham, T.T.; Izokh, A.E.; Pham, P.L.T.; Duangpaseuth, S.; Soulintone, O. Petrology of the Permian-Triassic granitoids in Northwest Vietnam and their relation to the amalgamation of the Indochina and Sino-Vietnam composite terranes. *Vietnam J. Earth Sci.* **2022**, *44*, 343–368. [CrossRef] [PubMed]
58. Pham, L.T. A comparative study on different methods for calculating gravity effect of an uneven layer: Application to computation of Bouguer gravity anomaly in the East Vietnam Sea and adjacent areas. *VNU J. Sci. Math Phys.* **2020**, *36*, 106–114.

minerals

Article

Impact of the Regional Pai-Khoi-Altai Strike-Slip Zone on the Localization of Hydrocarbon Fields in Pre-Jurassic Units of West Siberia

Aleksey Egorov, Vladimir Antonchik , Natalia Senchina * , Igor Movchan and Maria Oreshkova

Geophysics Department, Saint Petersburg Mining University, St. Petersburg 199106, Russia; egorov_as@pers.spmi.ru (A.E.); v.antonchik@mail.ru (V.A.); movchan_ib@pers.spmi.ru (I.M.)
* Correspondence: senchina_np@pers.spmi.ru; Tel.: +7-9119013891

Abstract: The paper presents the results of a geological interpretation using gravity, magnetic, and seismic data to understand the oil and gas potential of pre-Jurassic sedimentary intervals and basement in the central West Siberia basin. The 200 km long Pai-Khoi-Altai strike-slip zone was investigated. Reconstruction based on a data complex indicate the right-lateral kinematics of the principal strike-slip faults and possible fault inversion. The study evaluated the spatial and genetic relationship between the conditions for hydrocarbon trap development and the strike-slip fault systems, such as "flower structures". Strike-slip geometry and kinematics are confirmed based on 2D and 3D seismic data. Geological and geophysical criteria are used to forecast localization of hydrocarbon fields. Predictive zones are elongated in several different directions and have a different distribution pattern in the blocks separated by principal strike-slip faults, confirming its significance as a controlling factor for the hydrocarbon potential of the region's structures.

Keywords: strike-slip; hydrocarbons; geophysics; basement; West Siberia oil and gas province; geophysical data interpretation

Citation: Egorov, A.; Antonchik, V.; Senchina, N.; Movchan, I.; Oreshkova, M. Impact of the Regional Pai-Khoi-Altai Strike-Slip Zone on the Localization of Hydrocarbon Fields in Pre-Jurassic Units of West Siberia. *Minerals* **2023**, *13*, 1511. https://doi.org/10.3390/min13121511

Academic Editors: Luan Thanh Pham, Saulo Pomponet Oliveira and Le Van Anh Cuong

Received: 6 October 2023
Revised: 27 November 2023
Accepted: 28 November 2023
Published: 30 November 2023

1. Introduction

The West Siberia basin provides approximately 70% of Russia's oil production and over 90% of its gas production [1]. The region is characterized by intensive exploitation at a depth of 2–4 km, consisting of Jurassic and Cretaceous deposits covered by younger sediments. This well-studied interval is gradually depleting, with experts noting a declining trend in oil and gas reserves, while there is no anticipated decrease in demand for petroleum products [2,3].

Relatively less explored but still petroleum- and gas-bearing are the pre-Jurassic strata, including the so-called "intermediate structural layer" and the weathered crust of the consolidated basement. According to various estimates, 3 to 15% of the world's proven oil reserves are localized in the basement rocks. However, the volume of exploratory drilling to reach this level is an order of magnitude less than in the sedimentary cover, and the discovery of deposits of this type often happens by chance [4].

In the pre-Jurassic complex of Western Siberia, deposits have been identified at dozens of fields with oil reserves of about 0.2 billion tons (for comparison, current recoverable reserves from all horizons of Western Siberia are nearly 18 billion tons) [5,6]. The prospects for new discoveries in such deposits are significant, and the pre-Jurassic complex deposits are considered to be the principal source for locating large reserves in the Western Siberia region [7].

The West Siberia petroleum and gas basin is of interest to leading Russian geologists such as V.S. Surkov [8], who described the structure of the Western Siberia basement, and A.E. Kontorovich [9], who studied the oil and gas potential of the region. G.A. Lobova [10] wondered about the oil and gas content of the pre-Jurassic rock, A.I. Timurziev [4,11] described fault structure, S.F. Khafizov [12] assessed the hydrocarbon potential of the province, which made a great contribution to the study of the area. While the prospects for discovering oil deposits in the basement rocks of the West Siberia petroleum and gas province are highly rated, the factors controlling the formation of deposits in pre-Jurassic-age rocks are relatively poorly understood to date [13]. This is likely the reason for the lack of new discoveries at deep horizons in the sedimentary cover, in the "intermediate structural layer," and in the upper part of the basement.

According to the results of scientific research [10,12,14], the localization of hydrocarbon deposits in pre-Jurassic complex rocks is influenced by the following:

- The presence, thickness, and composition of "intermediate structural layer" rocks, including the availability of source rocks.
- The development of regional strike-slip dislocations, which create extensive zones of deformation and increased fracturing in the overlying sedimentary cover, controlled by "flower structures" [14].
- Triassic rifting and the development of volcanic–sedimentary deposits.
- The presence of uplifted basement surface features.
- The development of weathered crusts in the basement.

However, oil and gas extraction from crystalline basement rocks presents significant challenges and risks. These risks include geological complexities, such as fractured and heterogeneous formations, potential for induced seismicity, and difficulties in wellbore stability. In part, the risk is associated with the proximity and activity of faults.

Our study examines the deep structural characteristics and the impact on the hydrocarbon potential of the Pai-Khoi-Altai strike-slip zone region. The purpose of the work is to confirm the presence of a strike-slip zone at the test site using a set of geophysical data and to assess its possible impact on oil and gas potential.

The significance of strike-slip dislocations in predicting oil and gas deposits has been noted by researchers such as G.G. Gogonenkov [15], N.V. Nassonova [16], A.I. Timurziev [11], O.A. Smirnov [17], V.P. Igoshkin [18], V.S. Surkov [19], and others. The manifestation of strike-slip tectonics in the West Siberia sedimentary basin has been identified at both regional levels (based on potential field data and 2D seismic surveys) and large-scale levels (based on 3D seismic surveys). Such structures are often complex in structure and have low amplitude, so their identification requires special processing of seismic data, a good example of which is spectral decomposition of data [20,21]. It has been established that strike-slip movement appears in the form of echelon folds, duplex faults (in plan view) and "flower structures" (in cross-section). Publications mention the influence of the "piston" effect [16], which pressurizes fluids and operates during compression and extension, as well as the formation of "pull-apart" basins [11,22] with developed sedimentary units and the formation of curtain folds and faults, complicating the structural framework of the deposits.

In the current work, using one specific area as an example, we examine the relationship between the formation of deep oil and gas deposits and the positions of principal strike-slip and splaying faults. We started with the assumption that strike-slip zones largely control the formation of fluid-permeable zones within the areas of main and splaying faults. The study area chosen for research is located in the central part of the West Siberia petroleum and gas province and includes map sheets P-43 and P-44, which are intersected by the regional Pai-Khoi-Altai strike-slip zone from southeast to northwest (see Figure 1) [23].

Research has revealed that the Hercynian folded basement has a heterogeneous composition, consisting of metamorphic and variably metamorphosed sedimentary rocks (clay shales, ortho- and metashales, and carbonate rocks) that have been disrupted by intrusions of both acidic and mafic composition. During the Early Mesozoic, uplift movements predominated over a large part of the area, resulting in significant erosion of the pre-Mesozoic strata and the formation of extensive weathering crusts. In the Early Triassic, rift zones began to form and develop, which were filled with sedimentary–volcanogenic deposits. The accumulation of rocks was accompanied by folding and the development of faulted (including strike-slip) dislocations, driven by stresses along the boundaries of adjacent folded areas. Simultaneously with the deepening of rift basins, there were multiple shifts between basin and continental facies [24]. A marine regime was established in most of the region during the Late Jurassic, coinciding with the deposition of sandy–silt–clay sediments. At the end of the Jurassic period, carbonate-rich and bituminous clay silts were deposited in the marine basin [25,26]. During the same period, active movements occurred along the regional Pai-Khoi-Altai strike-slip zone.

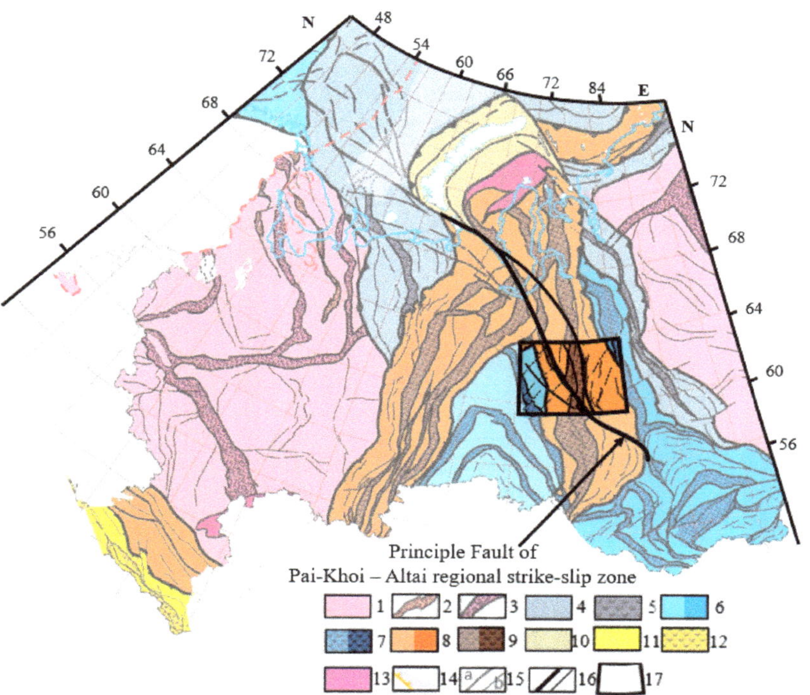

Figure 1. Location map showing the Pai-Khoi-Altai regional strike-slip zone in the West Siberia basin. Symbols: 1–3—ancient platforms; 4–5—Baikalian folded areas and interblock suture zones; 6–7—Caledonian folded; 8–9—Hercynian folded area; 10—Cimmerian folded areas; 11–12—Alpine folded areas; 13—oceanic areas; 14—continental slopes and deep-sea depressions; 15—faults; 16—faults of the Pai-Khoi-Altai strike-slip zone; 17—research site.

2. Hydrocarbon Potential of the Pre-Jurassic Units in the Study Area

Within the considered area, several hydrocarbon-bearing complexes (HBC) are distinguished: the pre-Jurassic (mainly Paleozoic), Jurassic (Tyumen, Vasyugan and Bazhenov), boundary time interval (Upper Jurassic to Lower Cretaceous), and Lower Cretaceous (Figure 2). Most of them are terrigenous (sandstone), but there are accumulations of hydrocarbons in the tops of the Triassic volcanic rocks where their porosity reaches 15%. The major source rocks are unnamed Triassic organic-rich sediments; Togur-Tyumen formation (J1-2) (coals in shales and sandstones); Bazhenov (J3-K1) organic-rich siliceous shale; and Pokur formation (K1-2). In the Caledonian part of the study area, there is a hydrocarbon-bearing "intermediate structural layer" (ISL) located between the consolidated basement and the platform cover. This layer is primarily composed of Late Paleozoic sediments and weathering crusts, partially re-deposited, with zones of organic matter (OM) generation [17,27]. This complex is poorly studied over most of the area due to its deep burial, and it has only been encountered in a limited number of wells.

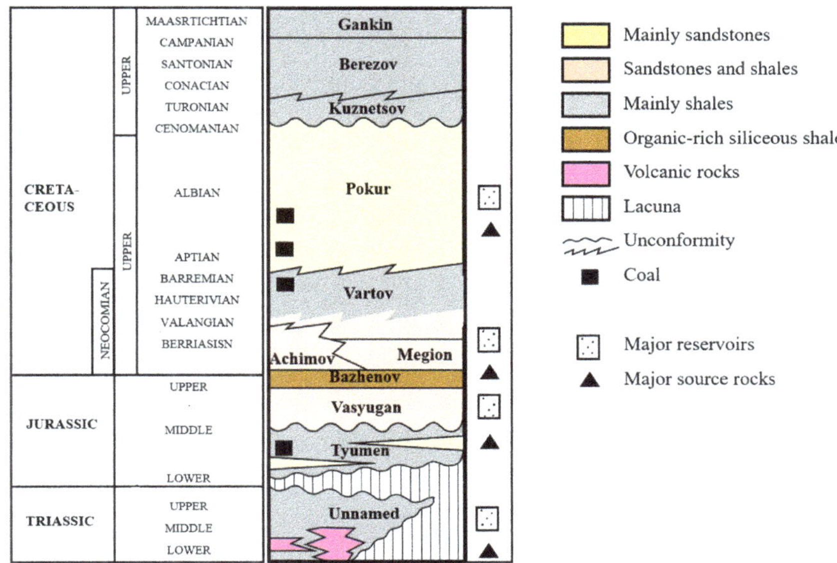

Figure 2. Stratigraphic position of major oil source beds and reservoirs.

In Paleozoic rocks, the following oil deposits have been explored and studied (Figure 3): Severo-Varieganskoye field, where oil accumulations are found in fractured volcanites and shales, as well as in their weathering crusts. Their structure is poorly studied [24].

The Sovetskoye field, reservoir M1 (Paleozoic), consists of loose weathered volcanites of the basic composition and fractured organogenic limestones. The pre-Jurassic HBC here is encountered in five wells at depths ranging from 2702 to 2776 m. Commercial oil flow is obtained from weathered fractured limestones [24].

The Kotygyeganskoye field (Vasyuganskaya oil- and gas-bearing region) is where the oil reservoir is located in fractured–cavernous Paleozoic collectors represented by dolomite breccia [25].

Several fields have oil reservoirs in pre-Jurassic rocks, including the unique Kharampur field, large fields like Bakhilovskoye, Varyngskoye, Verkhnekolik-Eganskoye, as well as the medium-sized and small Festivalnoye and Severnoye fields [25].

The thickness of the intermediate layer, based on the limited number of wells studied, is estimated to be highly variable, averaging several hundred meters [28,29]. Due to its strong metamorphism, ISL was not considered a promising target for oil exploration until the 1970s. However, subsequent assessments have shown that the ISL, in terms of sedimentary fill volumes, is "comparable to the volume of the Mesozoic-Cenozoic cover of the plate" and could be a significant source of organic matter [30]. High prospects for oil and gas in the actual intermediate structural layer are estimated [10] based on the study of oils, bitumens, and organic matter dispersed in rocks and underground waters [31]. Moreover, organic matter could have migrated into overlying horizons through subvertical fluid-permeable zones formed by strike-slips in an anoxic environment [32].

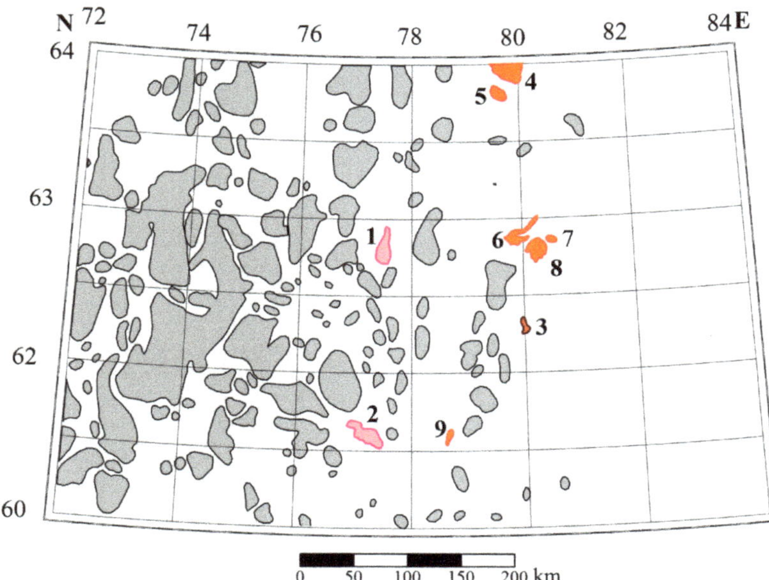

Figure 3. Oil and gas field near the study area. Grey color indicates oil and gas fields; red shades indicates fields with pre–Jurassic reservoirs. Fields are marked with numbers: 1—Severo-Varieganskoye (reservoir—fractured volcanic and basement, pink); 2—Sovetskoye (weathered mafic volcanics and fractured limestones, pink); 3—Kotygyeganskoe (dolomitic breccia of Paleozoic age, dark red). Pre-Jurassic fields with oil and gas pools (orange): 4—Kharampur; 5—Festivalnoye; 6—Bakhilovskoye; 7—Varyngskoye; 8—Verkhnekolik-Yeganskoye; 9—Severnoye field.

3. Characteristics of the Regional Pai-Khoi-Altai Strike-Slip System

The importance of evaluating the effects of tectonic disruptions on the rocks in the lower part of the platform cover and the geosyncline basement is exceptionally high. There are fundamental prerequisites for oil and gas potential in the pre-Jurassic rocks in this area: the presence of oil-source rocks in favorable conditions for hydrocarbon "maturation" and the existence of a cap rock. An important element is the presence and positioning of reservoirs, which could be located in areas where hydrothermal alteration processes occur due to tectonic activity [21,33]. According to various authors, deep-seated tectonic processes led to leaching of feldspar from acidic rocks, karst formation in carbonate rocks, the development of caverns and fractures, and the leaching of minerals from basic lavas (basalts), resulting in the formation of porous and fractured reservoirs [21,34,35]. Traps in such rocks are typically located in structural highs of the basement (with more developed weathering crust), in areas of Triassic rock exposure on basement uplifts [7], and near fault disruptions that create collectors and tectonic-screened traps [36].

To gain a better understanding of the patterns of deposit localization in areas influenced by the principal strike-slip fault and surrounding feathering dislocations, a tectonic reconstruction and detailed analysis of the placement of disjunctive dislocations are required. To describe typical deformations associated with strike-slip faults, the concept of "flower structures," proposed by Harding T.P. in the early 1970s in the process of seismic data analysis [37], is used. Strike-slip dislocations are characterized by an almost vertical narrow zone of deformation at deep levels of the section and an expansion of the dislocation area approaching the surface (see Figure 4). Various-scale strike-slip zones exhibit complex morphology, intense disruption of host rocks, and a wide stratigraphic range of oil and gas potential due to the high fluid permeability of "through" subvertical faults.

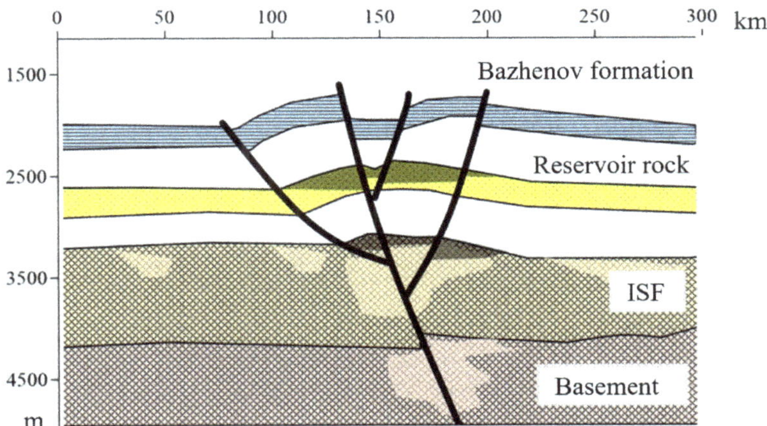

Figure 4. A conceptual diagram of the "flower structure" and fields associated with strike-slip deformations.

The formation of "flower structures" may disrupt the integrity of the reservoir with the creation of tectonically screened traps. Deposits in zones with the development of such structures will have a complex blocky or lens-like character, likely without a single water-oil contact. Furthermore, the actual development of tectonic faults results in the formation of additional porous space in reservoirs of various types [31,38], including in the weathering crust of the basement and weakly metamorphosed sediments in areas adjacent to fault disruptions and fluid alteration zones.

The presence of regional strike-slip faults in Western Siberia was noted by Ivanov K.S. et al. in 2005 [39], based on deep drilling data and potential field interpretation results. In 2007, K.O. Sobornov [40] came to the same conclusion based on regional seismic survey data. Independently, the presence of a strike-slip fault zone was suggested by Egorov A.S. and Antonchik V.I. based on comprehensive studies of the deep structure of the Earth's crust [23,41,42]. Later, the strike-slip fault zone stretching from the Pai-Khoi-Novozemelsk fold area to the Altai-Sayan fold area was studied by various authors [16,40], and it was given various names, such as the "Trans-Eurasian fault" or "Trans-Eurasian left-lateral fault" [43], the "Main strike-slip zone of Western Siberia" [40], "left-lateral fault system" [18] and others. The kinematics of the principle fault of the strike-slip zone is the subject of numerous discussions. Both left- and right-lateral kinematics are assumed. In certain regions, the distribution of a system of Neogene-Quaternary horizontal strike-slips in Paleozoic basement was observed, resulting in complex plicative and disjunctive deformations of the sedimentary cover [15] and perhaps even the fault inversion.

The location of the principal strike-slip fault varies significantly among different authors. In our study, the refined position of the principal and adjacent strike-slip dislocations is presented within an area up to 200 km in width. The structure of tectonic deformations heterogeneities of the basement and sedimentary cover of the West Siberia geosyncline are well manifested in potential fields. Zones of Jurassic sedimentary filling of consolidated basement depressions are reflected in gravimetric and magnetic data, as sedimentary rocks have lower density and magnetization compared to most consolidated basement rocks. Accordingly, the accumulation of sedimentary deposits in troughs during the Mesozoic era should be manifested as negative anomalies in potential fields. Rift structures, on the other hand, where sediments are complemented by dense and magnetized volcanics, manifest as positive anomalies in potential fields (Figure 5).

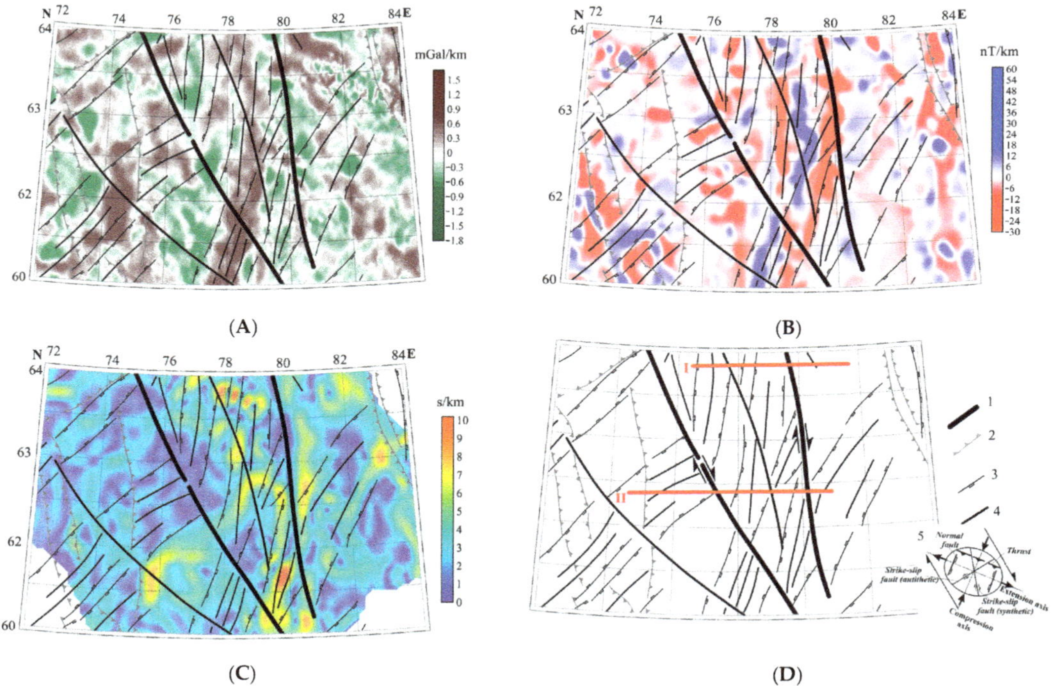

Figure 5. Maps of the vertical derivative of the gravitational field in Bouguer reduction (**A**), the vertical derivative of the anomalous magnetic field (**B**), the horizontal gradient of the isochron reflecting the "B" horizon (**C**), and the results of tectonic decoding (**D**). 1—major faults; 2—fold boundaries; 3—boundaries of horst-graben structures; 4—other fault disruptions; 5—strain ellipsoid. Red lines indicate the positions of seismic sections presented in Figure 6.

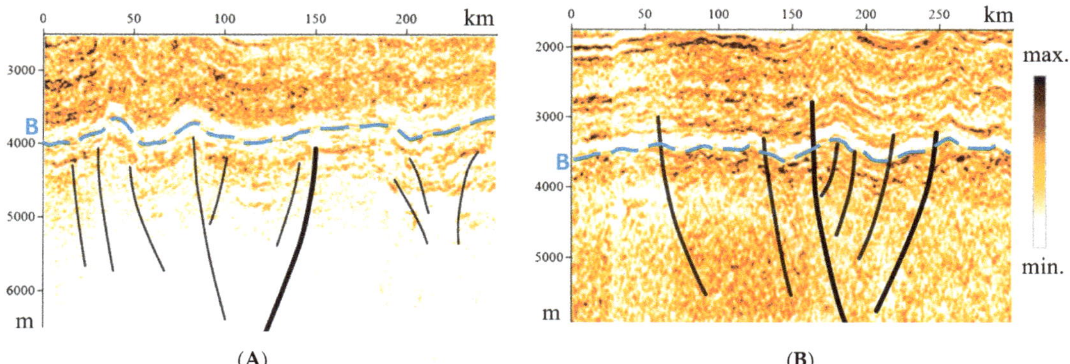

Figure 6. Examples of seismic sections along lines I (**A**) and II (**B**) in the indicators of reflexivity along latitudinal profiles I and II. The principal fault is shown by a bold line and feathering dislocations by thinner lines. The letter B and the dashed line correspond to the Bazhenov formation. See position of the lines in Figure 5D.

The analysis of the localization patterns of the identified principal strike-slip faults and the feathering rift dislocations allows us to draw conclusions about the kinematics of the principal strike-slip fault zone under consideration. To solve this problem, the "strain ellipsoid" model was applied, according to which the total stress in the strike-slip zone is considered to be the result of the superposition of tension and compression stress fields developing along the shear zone [44]. To decipher the deformations mapped on the research area, the right-lateral strike-slip model is most confidently suitable. In this case, stretching deformations (rifts) develop along the short axis of the ellipsoid at an angle of about +45° to the shear axis (Figure 5D).

However, it is necessary to explain the differences in the orientation of the rift structures in the three zones of the research area: the steepest angles relative to principal strike-slip faults are noted in the southwestern and northeastern parts of the research area; the most gentle are in the central zone located between the two principal strike-slip faults. Such picture is explained in the same work by R.G Park [44], in which the author, with reference to earlier works, showed that under conditions of general compression of the strike-slip zone, rifts are oriented at a steeper angle (60° or more) to the strike-slip axis; under extension conditions, the slope of the rifts decreases to 25–30°. This zonality can be used as a predictive sign of the localization of hydrocarbon traps.

Strike-slip structures are revealed in sections of the West Siberia sedimentary basin at different scales. For example, in 3D seismic survey data, strike-slips extending tens of kilometers form systems of disruptions within the Mesozoic layers [15]. When interpreting data from regional 2D seismic survey profiles at deeper levels of the section, the "flower structure" is less clearly observed due to the chaotic distribution of the wave field [45]. Figure 6 schematically illustrates the manifestation of the principal and adjacent feathering dislocations of the strike-slip zone.

Results of the analysis of geological–geophysical data and previous studies allow us to characterize the strike-slip fault as a long-lived structure, formed during the pre-rift stage of the region's evolution and continuing its tectonic activity during the post-rift stage of sediment accumulation in the West Siberia basin [40]. From our point of view, the beginning of the strike-slip formation should be correlated with the final stage of Hercynian orogeny, and active tectonic movements probably continued until the end of the Jurassic. Our research indicates that dislocations associated with the Pai-Khoi-Altai regional strike-slip are primarily manifested at the Mesozoic level of the sedimentary cover and in the consolidated basement. The presence of strike-slip dislocations is almost absent in the layers located above the Bazhenov horizon. Therefore, at the end of the Mesozoic and Cenozoic eras, the considered structure probably did not exhibit activity. Nevertheless, traces of the strike-slip fault zone's influence may be found in the folded structures of overlying horizons.

One of the clear confirmations of the position and structure of the regional strike-slip structure is in 3D seismic data, presented fragmentarily within small areas at the polygon. Data at one of the oil fields, studied by 3D seismic exploration at the level of the reflecting horizon B, indicate that the dominant strike-slip zone of the north–northwest direction. As practice shows [46,47], one of the best ways to demonstrate 3D seismic data is spectral decomposition, so this approach was used for the visualization of the Bazhenov formation surface in Petrel software. Taking into account orientation of en-echelon folds and faults, we can assume right-lateral kinematics. The section along line III confirms the presence and intensity of the flower structure (Figure 7).

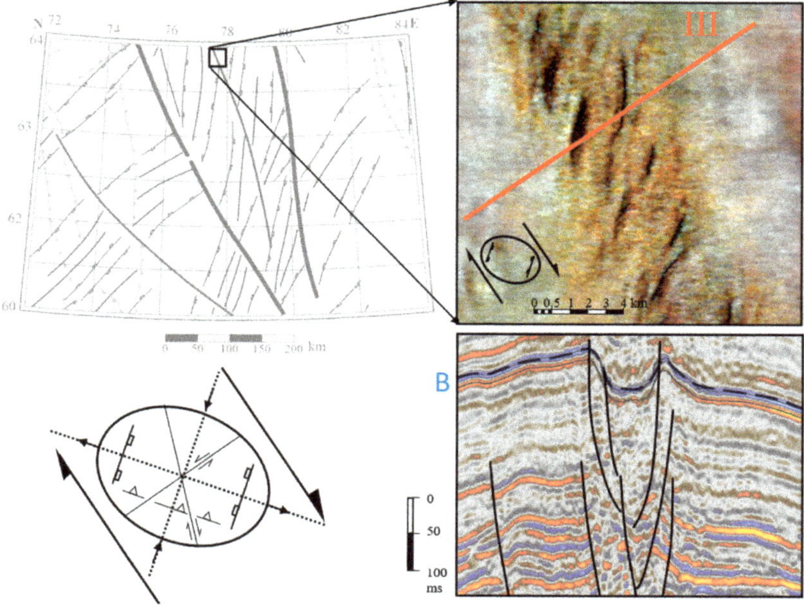

Figure 7. Results of spectral decomposition of 3D seismic exploration data along horizon B in the part of oil field at the study area with the position of the deformation ellipsoid and section along line III.

4. Predictive Assessment of the Study Area Based on Geological, Structural, and Geophysical Data

The study area is characterized by a high density location of oil and gas deposits and, consequently, the area is highly studied by geological and geophysical methods [48]. Third-generation of state geological maps (sheets P-43, P-44) are available online [49]. Three profiles of deep seismic sounding («Bitum», «Kimberlite», and «Quartz») are laid here, as well as a series of regional 2D seismic (CDP) profiles. Within individual deposits and prospective zones, numerous 3D seismic survey areas are localized. Digital gravity and magnetic survey data are available on a scale of 1:1,000,000 [50,51].

Rift structures served as zones of organic matter accumulation and provided heating of these layers to achieve a high degree of catagenesis for oil maturity [52], ensuring their migration capability. Several deposits are localized on the periphery of rifts [53], which makes it reasonable to use gravity and magnetic data as indirect predictive factors. Intrusions of various compositions are distinctly manifested in potential fields as isometric anomalies of different signs and amplitudes.

To study the heterogeneities of the sedimentary cover and conduct predictive assessments for oil and gas, several techniques are applied, such as the method of gravimetric detection of oil and gas (GDOG) [54], the method for identifying "deposit-type" anomalies (DTA) [55] and others. However, direct hydrocarbon reservoir prediction methods based on potential fields generally do not demonstrate high effectiveness.

The comprehensive interpretation method applied in this work, developed by A.S. Dolgal [56], consists of the following procedures: filtering (transformation), zoning (classification), combining the results of methodical interpretation, identification of prospective zones (forecast), and modeling.

Formalized forecasting within the study area is performed using a pattern recognition algorithm that utilizes geological-structural and geophysical parameters available in digital format as factors for training.

The optimal set of geophysical transformations for predictive assessment in the study area includes calculating the horizontal gradient magnitude to enhance the contrast of block boundaries and fault dislocations; calculating the vertical derivative of the field for contrasting the manifestations of local anomaly-forming objects (Oasis Montaj software), which can assist in mapping blocks and sutural zones, fault dislocations, basement protrusions, and granitic massifs; mapping "blocks" identified on gravity and magnetic fields [55], considering uniform amplitude–frequency characteristics of the fields; and analyzing the relief morphology of the basement [57]. Horizontal gradient is the simplest transformation for identifying interblock zones based on geophysical data. In the future, it may be interesting to use improved versions of the horizontal gradient (Enhanced Horizontal Gradient Amplitude (EHGA) [58,59] or Enhanced Total Gradient (ETG) [60]. Their use will help more accurately determine the position of faults—elements of the strike-slip zone.

In addition to tectonic parameters, we introduced the "horizontal gradient module of the isochron reflecting the "B" horizon" as a predictive factor (which can also be improved using various techniques [59]). This corresponds to the angle of inclination of the Bazhenov suite's surface. There is a visual correlation between the hydrocarbon reservoirs of the Jurassic basement and the areas with significant tilting of this surface [61,62]. This can be explained by the impact of the strike-slip fault on the geometry of the Bazhenov suite's roof, leading to the formation of specific traps.

As "training objects", we selected deposits with reservoirs located at various depths, including those in rocks of Jurassic age (deposits 3–9 in Figure 3). The largest of the selected "training objects" is the Kharampur deposit (see Figure 3, deposit 4). The basement here is exposed to a depth of about 4 km, where it is represented by basalts; sediments of the Lower Jurassic age lie unconformably above. Within the deposit, 28 reservoirs of the sheet, massive, and lithological barrier types have been identified, primarily in sandstones. Due to tectonic disruptions, productive layers are fragmented into separate reservoirs with sharp heterogeneity, variability, and fracturing of the reservoirs [63]. On the Upper Kolik-Egan oil and gas condensate deposit (Figure 3, deposit 8), the Paleozoic basement is penetrated by wells to a depth of 3.2 km. It consists of shales, limestones (Lower Carboniferous age), and sedimentary terrigenous formations. In the productive part of the sedimentary sequence, the deposit has 62 layers [35]. The oil reservoir, located on an erosional–tectonic protrusion of the Paleozoic basement, is exposed on the Festival Square (Figure 3, deposit 5). The host rocks are represented by sedimentary and volcanic formations [64]. The heterogeneity of the reference sample, due to the different geological-structural positions of the reference hydrocarbon deposits, necessitates the use of a combination of forecasting methods and a large number of criteria.

The spacing of linear, contoured, and digital predictive features, formed on the basis of a complex of geological and geophysical data, is presented in Table 1. Within the pattern recognition technology, we carried out the following: the formation and mutual alignment of data matrix nodes; the selection of the optimal interpolation procedure with averaging of the final areal distributions; a detailed analysis of the morphology of primary features; the formalization of initially non-digital features; and, specifically, object recognition with training. To implement the latter, a discriminant analysis in an author's modification was applied, aimed at considering a small neighborhood of an individual point in the reference sample as an independent reference object in a multidimensional feature space. The result of the analysis is predictive aureoles distributed over the study area and verified using the K-means method [65] realized in proprietary software.

Joint processing, implemented in the multidimensional space of the mentioned geological–structural and geophysical features, aims to create predictive aureoles of potential object detection, similar to hydrocarbon reference objects. The result of using the modified discriminant analysis in conjunction with the K-means method is generalized by us in the form of a scalar field of the probability distribution of object detection, similar to hydrocarbon reference objects (Figure 8). It can be seen that in the probability value range exceeding 0.5 (highlighted with dark blue shading), the areas extend beyond the boundaries of reference objects, forming the most representative anomalies in the areas of junction of horst–graben boundaries and principal fault zones.

Table 1. Table of structural geological and geophysical features involved in the task of pattern recognition with training.

Nos.	Name of Feature	Type of Feature	The Form of the Attribute Measurability
	Geological and structural parameters		
1	Pai-Khoi-Altai strike-slip fault (eastern principal fault)	linear	distance to the fault (km)
2	Pai-Khoi-Altai strike-slip fault (western principal fault)	linear	distance to the fault (km)
3	Near-strike-slip rifts (grabens)	contour	binary (yes-no)
4	Near-strike-slip horsts	contour	binary (yes-no)
5	Suture zones	linear	distance to the suture zone (km)
6	Relief	metrized	absolute elevations, m
	Geophysical parameters		
7	Magnetic field	metrized	field level (nT)
8	Gravitational field	metrized	field level (mGal)
9	Vertical derivative of magnetic field	metrized	values of field transformants (nTl/m)
10	Vertical derivative of gravitational field	metrized	values of field transformants (mGal/m)
11	Modulus of horizontal gradient of magnetic field	metrized	values of field transformants (nT/m)
12	Modulus of horizontal gradient of gravitational field	metrized	values of field transformants (mGal/m)
13	"Geoblocks" of magnetic field	metrized	conventional unit
14	"Geoblocks" of gravitational field	metrized	conventional unit
15	Horizontal gradient of the map isochron of reflecting horizon B	metrized	proportional to the inclination angle of the surface of the Bazhenov Formation (s/km)

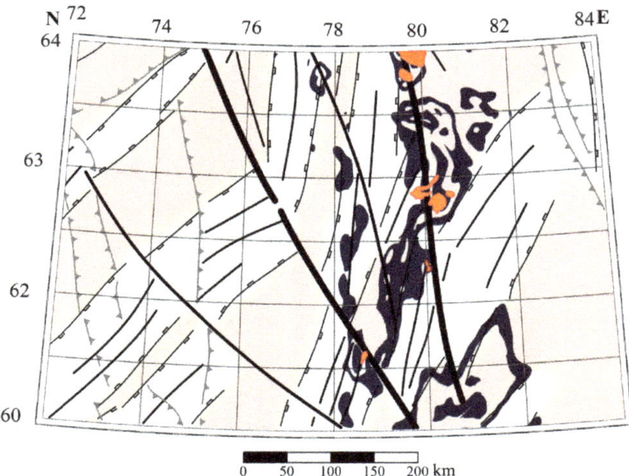

Figure 8. Forecast areas (blue color) calculated on the basis of the full set of features with applied elements of tectonic interpretation (see symbols in Figure 5D) and the contours of reference deposits (see symbols in Figure 3). Cream color emphasizes horst areas; grabens are marked in white.

As evident, the location of reference deposits aligns with the nodes of intersection of the major fault with adjacent extensional structures. When analyzing the interpretation results, we pay attention to the specificity of the distribution of contours of hydrocarbon reference objects, forming groups with a regular spatial step, manifested both along the mentioned fault zone and in its vicinity. Quasi-periodic systems are a manifestation of wave structuring of the geological environment and find application in forecasting deposits of both ore [66,67], and oil and gas. Furthermore, the belt of increased probability of detecting deposits similar to reference ones marks known deposits with local areas of increased morphological variability. Predictive aureoles are elongated in several different directions and have a different distribution pattern in the blocks separated by principal dislocations, confirming the significance of the strike-slip fault as a controlling factor for the hydrocarbon potential of the region's structures. Among the extensional structures, horsts tend to gravitate toward reference objects to the greatest extent.

5. Conclusions

The Pai-Khoi-Altai strike-slip zone covers a vast area and manifests itself as a system of two principal strike-slip dislocations and numerous adjacent horsts and grabens. Fundamental importance has the nature of feathering dislocations; our detailed reconstructions indicate right-lateral kinematics of the principal strike-slip faults (Figures 6 and 7). In addition, this zone extends to the Altai-Sayan folded area, the dislocations of which, including the Kuznetsk Alatau, have right-lateral kinematics. According to seismic survey data, strike-slip dislocations are manifested at depths below the Bazhenov horizon; above it, predominantly fold structures are observed, which can also form traps for hydrocarbons, increasing the significance of the studied structure.

The obtained predictive estimates indicate the high prospects of areas adjacent to the eastern major strike-slip fault zone and adjacent horst structures. This is explained by the probable presence of tectonically weakened fluid-permeable zones; hydrocarbon traps in plicatic dislocations; sources of hydrocarbons in the lower horizons of the sedimentary cover; and favorable conditions for the maturation of organic matter and hydrocarbon generation.

In this work, a methodological approach has been tested, allowing the application of formalized tectonic and geodynamic criteria in automated forecasting of prospective deep-seated hydrocarbon deposits in the study area, including the application of neural network technologies [64,67]. Seismic section data have been applied to verify the identified tectonic elements and to confirm the configuration of the regional Pai-Khoi-Altai strike-slip zone. This direction using specially designed seismic attributes seems useful for this purpose [68,69]. In the future, it is planned to use the steerable pyramid attribute [67] for identifying the weak strike-slip faults and to test the innovate approach presented in papers [70,71] with a detailed assessment of pressure and direction of fluid migration.

Based on the results of geological interpretation of geophysical data, a geological-structural basis has been developed, the formalized elements of which have been applied as factors in object recognition with training on reference deposits. Surfer, Oasis Montaj, Petrel and proprietary programs were used to implement the work. Predictive modeling based on the analysis of deep structure and tectonic dislocations of various kinematics is a justified tool in regions with promising lower structural strata for oil and gas. From the analysis conducted, it is evident that reference deposits are located in those areas of the regional strike-slip zone adjacent to which extensional horst structures are present. Studying the geological processes of accumulation, both in the traditional organic paradigm and from the perspective of an inorganic "mantle" source of oil and gas, along with an analysis of the geophysical manifestations of hydrocarbon criteria, may be the key to discovering new deep-seated hydrocarbon deposits in Western Siberia.

Author Contributions: A.E. proposed the concept of work, methodology, review and editing; V.A. conducted the literature review, software and resources support; N.S. prepared original draft, calculated the transformations of the fields; I.M. fulfilled the forecast, and formal analysis; M.O. corrected the text of the article. All authors have read and agreed to the published version of the manuscript.

Funding: This research received no external funding.

Data Availability Statement: The data used in this work are available on request to the corresponding author due to privacy restrictions.

Acknowledgments: The authors thank the employees of the Geophysics Department of the Mining University for participation in discussions of the work.

Conflicts of Interest: The authors declare no conflict of interest.

References

1. Efficient Use of the Resource and Production Potential of Western Siberia in the Implementation of Russia's Energy Strategy until 2020. Available online: https://neftegaz.ru/science/development/332532-effektivnoe-use-resursno-proizvodstvennogo-potentsiala-zapadnoy-sibiri-v-realizatsii-energe/ (accessed on 19 September 2023).
2. Litvinenko, V. The role of hydrocarbons in the global energy agenda: The focus on liquefied natural gas. *Resources* **2020**, *9*, 59. [CrossRef]
3. Litvinenko, V.S.; Petrov, E.I.; Vasilevskaya, D.V.; Yakovenko, A.V.; Naumov, I.A.; Ratnikov, M.A. Assessment of the role of the state in the management of mineral resources. *J. Min. Inst.* **2023**, *259*, 95–111. [CrossRef]
4. Timurziev, A.I. Oil in the basal complex of the Western Siberia: Reality and alternatives. *Gorn. Vedom.* **2016**, *5–6*, 100–118.
5. Oil production in Western Siberia: Reloaded. Available online: https://vygon.consulting/products/issue-1400/ (accessed on 19 September 2023).
6. Rogachev, M.K.; Mukhametshin, V.V.; Kuleshova, L.S. Improving the efficiency of using resource base of liquid hydrocarbons in Jurassic deposits of Western Siberia. *J. Min. Inst.* **2019**, *240*, 711. [CrossRef]
7. Timurziev, A.I. Fundamental oil of Western Siberia: Reality and alternatives. *Min. Her.* **2016**, *5–6*, 100–118. (In Russian)
8. Surkov, B.C.; Trofimuk, A.A.; Zhero, O.G. Triassic rift system of the West Siberia plate, its influence on the structure and oil and gas content of the platform Mesozoic-Cenozoic cover. *Geol. Geophys.* **1982**, *8*, 3–15. (In Russian)
9. Kontorovich, A.E.; Zolotova, O.V.; Ryzhkova, S.V. Methodology for rational planning of geological exploration works in subsoil areas intended for licensing in the southern regions of Western Siberia. *Geo-Sib* **2007**, *5*, 148–151.
10. Lobova, G.A.; Isaev, V.I.; Kuzmenkov, S.G.; Luneva, T.E.; Osipova, E.N. Oil and Gas Potential of Weathering and Paleozoic Reservoirs in the South-East of Western Siberia (Forecasting Hard-to-Recover Reserves). *Geophys. J.* **2018**, *40*, 73–106. (In Russian)
11. Timurziev, A.I. Recent strike-slip tectonics of sedimentary basins: Tectonophysical and fluid-dynamic aspects (in connection with oil and gas potential) part 1. *Deep Oil* **2013**, *1*, 561–605. (In Russian)
12. Khafizov, S.; Syngaevsky, P.; Dolson, J.C. The West Siberia Super Basin: The largest and most prolific hydrocarbon basin in the world. *AAPG Bull.* **2022**, *106*, 517–572. [CrossRef]
13. Zhdaneev, O.V. Technological sovereignty of the Russian Federation fuel and energy complex. *J. Min. Inst.* **2022**, *258*, 1061–1070. [CrossRef]
14. Kotikov, D.A.; Shabarov, A.N.; Tsirel, S.V. Connecting seismic event distribution and tectonic structure of rock mass. *Gorn. Zhurnal* **2020**, 28–32. [CrossRef]
15. Gogonenkov, G.N.; Timurziev, A.I. Shear deformations in the cover of the West Siberia Plate and their role in the exploration and development of oil and gas fields. *Geol. Geophys.* **2010**, *51*, 384–400. (In Russian)
16. Nassonova, N.V.; Romancheev, M.A. Geodynamic control of oil and gas content by shear dislocations in the east of Western Siberia. *Geol. Oil Gas* **2011**, *4*, 8–14. (In Russian)
17. Smirnov, O.A.; Lukashov, A.V.; Nedosedkin, A.S.; Moiseev, S.A. Stratigraphic pinch-out zones in riphean deposits as promising exploration targets for expanding gazprom's mineral resources base in the central and western parts of the Siberia platform. *GeoBaikal* **2018**, *2018*, 1–5. [CrossRef]
18. Igoshkin, V.J.; Dolson, J.C.; Sidorov, D.; Bakuev, O.; Herbert, R. *New Interpretations of the Evolution of the West Siberia Basin, Russia: Implications for Exploration*; American Association of Petroleum Geologists: Tulsa, OK, USA, 2008; pp. 1–35.
19. Surkov, V.S.; Zhero, O.G. *Foundation and Development of Platforms Change Cover of the West Siberia Plate*; Nedra: Moskow, Russia, 1984; 143p. (In Russian)
20. Naseer, M.T. Spectral decomposition-based quantitative inverted velocity dynamical simulations of early cretaceous shaly-sandstone natural gas system, Indus Basin, Pakistan: Implications for low-velocity anomalous zones for gas exploration. *Nat. Resour. Res.* **2023**, *32*, 1135–1146. [CrossRef]
21. Naseer, M.T. Seismic attributes and quantitative stratigraphic simulation' application for imaging the thin-bedded incised valley stratigraphic traps of Cretaceous sedimentary fairway, Pakistan. *Mar. Pet. Geol.* **2021**, *134*, 105336. [CrossRef]
22. Razmanova, S.V.; Andrukhova, O.V. Oilfield service companies as part of the economy digitalization: Assessment of the prospects for innovative development. *J. Min. Inst.* **2020**, *244*, 482–492. [CrossRef]
23. Egorov, A.S. *Deep Structure and Geodynamics of the Lithosphere of Northern Eurasia according to the Results of Geol.-Geofiz*; Modeling along the Geotraverses of Russia: St. Petersburg, Russia, 2004; pp. 13–28. (In Russian)
24. Astapov, A.P.; Braduchan, Y.V.; Borovsky, V.V.; Voronin, A.S.; Gorelina, T.E.; Kovrigina, E.K.; Lebedeva, E.A.; Markina, T.V.; Matyushkov, A.D.; Rubin, L.I.; et al. *State Geological Map of the Russian Federation. Scale 1: 1,000,000 (Third Generation). Series West Siberia. Sheet P43—Surgut. Explanatory Note*; VSEGEI Map Factory: St. Petersburg, Russia, 2012; 342p.
25. Faibusovich, Y.E.; Voronin, A.S.; Markina, T.V.; Rubin, L.I.; Chekanov, V.I. *State Geological Map of the Russian Federation, Scale 1: 1,000,000 (Third Generation). Series West Siberia. Sheet P-44—P. Wah. Explanatory note*; VSEGEI Publishing House: St. Petersburg, Russia, 2020; 192p.
26. Gusev, E.A. Results and prospects of geological mapping of the Arctic shelf of Russia. *J. Min. Inst.* **2022**, *255*, 290–298. [CrossRef]
27. Punanova, S.A.; Shuster, V.L. A new look at the prospects for oil and gas potential of deep-seated pre-Jurassic deposits in Western Siberia. *Georesources* **2018**, *20*, 67–80. [CrossRef]

28. Ulmishek, G.F. Petroleum Geology and Resources of the West Siberia basin, Russia. U.S. Geological Survey Bulletin 2201-G. 2003. Available online: https://pubs.usgs.gov/bul/2201/G/ (accessed on 19 September 2023).
29. Stupakova, A.V.; Sokolov, A.V.; Soboleva, E.V.; Kurasov, I.A.; Bordyug, E.V.; Kiryukhina, T.A. Geological study and oil and gas potential of the Paleozoic deposits of Western Siberia. *Georesources* **2015**, *2*, 63–76. (In Russian) [CrossRef]
30. Vazhenina, O.A.; Trigub, A.V. Forecast of pre-Jurassic basement rock filling with hydrocarbons based on oil and gas systems modeling (Western Siberia). *Geol. Oil Gas* **2018**, *4*, 39–51. (In Russian)
31. Big Oil of Siberia. Available online: https://www.igm.nsc.ru/images/history/book/21_igig_total_245-260.pdf (accessed on 19 September 2023).
32. Ageev, A.S.; Ilalova, R.K.; Duryagina, A.M.; Talovina, I.V. A link between spatial distribution of the active tectonic dislocation and groundwater water resources in the Baikal-Stanovaya shear zone. *Min. Inf. Anal. Bull.* **2019**, *5*, 173–180. [CrossRef]
33. Sohrabi, A.; Nadimi, A.; Talovina, I.V.; Safae, I.H. Structural model and tectonic evolution of the fault system in the southern part of the Khur area, Central Iran. *Notes Min. Inst.* **2019**, *236*, 142–152. [CrossRef]
34. Khisamutdinova, A.I.; Alekseeva, P.A.; Romanchuk, A.F.; Miroshnichenko, D.E.; Kerusov, I.N.; Bakulin, A.A.; Kucheryavenko, D.S.; Kiryanova, T.N. The structure and forecast of reservoirs of the pre-Jurassic complex in one of the sections of the Elizarovsky trough of the Frolov megadepression according to 3d seismic data in the absence of well information. *Oil Gas Geology. Theory Pract.* **2021**, *16*, 1–17. [CrossRef]
35. Talovina, I.V.; Krikun, N.S.; Yurchenko, Y.Y.; Ageev, A.S. Remote sensing techniques in the study of structural and geotectonic features of Iturup island (the Kuril islands). *Notes Min. Inst.* **2022**, *254*, 158–172. [CrossRef]
36. Shuster, V.L.; Punanova, S.A. Prospects for the oil and gas potential of deep Jurassic and pre-Jurassic deposits in the north of Western Siberia in unconventional traps. *Georesources* **2021**, *23*, 30–41. [CrossRef]
37. Huang, L.; Liu, C. Three Types of Flower Structures in a Divergent-Wrench Fault Zone: The Third Type of Flower Structures. *J. Geophys. Res. Solid Earth* **2017**, *122*. [CrossRef]
38. Lavrova, N.V. To the question of the evolution of deformation zones in platform conditions on the example of the Kungur Ice Cave (Urals). *Notes Min. Inst.* **2020**, *243*, 279–284. [CrossRef]
39. Ivanov, K.S.; Fedorov, Y.N.; Kormiltsev, V.V. Shear system in the basement of the West Siberia megabasin. *Dokl. Akad. Nauk.* **2005**, *405*, 371–375. (In Russian)
40. Smirnov, O.A.; Borodkin, V.N.; Lukashov, A.V.; Plavnik, A.G.; Trusov, A.I. Regional model of riftogenesis and structural-tectonic zoning of the north of Western Siberia and the South Kara syneclise according to the complex of geological and geophysical studies. *Neftegazov. Geol. Theory Pract.* **2022**, *17*, 1. Available online: http://www.ngtp.ru/rub/2022/1_2022.html (accessed on 19 September 2023).
41. Egorov, A.S.; Ageev, A.S. Tectonic zoning and sequence of formation of consolidated crust of Northern Eurasia and adjacent shelf. In Proceedings of the Materials of the LIV Tectonic Meeting Tectonics and Geodynamics of the Earth's Crust and Mantle: Fundamental Problems, Moscow, Russia, 31 January–4 February 2023; pp. 155–160.
42. Egorov, A.S.; Smirnov, O.E.; Vinokurov, I.Y.; Kalenich, A.P. Similarities and differences in the structure of the Ural and Paikhoi-Novaya Zemlya folded regions. *Notes Min. Inst.* **2013**, *200*, 34–42.
43. Sobornov, K.O. Late Paleozoic-Cenozoic structural development and oil content of Western Siberia. In Proceedings of the International Conference of Geophysicists and Geologists, Tyumen, Russia, 4–7 December 2007; pp. 11–14. (In Russian).
44. Park, R.G. *Geological Structures and Moving Plates*; Blackie: Glasgow, UK; London, UK, 1988; 337p.
45. Mingaleva, T.; Gorelik, G.; Egorov, A.; Gulin, V. Correction of depth-velocity models by gravity prospecting for hard-to-reach areas of the shelf zone. *MIAB Min. Inf. Anal. Bull.* **2022**, 77–86. [CrossRef]
46. Naseer, M.T. Appraisal of tectonically-influenced lowstand prograding clinoform sedimentary fairways of Early Cretaceous Sember deltaic sequences, Pakistan using dynamical reservoir simulations: Implications for natural gas exploration. *Mar. Pet. Geol.* **2023**, *151*, 106166. [CrossRef]
47. Naseer, M.T.; Naseem, S.; Shah, M.A. Simulating the stratigraphy of meandering channels and point bars of Cretaceous system using spectral decomposition tool, Southwest Pakistan: Implications for petroleum exploration. *J. Pet. Sci. Eng.* **2022**, *212*, 110201. [CrossRef]
48. Zhdaneev, O.V.; Zaitsev, A.V.; Lobankov, V.M. Metrological support of equipment for geophysical research. *Notes Min. Inst.* **2020**, *246*, 667–677. [CrossRef]
49. Digital Catalog of the State Geological Maps of the Russian Federation scale 1:1,000,000 (Third Generation). Available online: https://www.vsegei.ru/ru/info/pub_ggk1000-3/ (accessed on 19 September 2023).
50. The World Magnetic Model. Available online: https://www.ngdc.noaa.gov/geomag/WMM/ (accessed on 19 September 2023).
51. WGM2012 Global Model. Available online: https://bgi.obs-mip.fr/data-products/grids-and-models/wgm2012-global-model/ (accessed on 19 September 2023).
52. Marin, Y.B. On mineralogical studies and the use of mineralogical information in solving petro- and ore genesis problems. *Geol. Ore Depos.* **2021**, *63*, 625–633. [CrossRef]
53. Khaliulin, I.I.; Shelikhov, A.P.; Yaitsky, N.N. Analysis of the relationship between anomalies of potential fields and the structural framework of the sedimentary cover of Western Siberia. In *Questions of Theory and Practice of Geological Interpretation of Geophysical Fields: Materials of the 47th Session of the International Scientific Seminar of D. G. Uspensk -V. N. Strakhova*; VSU: Voronezh, Russia, 2020; pp. 288–290. (In Russian)

54. Mikhailov, I.N.; Veselov, K.E. Results and place of gravity exploration in the problem of direct forecasting of oil and gas. *Appl. Geophys.* **1989**, *120*, 147–153. (In Russian)
55. Kulikov, G.N.; Mavrichev, V.G. Aeromagnetic exploration for oil and gas. *Geophysics* **1995**, *2*, 37–42. (In Russian)
56. Dolgal, A.S. Computer Technologies for the Interpretation of Gravitational and Magnetic Fields in Mountainous Terrain. Ph.D. Thesis, Russian Academy of Sciences, Moscow, Russia, 2002; 343p. (In Russian).
57. Ananyev, V.V.; Grigoryev, G.S.; Gorelik, G.D. Physical modeling of the Bazhenov formation in combination with CSEM and seismic methods. *E3S Web Conf.* **2021**, *266*, 07003. [CrossRef]
58. Stephen, E., II; Mohammed, E.A.; Ogiji-Idaga, A.; Anthony, A.; Luan, P.; Kamal, A.; David, G.-O.; Ubong, B.; Fnais, M. Application of the enhanced horizontal gradient amplitude (EHGA) filter in mapping of geological structures involving magnetic data in Southeast Nigeria. *J. King Saud Univ. Sci.* **2022**, *34*, 102288. [CrossRef]
59. Luan, P.; Erdinc, O.; Özkan, K.; Phan, T.; Dat, N.; Quynh, V.; Sang, L.; Do, T. Determination of subsurface lineaments in the Hoang Sa islands using enhanced methods of gravity total horizontal gradient. *Vietnam J. Earth Sci.* **2022**, *44*, 1–15. [CrossRef]
60. Prasad, K.; Luan, P.; Anand, S.; Ahmed, M.E.; Kamal, A.; Fnais, M.; David, G.-O. A Novel Enhanced Total Gradient (ETG) for Interpretation of Magnetic Data. *Minerals* **2022**, *12*, 1468. [CrossRef]
61. Mingaleva, T.A.; Shakuro, S.V.; Egorov, A.S. Analysis of studies, observed results of geophysical surveys in areas contaminated with LNAPL. *Russ. J. Earth Sci.* **2023**, *1*, 1–16. [CrossRef]
62. Oil and Gas Potential of Pre-Jurassic Rock Complexes in Western Siberia. Available online: https://geonedra.ru/2020/neftegazonosnost-doyurskogo-kompleksa/ (accessed on 19 September 2023).
63. Reservoirs of Hydrocarbons in Erosional-Tectonic Protrusions of Pre-Jurassic Rocks in the Southeastern Part of the Western Siberia Plate. Available online: https://neftegaz.ru/science/development/332012-rezervuary-uglevodorodov-v-erozionno-tektonicheskikh-vystupakh-doyurskikh-porod-yugo-vostochnoy-chas/ (accessed on 19 September 2023).
64. Movchan, I.V.; Yakovleva, A.A.; Frid, V.; Movchan, A.B.; Shaygallyamova, Z.I. Modeling of seismic assessment for large geological systems. *Philos. Trans. R. Soc. A Math. Phys. Eng. Sci.* **2022**, *380*. [CrossRef]
65. Mokeev, V.V.; Tomilov, S.V. On the efficiency of image analysis and recognition using the principal component method and linear discriminant analysis. In *Vestnik SUSU, vol. 13, no. 3. Series "Computer Technologies, Control, Radio Electronics"*; South Ural State University: Chelyabinsk, Russia, 2013; pp. 61–70.
66. Ageev, A.; Egorov, A.; Krikun, N. The principal characterized features of Earth's crust within regional strike-slip zones. In *Advances in Raw Material Industries for Sustainable Development Goals*; CRC Press: London, UK, 2020; pp. 78–83.
67. Kochnev, A.A.; Kozyrev, N.D.; Kochneva, O.E.; Galkin, S.V. Development of a comprehensive methodology for the forecast of effectiveness of geological and technical measures based on machine learning algorithms. *Georesursy* **2020**, *22*, 79–86. [CrossRef]
68. Tang, Q.; Tang, S.; Luo, B.; Luo, X.; Feng, L.; Li, S.; Wu, G. Seismic Description of Deep Strike-Slip Fault Damage Zone by Steerable Pyramid Method in the Sichuan Basin, China. *Energies* **2022**, *15*, 8131. [CrossRef]
69. Naseer, M.T. Seismic attributes and reservoir simulation' application to image the shallow-marine reservoirs of Middle-Eocene carbonates, SW Pakistan. *J. Pet. Sci. Eng.* **2020**, *195*, 107711. [CrossRef]
70. Ze, T.; He, Z.; Alves, T.M.; Guo, X.; Gao, J.; He, S.; Zhao, W. Structural inheritance and its control on overpressure preservation in mature sedimentary basins (Dongying depression, Bohai Bay Basin, China). *Mar. Pet. Geol.* **2022**, *137*, 105504. [CrossRef]
71. Jing, S.; Alves, T.M.; Omosanya, K.O.; Hales, T.C.; Ze, T. Tectonic evolution of strike-slip zones on continental margins and their impact on the development of submarine landslides (Storegga Slide, northeast Atlantic). *GSA Bull.* **2020**, *132*, 2397–2414. [CrossRef]

 *minerals*

Article

Electrical Prospecting of Gold Mineralization in Exhalites of the Digo-Digo VMS Occurrence, Central Brazil

Pedro Augusto Costa do Amaral [1], Welitom Rodrigues Borges [1,*], Catarina Labouré Bemfica Toledo [1], Adalene Moreira Silva [1], Hygor Viana de Godoy [1] and Marcelo Henrique Leão Santos [2]

[1] Institute of Geosciences, University of Brasilia (UnB), Darcy Ribeiro Campus, Brasília 70910-900, DF, Brazil; pedro.amaral.unb@gmail.com (P.A.C.d.A.); catarinatoledo@unb.br (C.L.B.T.); adalene@unb.br (A.M.S.); hygorv@gmail.com (H.V.d.G.)

[2] Faculty of Planaltina, University of Brasilia (UnB), Planaltina 73345-010, DF, Brazil; marcelo.leao@unb.br

* Correspondence: welitom@unb.br; Tel.: +55-619-9646-9090

Abstract: The greenstone belts of the Crixás-Goiás Domain are economically important due to significant epigenetic gold deposits and the potential for under-researched syngenetic deposits. The gold occurrences associated with the volcanogenic massive sulfide (VMS) deposits in the region are documented only in the volcanoclastic rocks of the Digo-Digo Formation, Serra de Santa Rita greenstone belt. The objective of this work is to discuss the efficiency of the induced polarization methods in the time and frequency domains for differentiating and identifying potentially mineralized zones in the exhalites associated with the VMS-type gold of the Digo-Digo Formation. Data were acquired using a multielectrode resistivity meter with the dipole–dipole array and 10 m spacing between electrodes, as well as different current injection times (250, 1000, and 2000 ms). After the electrical data processing and inversion, the sections were integrated into ternary red-green-blue and cyan-magenta-yellow models to highlight areas of high chargeability, low resistivity, and high metal factor (frequency domain) and, thus, map the higher potential zones to host polarizable metallic minerals. The geological–geophysical model elaborated from the correlation of electrical and surface geological data allowed us to identify four anomalous areas related to potential mineralized zones. The geological data confirm that two targets are associated with the geological contacts between metamafic and intermediate metavolcanic units and the exhalative horizon. One of the targets coincides with a sulfide-rich exhalative horizon (VMS), while the last target occurs in the occurrence area of metaultramafic rocks, where gold mineralization occurrences have not been previously described, being a promising target for future investigations.

Keywords: electrical prospecting; VMS deposits; Digo-Digo formation

Citation: do Amaral, P.A.C.; Borges, W.R.; Toledo, C.L.B.; Silva, A.M.; de Godoy, H.V.; Leão Santos, M.H. Electrical Prospecting of Gold Mineralization in Exhalites of the Digo-Digo VMS Occurrence, Central Brazil. *Minerals* **2023**, *13*, 1483. https://doi.org/10.3390/min13121483

Academic Editors: Stanisław Mazur and Amin Beiranvand Pour

Received: 21 June 2023
Revised: 11 October 2023
Accepted: 17 October 2023
Published: 24 November 2023

1. Introduction

The deposits of volcanogenic massive sulfide (VMS) usually occur as polymetallic massive sulfide (>40%) lenses rich in base (Cu, Zn, Pb) and precious (Au and Ag) metals. VMS is generated from the focused discharge of hot, metal-rich hydrothermal fluids associated with hydrothermal ocean floor convection. For this reason, VMS deposits are generally classified as "Exhalative" deposits. These deposits can also form as either exhaling stratiform or replacement bodies, and commonly have stockwork-type mineralizations associated with the proximal footwall [1].

There is a strong contrast in physical properties between exhalative horizons and their host rocks due to the presence of large chalcopyrite and pyrrhotite amounts [2–4]. Such properties include density, magnetic intensity and susceptibility, gravity, electrical resistance, and acoustical velocity [5]. Due to these different physical properties, gravimetry, self-potential, magnetics, electromagnetics, and induced polarization methods have been successfully used for prospecting VMS deposits [5–20].

Recently, electrical resistivity and induced polarization surveys have been successfully applied to generate polymetallic targets in the Skellefte mineral district, northern Sweden, where more than 80 VMS deposits have already been discovered [11,21]. The efficiency of applying electrical methods to detect VMS mineralization has been previously attested in works where data acquisition was performed above already known buried ore bodies through drilling [7,21].

In Brazil, the records of VMS-type deposits are scarce [22–29], possibly due to the lack of detailed geological mapping and geophysical surveys in areas with geological potential for these deposits to occur.

The objective of this work is to show the efficiency of the induced polarization in the time and frequency domains for differentiating and identifying potentially mineralized zones in exhalites associated with the VMS-type gold in the Digo-Digo Formation, Serra de Santa Rita greenstone belt, Goiás, Brazil.

2. Area Geology

2.1. Crixás-Goiás Domain

The Crixás-Goiás Domain (CGD) [30], formerly the Archean-Paleoproterozoic Terrain of Goiás [31,32], is located in the central–west portion of Goiás, and is part of the Tocantins Province. The province represents a large Brasiliano/Pan-African orogen of the South American Platform formed by a collision between the Amazonian, Paranapanema, and São Francisco/Congo cratons during the Brasiliano Orogeny, which led to the amalgamation of the West Gondwana (Figure 1).

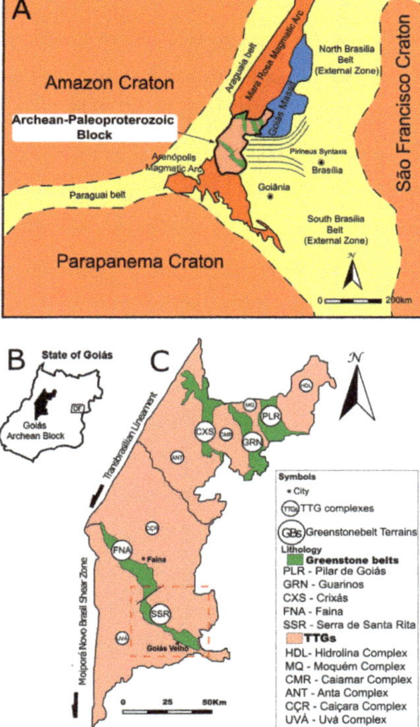

Figure 1. (**A**) Geotectonic scenario and main subdivisions of the Tocantins Province, Central Brazil, emphasizing the Crixás-Goiás Domain in the Brasília Belt (modified from 31). (**B**) Map of Goiás showing the CGD. (**C**) CGD limits and subdivisions; red polygon represents the area of Figure 2 (modified from 35).

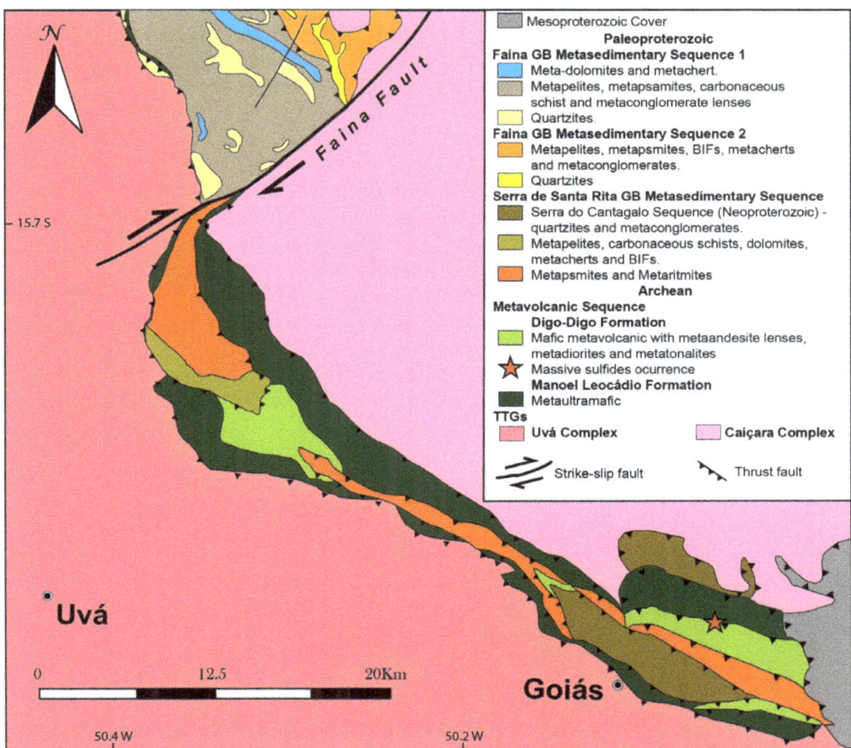

Figure 2. Geological map of the Serra de Santa Rita greenstone belt (modified from 38); the study area corresponds to the massive sulfide occurrence on the map (red star).

The CGD was interpreted as an allochthonous microplate amalgamated on the western margin of the Brasília Belt, in the last evolution stages of the Neoproterozoic orogen [32]. The CGD extends for about 18,000 km², consisting of approximately 80% of tonalite–trondhjemite–gneisses-type complexes (TTGs) and 20% of narrow greenstone belts terrains with Archean to Paleoproterozoic ages [33,34].

The granite–gneiss terrains correspond to six main complexes. In the northern portion, Anta, Caiamar, Moquém, and Hidrolina are found, whereas in the south of the block are the Caiçara and Uvá complexes. These complexes have different structural frameworks, ages, and rock associations [35]. The greenstone belts correspond to five narrow and irregular belts. In the north part of CGD are located the greenstone belts of Crixás, Pilar, and Guarinos, while in the south portion are situated the Faina and Serra de Santa Rita greenstone belts.

2.2. Serra de Santa Rita Greenstone Belt

In the southern portion of the Crixás-Goiás Domain, the Faina and Serra de Santa Rita greenstone belts are juxtaposed along an N 30° E dextral fault (Figure 2), extending approximately 100 km in length and up to 7 km in width [36,37]. The two greenstone belts have similar basal sequences, composed of volcanic sequences, but the upper sedimentary records are contrasting and represent different depositional regimes with a probable Paleoproterozoic age [35,36,38,39].

The basal sequence of the Serra de Santa Rita greenstone belt consists of volcanic rocks of ultramafic composition, which characterize the Manoel Leocádio Formation, followed by mafic and intermediate volcanic rocks of the Digo-Digo Formation [36]. The basal volcanic sequence was dated in 2.96 and 2.92 Ga [38] and covered by a sedimentary sequence formed by carbonaceous phyllites, with intercalations of metachert, calc-shists, banded iron formations, and metagraywacke composing a typical turbiditic sequence [38].

2.3. Digo-Digo Formation

The base of the Digo-Digo Formation consists of mafic metavolcanic rocks represented by amphibole schists with varying proportions of actinolite, albite, epidote, and quartz, intercalated with thin layers of feldspathic metatuffs, talc schists, carbonaceous metapelites, and metacherts [36]. The upper portion of the Digo-Digo Formation consists of sericite–chlorite–quartz schists with relicts of original pyroclastic textures ranging from recrystallized ash to coarse tuffs, including abundant layers with lapilli-sized fragments. The composition of the felsic tuffs varies from predominantly dacitic to rhyolitic [36]. The volcanoclastic rock granulometry tends to increase towards metric exhalative horizons composed of metachert banded with carbonaceous phyllite, rich in pyrite and gold (0.5 to 2 ppm), or intervals of massive pyrite [35,40]. U-Pb dating in zircons of andesitic metavolcanic rocks of the Digo-Digo formation indicates an age of 2.97 Ga [41].

2.4. Local Geology

The study area is located at the confluence of Digo-Digo Creek and the Vermelho River, where the upper portion of the Digo-Digo Formation outcrops and massive sulfide layers are exposed. To characterize the main geological units and structures, the area was mapped at a 1:10,000 scale (Figure 3A).

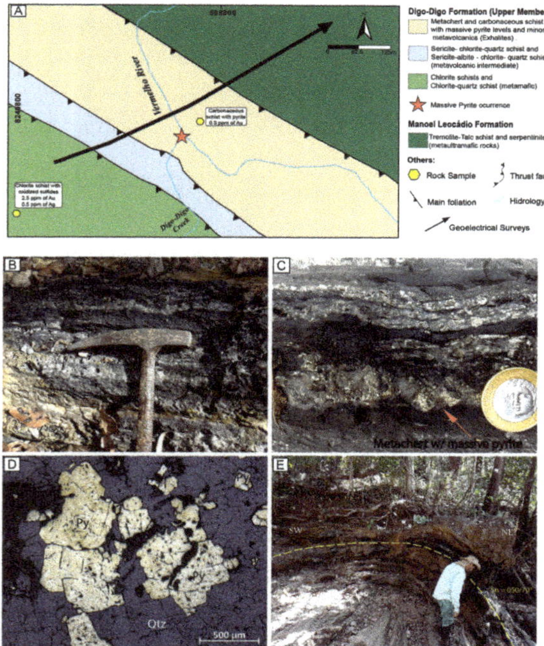

Figure 3. (**A**) Detailed geological map of the confluence region of the Digo-Digo Stream with the Vermelho River. (**B**) Carbonaceous metachert layer with centimetric scale massive sulfides. (**C**) Detail of Figure 3(**B**), with bands/horizons of coarse-grained pyrite. (**D**) Polished section of metachert with pyrite aggregates showing modal composition of 35%–45% pyrite and 45%–55% quartz. (**E**) Carbonaceus metachert layer folded.

In the area, the Digo-Digo Formation is characterized by three distinct stratigraphic units: (i) a mafic metavolcanic unit; (ii) an intermediate metavolcanic unit; and (iii) an exhalative unit (see Figure 3A). The boundaries between these units are delineated by NW-verging thrust faults. Within all these units, the primary tectonic foliation trends in the northwest direction and typically dips between 300 and 700 degrees to the southwest.

The mafic metavolcanic unit outcrops in the southwest portion of the area and comprises mafic schists and amphibolites. The mafic schists consist of varying proportions of chlorite, epidote, actinolite, and hornblende, sometimes with the presence of magnetite, disseminated sulfides, and boxwork textures. The amphibolites are composed of hornblende, plagioclase, ±epidote, ±chlorite, and exhibit tectonic foliation marked by the preferential orientation of hornblende. The metamafic rocks exhibit a transitional geochemical signature, enriched with light rare earth elements and pronounced negative anomalies of Nb and Ti, similar to the patterns of oceanic arc basalts. Additionally, the amphibolites have a high Nb content, showing similarities with Nb-enriched basalts generated in modern subduction zone environments.

The intermediate metavolcanic unit is composed of metandesites, metatuffs, and rare subvolcanic bodies of metadiorite. Typically, discontinuous layers of metandesite are interlayered with metatuffs. In less deformed areas, the metandesites display a porphyritic texture, consisting of euhedral to subhedral phenocrysts of plagioclase (20 to 35%), immersed in a fine matrix (comprising 65 to 80% of the rock) composed of plagioclase, quartz, muscovite, biotite, and chlorite. When intensely deformed, the original texture is entirely obliterated, giving way to pervasive tectonic foliation.

Andesitic metatuffs levels occur intercalated with the metandesites and may represent different episodes of explosive eruption or, alternatively, duplicated and disrupted layers due to later tectonic processes. In less deformed regions, a diagnostic feature of the metatuffs is the high content of plagioclase crystals (70%–80%) surrounded by a fine-grained matrix (20%–30%).

Locally, bodies of diorite occur in association with intermediate and mafic volcanic rocks. U-Pb analyses on zircons from the metadiorite and metandesite indicate crystallization ages of 2.96 and 2.97 Ga [41]. The intermediate metavolcanic rocks exhibit calc-alkaline affinity with geochemical signatures similar to high-silica adakites [41]. The geochemical and isotopic signatures of the mafic and intermediate metavolcanic rocks suggest that the evolution of this supracrustal sequence is linked to the development of a Mesoarchean volcanic arc [38,41].

The exhalative horizon of the Digo-Digo Formation outcrops near the confluence of the Digo-Digo Creek with the Vermelho River. It is represented by intercalations of centimeter- to meter-scale layers of metavolcanics, metacherts, and carbonaceous phyllites. Massive sulfide lenses, with traces of gold, are hosted within this exhalative horizon (Figure 3B,C). Decimeter-scale lenticular levels of metachert are interlayered with chlorite schists. These rocks are generally composed of very fine-grained recrystallized quartz (~30 μm), possibly containing subordinate amounts of chlorite/muscovite and opaque minerals, as well as carbonaceous material. The sulfide lenses are hosted in carbonaceous metachert and comprise 30 to 50% pyrite, 40 to 70% quartz with muscovite and ilmenite as trace minerals.

The main outcrop of the exhalative horizon, which hosts the massive sulfide lenses, is strongly deformed and exhibits metric to decametric tight asymmetrical folds verging towards NE (Figure 3D). Apart from the notable gold content found in rock samples from the massive sulfide lenses, petrographic studies conducted in the exhalative horizons near the Vermelho River confluence have uncovered gold grains exceeding 5 μm in size. These grains are hosted within centimeter-scale sericite–albite–chlorite schists of the intermediate metavolcanic unit [41].

On the northern margin of Vermelho River, the exhalative horizon is in tectonic contact with the metaultramafic rocks of the Manoel Leocádio Formation, primarily represented by talc schists and talc–tremolite schists (Figure 3A).

3. Methodology

3.1. Data Acquisition

The electrical resistivity and chargeability data were acquired over a 550 m section where massive outcropping sulfide layers are exposed. Three sets of data were acquired in the same spatial positions due to the difficult access and crossing of the banks of the Vermelho River (Figure 3A). The SW-NE survey line was positioned perpendicularly to the main direction of the regional geological structures to reduce the effects of three-dimensional structures that cannot be accurately modeled via 2D inversions, thus causing distortions in the electrical resistivity models [29,42–44].

The apparent resistivity and chargeability data were acquired using the Syscal Pro Switch 72 multi-electrode resistor (Iris Instruments, Orléans, France; Figure 4A). Field acquisitions were performed using the time and frequency domain IP methods with dipole–dipole array and 10 m spacing electrodes. In the IP time mode, non-polarizable electrodes were used at the electrical potential recording points and stainless-steel electrodes at the electric current injection points.

Figure 4. (**A**) Resistivity meter Syscal Pro 72, prepared for acquiring chargeability measurements. (**B**) Stainless steel electrode connected to a multi-cable to perform resistivity measurements (250 and 1000 ms). (**C**) Survey line for measuring chargeability on the northeast bank of the Vermelho River. (**D**) Position of the Syscal Pro during resistivity data (250 and 1000 ms) acquisition on the southwest bank of the Vermelho River. (**E**) Spatial configuration of the electric survey line.

In the IP frequency, 250 and 1000 ms electrical resistivity recording periods were applied, using multi-electrode systems with stainless steel electrodes (Figure 4B). In both surveys, a saline water compound was added to the injection/current record and electrical potential points to reduce the high values of contact resistance.

The survey sections performed with multielectrode systems (250 and 1000 ms) reached 56 m depth, while the resistivity and chargeability (2000 ms) acquisition carried out simultaneously using common cables and porous potential electrodes reached only 32.5 m depth, limited by the maximum of 10 depth investigation levels of the equipment (Figure 4C–E).

3.2. Data Modeling

The electrical resistivity and chargeability data were reduced and filtered separately using the software Prosys 2.0 (Iris Instruments):

Electrical resistivity: (1) conversion to absolute values (the negative records correspond to negative electrical potential measurements that result from a reversal of electrical current); (2) insertion of topography at each electrode point; and (3) filtering of electrical resistivity outliers, high values (spikes), null values (caused by the lack of electrical current that sometimes occurred due to accidental disconnection between the electrodes and cables). Most of the outliers correspond to the Vermelho River passage places and where colluvial sediments (boulders) are present on the surface (Figure 5).

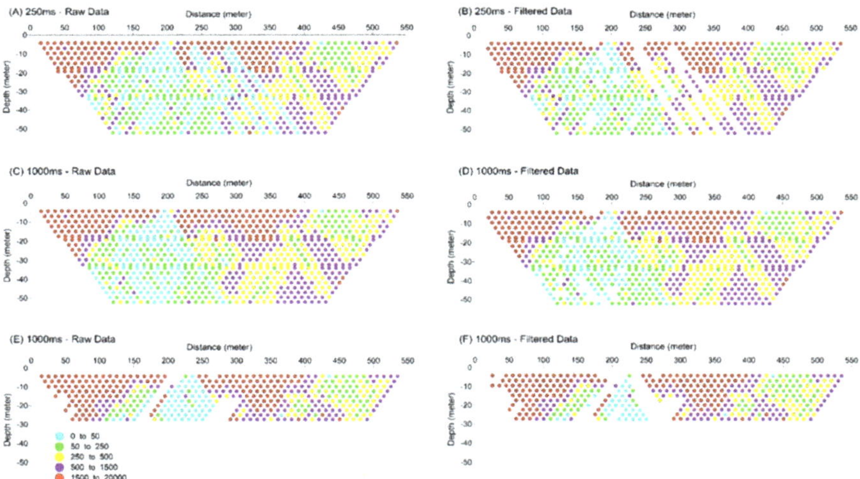

Figure 5. Distribution of the electrical resistivity measurement points in the Digo-Digo Stream region. Sections of raw and filtered data, obtained with sampling periods of 250, 1000, and 2000 ms. (**A**) 250 ms—raw data. (**B**) 250 ms—filtered data. (**C**) 1000 ms—raw data. (**D**) 1000 ms—filtered data. (**E**) 2000 ms—raw data. (**F**) 2000 ms—filtered data.

Chargeability: (1) removal of data where the electric potential decay curves did not follow an exponential behavior, that is, the temporal windows' decay curve behaved erratically without monotonous tendencies with patterns different from the assumed Cole–Cole model; (2) negative or null charge records; and (3) values greater than 100 mV/V so that the decay curve did not present perfect monotonous tendencies. Figure 6 shows the chargeability data recorded in the field (Figure 6A), and after data filtering (Figure 6B). Figure 6C shows the electric potential decay curves of each registered point, and in Figure 6D, these are shown after removing the incoherent curves (without exponential decay).

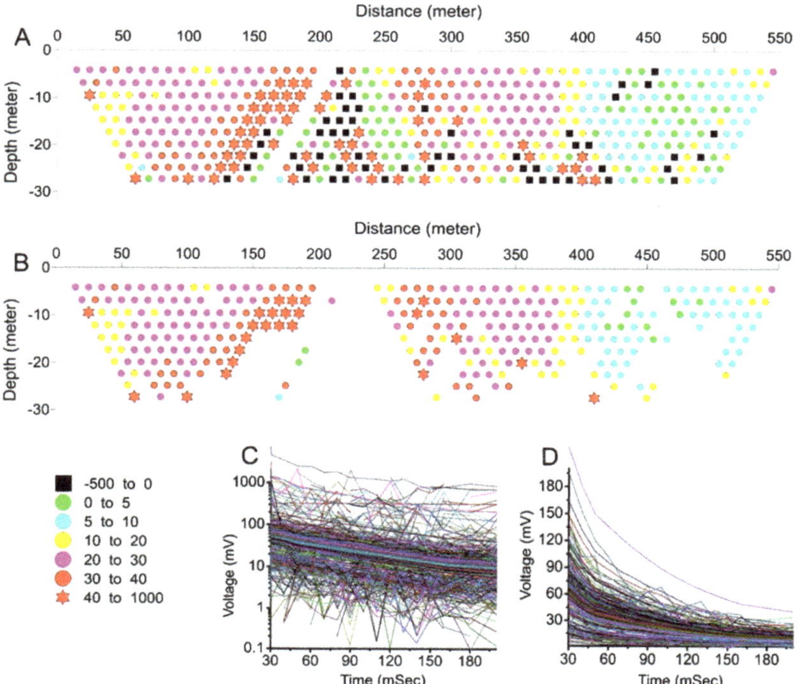

Figure 6. (**A**) Sampling points of chargeability data indicating the lack of sampling only at the Digo-Digo stream crossing—210 m. (**B**) Filtered sampling points. Decay curves of the electrical potential of raw data (**C**) and after filtering (**D**).

After filtering, the inverse modeling of 2D data was performed in the Res2Dinv inversion software [45], which uses least-squares inversion as a standard to minimize the square of the difference between the measured and calculated resistivities. The apparent resistivity data were inserted in the program and filtered for a second time using the exterminate bad datum points tool, followed by the application of the inversion routine.

In the frequency-domain-induced polarization, the apparent resistivity is measured at two different frequencies commonly varying between 0.1 and 10 Hz, while the result is presented as a percentage of effective frequency and/or metal factor [45]. The metal factor (MF) is a standardized parameter used mainly in the frequency-domain-induced polarization method [34]. Despite not measuring any exclusive physical property, this parameter is extensively used in mineral exploration as a data display technique [46]. Based on induced polarization results obtained in the field, ref. [34] suggests that the metal factor is a much better indicator of the mineralization quantity responsible for causing an IP anomaly greater than the frequency effects (PFE). The metal factor can be calculated for both frequency and time domain measurements. In the frequency domain, the metal factor (MF_f) value is calculated as in Equation (1) [45].

$$MF_f = 10^5 \left(\frac{\rho_{dc} - \rho_{ac}}{\rho_{ac}^2} \right) \tag{1}$$

where ρ_{dc} and ρ_{ac} are the resistivity values measured at low and high frequencies, respectively, expressed as ohms per meter. In the time domain, the metal factor is calculated as in Equation (2) [47,48].

$$MF_t = 10^3 \left(\frac{M}{\rho_{ac}} \right) \tag{2}$$

where the chargeability (M) is expressed in milliseconds [46].

The metal factor values were also calculated for the unfiltered data (only the topographic correction was performed) using the above equations in module (to avoid negative values). Subsequently, the calculated MF values were filtered to exclude the outliers, points with extremely high Root Mean Square Error (RMS) values.

For metal factor calculations, the topography was initially inserted in the raw data to correct the resistivity values. Subsequently, the MF calculation was performed using module Equation (2) to avoid negative results. Then, outlier MF values or those with exceedingly high RMS errors were eliminated, aiming to reduce the high variance in MF data and consequently reduce the RMS error values of the inversion models.

4. Results

4.1. Electrical Resistivity

Six main geoelectric zones with high lateral contrast are observed in the three electrical resistivity sections (Figure 7) as follows:

i. Thick layer of high resistivity positioned between 0 and 70 m, with values ranging from 1100 to 9000 ohm·m;

ii. Thick layer of low resistivity positioned between 100 and 215 m, with values ranging from 7 to 200 ohm·m underlying a thin superficial layer of high resistivity with values above 7000 ohm·m;

iii. Low resistivity zone position between 215 and 400 m, similar to (ii) but with slightly higher resistivity values ranging from 65 to 500 ohm·m;

iv. Thick zone of low resistivity associated with zones of medium resistivity ranging from 25 to 1000 ohm·m, which is better observed in the 250 ms section (Figure 6C), positioned between 400 and 550 m;

v. Surface zone at the top of the electrical profile with high resistivity, with values above 7000 ohm·m, extending laterally over almost every profile;

vi. Surface zone in lateral contact with zone (v), with low resistivities varying between 30 and 500 ohm·m, occurring between 410 and 480 m on the surface.

4.2. Chargeability

The chargeability model obtained after the inverse modeling converged with 7.5% RMS error after four iterations (Figure 8). In Section 2D, there are three main positive anomalies with values ranging from 45 to 100 mv/V between 100 and 360 m. The three main anomalies identified have a spatial association with zones (ii) and (iii) that have the lowest values of resistivity obtained in the survey on the exhalative horizon of the Digo-Digo Formation. In the IP time section, there is also a vertical anomaly with high chargeability at a depth of 24 m at a position of 400 m. This anomaly correlates with the high resistivity zone (3000 and 8000 ohm).

4.3. Metal Factor

After filtering the data, the metal factor calculated using the models (Figure 8) presented values between 0 and 35,500. Therefore, the RMS error calculated via RES2DINV reaches approximately 80% due to the high variation (1 to 35,500) in the normalized values resulting from the metal factor calculations.

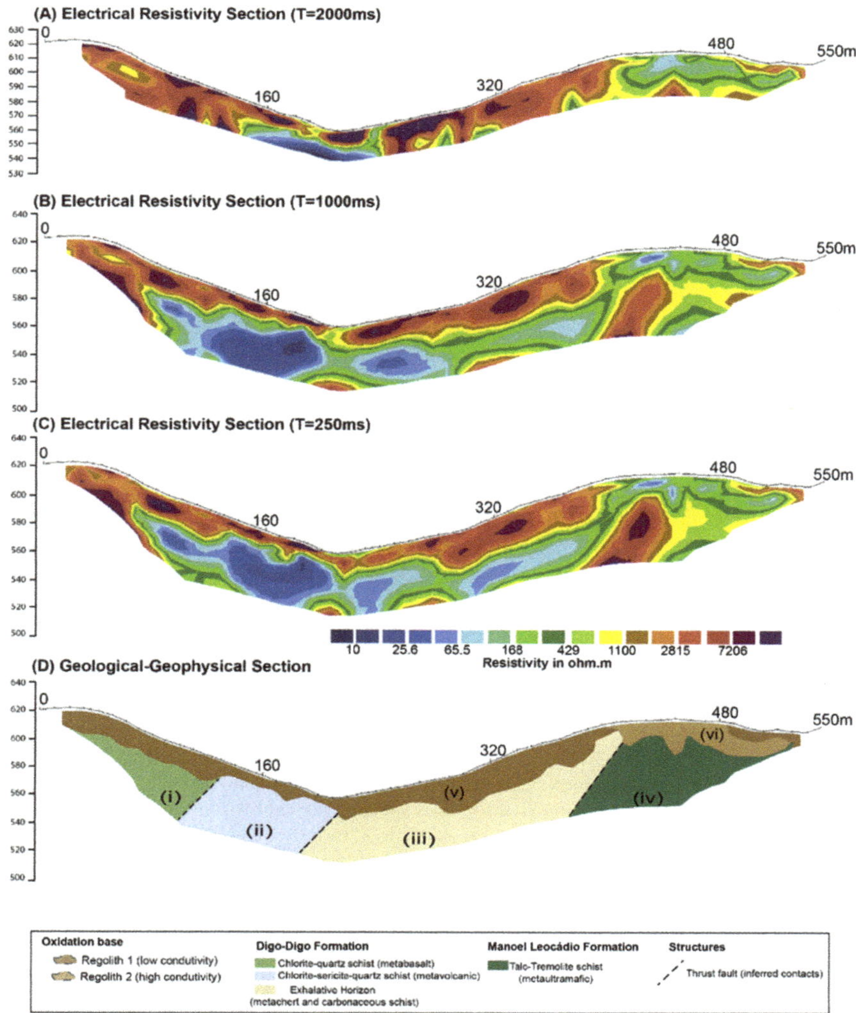

Figure 7. Electrical resistivity models obtained with current cycles of 2000 ms (**A**), 1000 ms (**B**), and 250 ms (**C**). (**D**) Geophysical–geological model of the 2D section elaborated from the correlation of the resistivity data with surface geological data.

The main positive anomalies observed in both metal factor inversions ranged between 20,500 and 35,500 and are spatially associated with the survey central portion, where the main IP anomalies corresponding to low resistivity areas in the electrical resistivity model also occur. In general, the anomalies in the MF models had lower values in the time domain compared to the frequency domain.

The MF time domain model showed three pinnacle-shaped anomalies, with mean values of 25,500 and greater than 30,000 at certain points in time. The MF frequency domain model presented two anomalies with values greater than 30,000, the anomalies continue in the central portion of the data (160–300 m) where values below 10,000 are observed, and the model geometry of the anomalies approximates a slightly distorted arched layer (Figure 8).

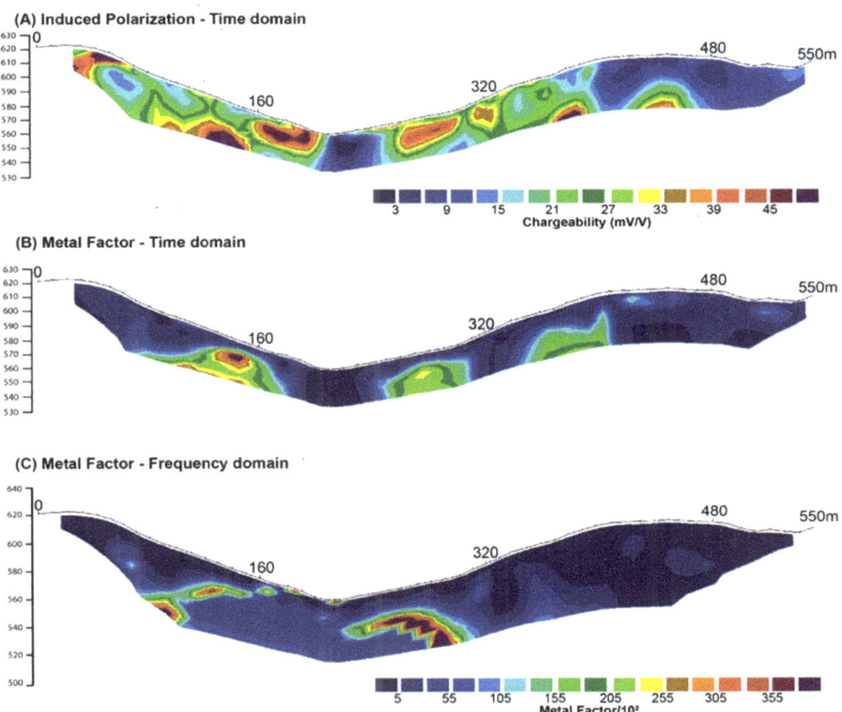

Figure 8. Electrical models: (**A**) chargeability (time domain IP), (**B**) time domain metal factor and (**C**) frequency domain metal factor.

4.4. Data Correlation

To determine the best subsurface locations that can potentially host mineralizations associated with massive and disseminated sulfides, the electric models were imported into the Oasis Montaj 9.1 software to correlate the chargeability, resistivity and metal factor data (frequency domain).

After that, new interpolations were performed via the Inverse Distance Weighting (IDW) method and the results showed acceptable correspondence with the models previously generated in the Res2Dinv software (version 3.71.118). This method assumes that the value attributed to a non-sampled point is the weighted average of the known neighboring values and that the weights between the non-sampled and sampled locations are inversely correlated [49]. The interpolation used cell size values of 2.7 m (approximately 1/2 of the spacing between points), and the values applied to the slope and power variables are, respectively, 1 and 2, which correspond to the software standard values.

Subsequently, the new grids were superimposed on the RGB (red-green-blue) and CMY (cyan-magenta-yellow) spectra, resulting in two ternary models. In these products, the zones with high chargeability as well as high metal factor and conductivity (low resistivity) are superimposed, highlighting the areas with the greatest potential for mineralization associated with the presence of metal sulfides.

In the ternary models (Figure 9), the purple colors in the RGB spectrum and greenish in the CMY indicate areas with greater potential for mineralization occurrence, while areas colored green (RGB) and purple/pink (CMY) suggest areas of high resistivity and low metal factor and chargeability values, indicating zones with properties of non-metallic materials and possible barren zones.

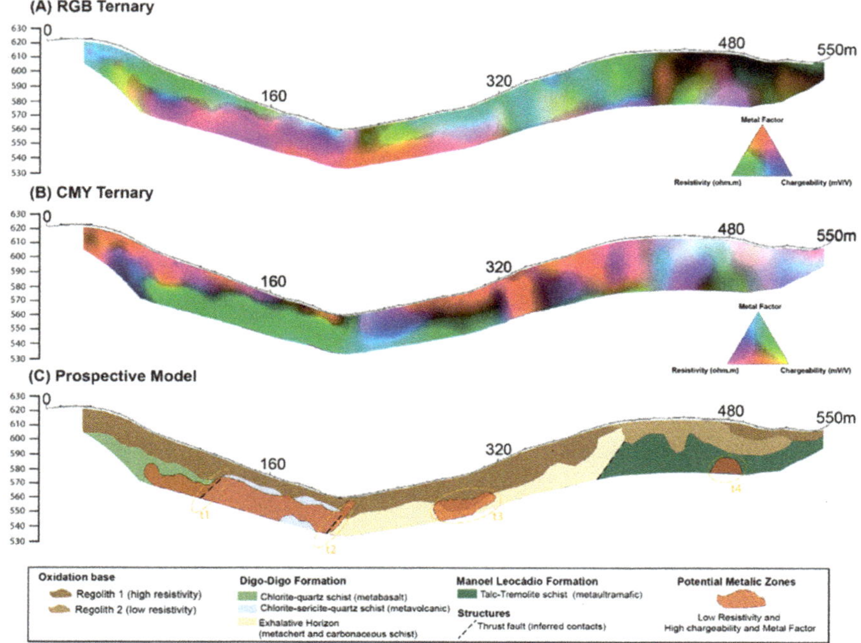

Figure 9. Ternary models resulting from the integration of 2D results of electrical resistivity, chargeability and metal factor (frequency domain). (**A**) Electric ternary (RGB); areas with greater potential to host metallic minerals are colored purple and pink. (**B**) Electric ternary (CMY); an area with high potential for mineralization is colored green. (**C**) Prospective model for the confluence region of the Digo-Digo stream with Vermelho River; areas with high potential to host metallic minerals are highlighted in red. More restricted target areas (t1, t2, t3 and t4) are identified by orange circles.

In ternaries, three zones with the characteristics with potential for sulfide mineralization occur between (1) 75 and 220 m, (2) 270 and 310 m, and (3) 470 and 490 m, starting at depths between 10 and 20 m (Figure 9).

5. Discussion and Conclusions

The geological information obtained on the surface is consistent with the respective lateral variations identified in the electrical resistivity sections.

i.   Zone (i) is spatially associated with chlorite–quartz schists from the Digo-Digo Formation;

ii.   Zone (ii) correlates spatially with intermediate metavolcanic rocks (sericite–chlorite–quartz schists);

iii.   Zone (iii) refers to the central areas of the survey, has the lowest resistivity values of the section, and can be spatially correlated to the exhalative horizon with disseminated to massive levels of pyrite and gold traces.

iv.   Zone (iv) occurs to the northeast portion of the Vermelho River, with different electrical responses and intermediate resistivity values (500–1100 ohms.m) and spatial correlation with the meta-ultrabasic rocks of the Manoel Leocádio Formation.

v.   The zones (v) and (vi) of the model correspond to two superficial layers (top of the profiles) with different resistivity values; zone (v) corresponds to a layer with resistivity values between 5000 and 8000 ohms.m that extends laterally over almost the entire profile, except between 400 and 480 m. These zones are interpreted as a product of rock weathering (regolith).

vi. Based on the electrical characteristics of the zones (i to vi), the geological data obtained on the surface, and their correlations, we propose an interpreted section with lithologies, structures, and coverage of the electrical resistivity models.

The model integrating the electrical resistivity, chargeability and metal factor (frequency domain) data generated by superimposing the models in the RGB and CMY spectra enables the determination of the zones with higher potential to host mineralizations associated with sulfides and, finally, the proposal of a prospective model of the studied section.

From the prospective model and the areas with potential for mineralization, four prospective targets were generated (t1, t2, t3 and t4):

- Targets t1 and t2 correspond to possible contact zones with the possible presence of metallic minerals; t1 is at the contact of chlorite–quartz schists (metabasalts) with the intermediate metavolcanic rocks, and t2 at the contact between the intermediate metavolcanic with the exhalative horizon; deposits and mineral occurrences are commonly associated with geological contact zones due to metal remobilizations, structural controls (shear zones), redox reactions and differences in buffering or permeability between rocks [50].
- The t3 target is hosted in the intercalations of metavolcanic, metacherts and carbonaceous phyllites with disseminated to massive sulfide levels containing gold traces, the target of the present study.
- Target t4 is hosted in the metaultramafic rocks of the Manoel Leocádio Formation, where previous studies described no associated mineralization; therefore, this target region has been selected for further investigation.

Some geological structures, such as the possible folding evidenced in the exposure of rocks from the exhalative horizon, could not be clearly identified in the resistivity and chargeability data due to the resolution of the acquired data. However, the MF frequency domain model shows that the zone with anomalous metal factor values presents a continuous geometry similar to the shape of an open fold of hectametric scale. If the geometry present in the MF model really corresponds to a geological structure, the fold observed in the field might be a parasitic fold that can be associated with a larger scale fold of the metal factor model.

Regarding the differences identified in the acquired data and results in the time and frequency domains, the following can be observed:

(1) The frequency domain IP data acquired using multi-electrodic systems obtained electrical resistivity measurements in 80% more points, reaching about 20 m deeper than in the time domain, and required only 50% of the acquisition time.

(2) The chargeability data had broader and more dispersed anomalies across the section, while the frequency domain metal factor data showed more restricted and structured anomalies, but still with good spatial correlation with chargeability and time domain metal factor data, mainly in the central portion associated with the exhalative horizon of the Digo-Digo Formation.

In conclusion, in cases of exploration projects that require data acquisition with greater depths and reduced costs, the frequency domain method using electrical resistivity meters with relatively low current transmissions was proven to be as efficient as the time domain data for detecting potential mineralized zones in the Digo-Digo Formation.

Author Contributions: Conceptualization, W.R.B., C.L.B.T. and P.A.C.d.A.; methodology, W.R.B. and P.A.C.d.A.; software, P.A.C.d.A. and W.R.B.; validation, P.A.C.d.A., W.R.B. and A.M.S.; formal analysis, W.R.B., P.A.C.d.A. and C.L.B.T.; investigation, W.R.B., P.A.C.d.A., C.L.B.T., A.M.S. and H.V.d.G.; writing—original draft preparation, P.A.C.d.A., W.R.B., C.L.B.T. and M.H.L.S.; writing—review and editing, P.A.C.d.A., W.R.B., C.L.B.T., A.M.S. and M.H.L.S.; visualization, W.R.B., P.A.C.d.A., C.L.B.T. and M.H.L.S.; supervision, W.R.B. and C.L.B.T.; project administration, C.L.B.T., A.M.S. and W.R.B. All authors have read and agreed to the published version of the manuscript.

Funding: This study was financed in part by the Coordenação de Aperfeiçoamento de Pessoal de Nível Superior—Brasil (CAPES)—Finance Code 001. The authors thank the Decanate of Research and Innovation and the Decanate of Graduate Studies of the University of Brasilia for the financial support via Public Notice DPI/ DPG n° 02/2023 for payment of the article publication fee.

Data Availability Statement: The authors confirm that the data supporting the findings of this study are available within the article.

Acknowledgments: We would like to thank the Orinoco Gold Ltd. and their staff for providing the drill core samples and supporting the field works and geophysical data acquisition. The first author thanks CAPES for his scholarship. C.L.B. Toledo and A.M. Silva thank the Conselho Nacional de Desenvolvimento Científico e Tecnologico (CNPq) for their respective research grants.

Conflicts of Interest: The authors declare no conflict of interest.

References

1. Galley, A.G.; Hannington, M.D.; Jonasson, I.R. Volcanogenic massive sulfide deposits. In *Mineral Deposits of Canada: A Synthesis of Major Deposit-Types, District Metallogeny, the Evolution of Geological Provinces, and Exploration Methods*; Special Publication; Geological Association of Canada, Mineral Deposits Division: Ottawa, ON, Canada, 2007; Volume 5, pp. 141–161.
2. Pelton, W.H.; Ward, S.H.; Hallof, P.G.; Sill, W.R.; Nelson, P.H. Mineral discrimination and removal of inductive coupling with multifrequency IP. *Geophysics* **1978**, *43*, 588–609. [CrossRef]
3. Palacky, G. Resistivity Characteristics of Geological Targets. In *Electromagnetic Methods in Applied Geophysics-Theory*; Nabighian, M., Ed.; Society of Exploration Geophysicists: Tulsa, OK, USA, 1987; pp. 53–129.
4. Clark, D.A.; Emerson, D.W.; Kerr, T.L. The use of electrical conductivity and magnetic susceptibility tensors in rock fabric studies. *Explor. Geophys.* **1988**, *19*, 244. [CrossRef]
5. Morgan, L.A. Geophysical Characteristics of Volcanogenic Massive Sulfide Deposits. In *Volcanogenic Massive Sulfide Occurrence Model*; Shanks, P., Thurston, R., Eds.; US Geological Survey: Reston, VA, USA, 2012; pp. 113–131.
6. Paterson, N.R.; Hallof, P.G. *Geophysical Exploration for Gold. Gold Metallogeny and Exploration*; Springer: Boston, MA, USA, 1993. [CrossRef]
7. Dentith, M.C.; Frankcombe, K.F.; Trench, A. Geophysical signatures of Western Australian mineral deposits: An overview. *Explor. Geophys.* **1994**, *25*, 103. [CrossRef]
8. McIntosh, S.M.; Gill, J.P.; Mountford, A.J. The geophysical response of the Las Cruces massive sulfide deposit. *Explor. Geophys.* **1999**, *30*, 123. [CrossRef]
9. Swiridiuk, P.; Close, B. Geophysical surveying over VMS deposits in Oman. *ASEG Ext. Abstr.* **2004**, 1–4. [CrossRef]
10. Hodges, G.; Chen, T.; Van Buren, R. HELITEM detects the Lalor VMS deposit. *Explor. Geophys.* **2016**, *47*, 285–289. [CrossRef]
11. Tavakoli, S.; Dehghannejad, M.; Juanatey, M.A.G.; Bauer, T.; Weihed, P.; Elming, S. Potential Field, Geoelectrical and Reflection Seismic Investigations for Massive Sulfide Exploration in the Skellefte Mining District, Northern Sweden. *Acta Geophys.* **2016**, *64*, 2171–2199. [CrossRef]
12. Newton, O.; Vowles, A. Geophysical Overview of Lalor VMS Deposit. In Proceedings of the Exploration 17: Sixth Decennial International Conference on Mineral Exploration, Toronto, ON, Canada, 21–25 October 2017; Tschirhart, V., Thomas, M.D., Eds.; Canadian Science Publishing: Ottawa, ON, Canada, 2017; pp. 619–635.
13. Ugalde, H.; Morris, W.A.; Van Staal, C. The Bathurst Mining Camp, New Brunswick: Data Integration, Geophysical Modelling and Implications for Exploration. *Can. J. Earth Sci.* **2018**, *56*, 433–451. [CrossRef]
14. Palacky, G.J.; Kadekaru, K. Effect of tropical weathering on electrical and electromagnetic measurements. *Geophysics* **1979**, *44*, 69–88. [CrossRef]
15. Claprood, M.; Chouteau, M.; Cheng, L.Z. Rapid detection and classification of airborne time-domain electromagnetic anomalies using weighted multi-linear regression. *Explor. Geophys.* **2008**, *39*, 164. [CrossRef]
16. Cox, L.H.; Wilson, G.A.; Zhdanov, M.S. 3D inversion of airborne electromagnetic data. *Geophysics* **2012**, *77*, WB59–WB69. [CrossRef]
17. Legault, J.M.; Plastow, G.; Zhao, S.; Bournas, N.; Prikhodko, A.; Orta, M. ZTEM and VTEM airborne EM and magnetic results over the Lalor copper-gold volcanogenic massive sulfide deposit region, near Snow Lake, Manitoba. *Interpretation* **2015**, *3*, SL83–SL94. [CrossRef]
18. Gibson, H.L.; Allen, R.L.; Riverin, G.; Lane, T.E. The VMS model: Advances and application to exploration targeting. In Proceedings of the Exploration 07: Fifth Decennial International Conference on Mineral Exploration, Toronto, ON, Canada, 9–12 September 2007; pp. 713–730.
19. Phillips, N.; Oldenburg, D.; Chen, J.; Li, Y.; Routh, P. Cost effectiveness of geophysical inversions in mineral exploration: Applications at San Nicolas. *Lead. Edge* **2001**, *20*, 1351–1360. [CrossRef]
20. Boszczuk, P.; Cheng, L.Z.; Hammouche, H.; Roy, P.; Lacroix, S.; Cheilletz, A. A 3D gravity data interpretation of the Matagami mining camp, Abitibi Subprovince, Superior Province, Québec, Canada: Application to VMS deposit exploration. *J. Appl. Geophys.* **2011**, *75*, 77–86. [CrossRef]

21. Tavakoli, S.; Bauer, T.E.; Rasmussen, T.M.; Weihed, P.; Elming, S. Deep massive sulfide exploration using 2D and 3D geoelectrical and induced polarization data in Skellefte mining district, northern Sweden. *Geophys. Prospect.* **2016**, *64*, 1602–1619. [CrossRef]
22. Araujo, S.M.; Fawcett, J.J.; Scott, S.D. Metamorphism of hydrothermally altered rocks in a volcanogenic massive sulfide deposit: The Palmeirópolis, Brazil, example. *Rev. Bras. Geociências* **1995**, *25*, 173–184. [CrossRef]
23. Neder, R.D.; Figueiredo, B.R.; Beaudry, C.; Collins, C.; Leite, J.A.D. The Expedito Massive Sulfide Deposit, Mato Grosso. *Rev. Bras. Geol.* **2000**, *30*, 222–225. [CrossRef]
24. Hartmann, L.A.; Delgado, I. Cratons and orogenic belts of the Brazilian Shield and their contained gold deposits. *Miner. Depos.* **2001**, *36*, 207–217. [CrossRef]
25. NetunoVillas, R.; Santos, M. Gold deposits of the Carajás mineral province: Deposit types and metallogenesis. *Miner. Depos.* **2001**, *36*, 300–331. [CrossRef]
26. Grainger, C.J.; Groves, D.I.; Tallarico, F.H.B.; Fletcher, I.R. Metallogenesis of the Carajás Mineral Province, Southern Amazon Craton, Brazil: Varying styles of Archean through Paleoproterozoic to Neoproterozoic base- and precious-metal mineralization. *Ore Geol. Rev.* **2008**, *33*, 451–489. [CrossRef]
27. Fruchting, A.; Boniatti, J.; Oliveira, G.; Oliveira, S.B.; Pires, P.F.R.; Henrique, E. Aplicação dos métodos eletromagnéticos aéreos e de polarização induzida espectral em mineralizações de cobre/zinco tipo VMS. In Proceedings of the 11th International Congress of the Brazilian Geophysical Society & EXPOGEF 2009, Bahia, Brazil, 24–28 August 2009. [CrossRef]
28. Teixeira, J.B.G.; Silva, M.G.; Misi, A.; Cruz, S.C.P.; Silva Sá, J.H. Geotectonic setting and metallogeny of the northern São Francisco craton, Bahia, Brazil. *J. S. Am. Earth Sci.* **2010**, *30*, 71–83. [CrossRef]
29. Couto, M.A., Jr.; Wosniak, R.; Marques, E.D.; Duque, T.; Carvalho, M.T.N. VTEM and Aeromagnetic Data Modeling Applied to Cu, Zn and Pb Prospection in Palmeirópolis Project, TO, Brazil. In Proceedings of the 15th International Congress of the Brazilian Geophysical Society, Rio de Janeiro, Brazil, 31 July–3 August 2017. [CrossRef]
30. Cordeiro, P.F.O.; Oliveira, C.G. The Goiás Massif: Implications for a pre-Columbia 2.2–2.0 Ga continent-wide amalgamation cycle in central Brazil. *Precambr. Res.* **2017**, *298*, 403–420. [CrossRef]
31. Jost, H.; Junior, F.C.; Fuck, R.A.; Dussin, I.A. Uvá complex, the oldest orthogneisses of the Archean-Paleoproterozoic terrane of central Brazil. *J. S. Am. Earth Sci.* **2013**, *47*, 201–212. [CrossRef]
32. Pimentel, M.M. The tectonic evolution of the Neoproterozoic Brasília Belt, central Brazil: A geochronological and isotopic approach. *Braz. J. Geol.* **2016**, *46*, 67–82. [CrossRef]
33. Jost, H.; Sial, A.N.; Benell, M.R.; Ferreira, V.P. *Carbon Isotopes in Dolostones of Goias State Greenstone Belts, Central Brazil*; Brazilian Geological Congress: Curitiba, Brazil, 2008; Volume 44, p. 687.
34. Jost, H.; Chemale, F., Jr.; Dussin, I.A.; Tassinari, C.C.G.; Martins, R. U-Pb zircon Paleoproterozoic age for the metasedimentary host rocks and gold mineralization of the Crixás greenstone belt, Goiás, Central Brazil. *Ore Geol. Rev.* **2010**, *37*, 127–139. [CrossRef]
35. Jost, H.; Carvalho, M.J.; Rodrigues, V.G.; Martins, R. Metalogênese dos greenstone belts de Goiás. In *Metalogênese das Províncias Tectônicas Brasileiras*; CPRM: Belo Horizonte, Brazil, 2014; pp. 141–168.
36. Resende, M.G.; Jost, H.; Osborne, G.A.; Mol, A.G. The stratigraphy of the Goiás and Faina Greenstone Belts, Central Brazil: A new proposal. *Rev. Bras. Geociências* **1998**, *28*, 77–94. [CrossRef]
37. Resende, M.G.; Jost, H.; Lima, B.E.M.; Teixeira, A.D.A. Proveniência e Idades modelos Sm/Nd das Rochas Siliciclásticas Arqueanas dos Greenstone Belts de Faina e Santa Rita, Goiás. *Rev. Bras. Geociências* **1999**, *29*, 281–290.
38. Borges, C.C.A.; Toledo, C.L.B.; Silva, A.M.; Junior, F.C.; Jost, H.; de Carvalho Lana, C. Geochemistry and isotopic signatures of metavolcanic and metaplutonic rocks of the Faina and Serra de Santa Rita greenstone belts, Central Brazil: Evidence for a Mesoarchean intraoceanic arc. *Precambrian Res.* **2017**, *292*, 350–377. [CrossRef]
39. Bogossian, J.; Hagemann, S.G.; Rodrigues, V.G.; Lobato, L.M.; Roberts, M. Hydrothermal alteration and mineralization in the Faina greenstone belt: Evidence from the Cascavel and Sertão orogenic gold deposits. *Ore Geol. Rev.* **2020**, *119*, 103293. [CrossRef]
40. Tomazzoli, E.R. Geologia, Petrologia, Deformação e Potencial Aurífero do Greenstone Belt de Goiás-GO. Master's Thesis, Universidade de Brasília, Brasília, Brazil, 1985; 206p.
41. Godoy, H.V. Petrografia e Geoquímica das Rochas Vulcânicas e Sedimentares da Formação Digo-Digo: Implicações Para Prospecção de Depósitos de Sulfetos Maciços Vulcanogênicos no Greenstone belt Serra de Santa Rita, GO. Master's Thesis, Universidade de Brasília, Brasília, Brazil, 2019; 70p. Available online: https://repositorio.unb.br/handle/10482/40498 (accessed on 18 November 2022).
42. Chambers, J.; Ogilvy, R.; Kuras, O.; Cripps, J.; Meldrum, P. 3D electrical imaging of known targets at a controlled environmental test site. *Environ. Geol.* **2002**, *41*, 690–704. [CrossRef]
43. Bentley, L.; Gharibi, M. Two-and three-dimensional electrical resistivity imaging at a heterogeneous remediation site. *Geophysics* **2004**, *69*, 674–680. [CrossRef]
44. Nimmer, R.E.; Osiensky, J.L.; Binley, A.M.; Williams, B.C. Three-dimensional effects causing artifacts in two-dimensional, cross-borehole, electrical imaging. *J. Hydrol.* **2008**, *359*, 59–70. [CrossRef]
45. Loke, M.H. *RES2DINVx64*, Version 4.09. Rapid 2D Resistivity & IP Inversion Using the Least-Square Method-Geoelectrical Imaging 2-D & 3D. Geotomo Software: Penang, Malaysia, 2019; 137p. Available online: https://www.geotomosoft.com/downloads.php(accessed on 18 November 2022).

46. Witherly, K.E.; Vyselaar, J. A Geophysical Case History of the Poplar Lake Copper-Molybdenum Deposit, Houston Area, British Columbia. In *Induced Polarization. Applications and Case Histories*; Ward, S.H., Ed.; Society of Exploration Geophysicists: Tulsa, OK, USA, 1990; pp. 304–324.

47. Madden, T.R.; Marshall, D.J. Induced polarization, a study of its causes. *Geophysics* **1959**, *24*, 790–816. [CrossRef]

48. Telford, W.M.; Geldart, L.P.; Sheriff, R.E. *Applied Geophysics*; Cambridge University Press: Cambridge, UK, 1990; 740p.

49. Lu, G.Y.; Wong, D.W. An adaptive inverse-distance weighting spatial interpolation technique. *Comput. Geosci.* **2008**, *34*, 1044–1055. [CrossRef]

50. Volkov, A.V. Types of Ore Material Remobilization. *Dokl. Earth Sci.* **2005**, *403*, 688–691.

Article

Enhancement of Potential Field Source Boundaries Using the Hyperbolic Domain (Gudermannian Function)

Ahmad Alvandi [1,*], Kejia Su [2,*], Hanbing Ai [3], Vahid E. Ardestani [1] and Chuan Lyu [2]

[1] Institute of Geophysics, University of Tehran, Tehran 14359-44411, Iran; ebrahimz@ut.ac.ir
[2] Research Institute No. 270, China National Nuclear Corporation (CNNC), Nanchang 330200, China; lvc270@163.com
[3] School of Geophysics and Geomatics, China University of Geosciences, Wuhan 430074, China; ahb_ecut@163.com
[*] Correspondence: aalvandi@ut.ac.ir (A.A.); sukejia2022@163.com (K.S.)

Abstract: Horizontal boundary identification of causative sources is an essential tool in potential field data interpretation due to the feasibility of automatically retrieving the boundary information of subsurface gravity or geomagnetic structures. Although many approaches have been proposed to address these issues, it is still a hot research topic for many researchers to derive novel methods or enhance existing techniques. We present two high-resolution edge detectors based on the Gudermannian function and the modifications of the second-order derivative of the field. The effectiveness of the newly proposed filters was initially tested on synthetic gravity anomalies and geomagnetic responses with different assumptions (2-D and 3-D; imposed and superimposed; noise-free and noise-contaminated). The obtained results verified that the two novel methods yield the capability of producing high-resolution, balanced amplitudes and accurate results for better imaging causative sources with different geometrical and geophysical properties, compared with the other nine representative edge enhancement techniques. Furthermore, the yielded results from the application of the two strategies to a real-world aeromagnetic data set measured from the Central Puget Lowland (C.P.L) of the United States and a gravity data set surveyed from the Jalal Abad area of Kerman province, Iran, with detailed comparative studies validated that the edges identified via the two methods are in good agreement with the major geological structures within the study areas and the determined lateral information using the tilt-depth, top-depth estimation method. These features make them valuable tools for solving edge detection problems.

Keywords: edge detection; Gudermannian function; GD_T and GD_H filters; potential field data

Citation: Alvandi, A.; Su, K.; Ai, H.; Ardestani, V.E.; Lyu, C. Enhancement of Potential Field Source Boundaries Using the Hyperbolic Domain (Gudermannian Function). *Minerals* **2023**, *13*, 1312. https://doi.org/10.3390/min13101312

Academic Editors: Emilio L. Pueyo, Luan Thanh Pham, Saulo Pomponet Oliveira and Le Van Anh Cuong

Received: 21 August 2023
Revised: 2 October 2023
Accepted: 6 October 2023
Published: 10 October 2023

1. Introduction

Imaging the horizontal boundaries of anomalous structures provides impressive visibility to determine lateral changes in gravity and magnetic data [1–10]. Acquired horizontal boundary information from buried causative sources has an essential role in modeling and interpreting gravity and magnetic data. Edge detection filters are constantly applied to delineate geologic structures in the form of faults, contacts, dykes, mineral deposits, and other tectonic features [5,6,10]. In the geophysical literature, many different filters have been presented to locate the geologic structures from measured data based on the horizontal and vertical derivatives of the gravity anomalies, the reduced-to-the-pole magnetic anomalies, or the ratios of these directional derivatives [9,11–20]. For comparison purposes, some conventional and standard edge detectors have been selected.

The total horizontal gradient (THG) filter of the potential field is given by Equation (1) [16]. THG is stable and popular for enhancing the edges of causative sources. The lateral boundaries of causative structures are determined by the peaks of the THG amplitude.

$$\text{THG} = \sqrt{\left(\frac{\partial M}{\partial x}\right)^2 + \left(\frac{\partial M}{\partial y}\right)^2} \tag{1}$$

where M is the gravity or the reduced-to-the-pole magnetic field, $\frac{\partial M}{\partial x}$ is the field gradient in the x direction, and $\frac{\partial M}{\partial y}$ is the field gradient in the y direction.

The analytical signal (AS) technique is another edge delineation filter proposed by [17]. The amplitude maxima are directed over the edge center; however, AS cannot balance the edges of various sources at different depths and would be dominated by larger amplitudes caused by shallow bodies [18]. The AS method is constructed as follows:

$$\text{AS} = \sqrt{\left(\frac{\partial M}{\partial x}\right)^2 + \left(\frac{\partial M}{\partial y}\right)^2 + \left(\frac{\partial M}{\partial z}\right)^2} \tag{2}$$

where $\frac{\partial M}{\partial z}$ is the measured field gradient in the z direction [17].

The tilt angle (TA) phase-based filter was proposed by [19], which is a normalized function of the THG filter. The TA edge detection method equation is given by

$$\text{TA} = \text{atan}\left(\frac{\frac{\partial M}{\partial z}}{\text{THG}}\right) \tag{3}$$

The lateral boundaries of subsurface sources are delineated by the zero crossing [19]. However, tilt angle edge detector usually generates blurry results, namely, TA suffers from the deficiency of yielding low-resolution outputs, especially in the case of thin and deep causative structures [7,9].

The total horizontal gradient of the tilt angle (TA-THG) filter was proposed by [20], which is calculated using Equation (4):

$$\text{TA-THG} = \sqrt{\left(\frac{\partial \text{TA}}{\partial x}\right)^2 + \left(\frac{\partial \text{TA}}{\partial y}\right)^2} \tag{4}$$

The edges of buried structures are recognized by the peaks of the TA-THG amplitude [20–23]. The TA-THG estimates diffuse results that are wider than true edges [5].

The TA of the THG (TAHG) method was introduced by [24]. This technique is popular for detecting the horizontal boundaries of buried structures, but TAHG is not suitable for discrimination between closely spaced bodies [13,14,23], which is defined as follows:

$$\text{TAHG} = \text{atan}\left(\frac{\frac{\partial \text{THG}}{\partial z}}{\sqrt{\left(\frac{\partial \text{THG}}{\partial x}\right)^2 + \left(\frac{\partial \text{THG}}{\partial y}\right)^2}}\right) \tag{5}$$

The TAHG filter has the feature of simultaneously maintaining both smaller and larger amplitudes. The edges of buried structures are retrieved by the maximum values of the processed results via TAHG [7,10,13,14,21–24].

The improved local phase (ILF) method was introduced by [25], which is constructed by calculating different order gradients of potential field data. The ILF is given by

$$\text{ILF} = \text{asin}\left(\frac{\text{THG}}{\sqrt{\left(\frac{\partial M}{\partial x}\right)^2 + \left(\frac{\partial M}{\partial y}\right)^2 + \left(\frac{\partial^2 M}{\partial x^2} + \frac{\partial^2 M}{\partial y^2}\right)^2}}\right) \tag{6}$$

where $\frac{\partial^2 M}{\partial x^2}$ and $\frac{\partial^2 M}{\partial y^2}$ are the second-order horizontal gradients of field. The ILF maximum values correspond to the horizontal boundaries of gravity and magnetic causative sources.

Moreover, the tilt angle of the first-order vertical gradient of the total horizontal gradient defines the THVH method [26]. The THVH filter is estimated using Equation (7):

$$\text{THVH} = \operatorname{atan}\left(\frac{-\left(\frac{\partial^2 \text{THG}}{\partial x^2} + \frac{\partial^2 \text{THG}}{\partial y^2} \right)}{\sqrt{\left(\frac{\partial \text{THV}}{\partial x} \right)^2 + \left(\frac{\partial \text{THV}}{\partial y} \right)^2}} \right) \tag{7}$$

where THV is the derivative of the THG filter in the z direction [26]. The THVH maximum amplitudes indicate the edges of buried sources. The THVH is capable of extracting more information from the potential field data, especially for superimposed bodies [26].

The TA of the balanced total horizontal gradient (TBHG) filter was proposed by [27]. The TBHG filter is expressed as follows:

$$\text{TBHG} = \operatorname{atan}\left(\frac{\frac{\partial \text{BTHG}}{\partial z}}{\sqrt{\left(\frac{\partial \text{BTHG}}{\partial x} \right)^2 + \left(\frac{\partial \text{BTHG}}{\partial y} \right)^2}} \right) \tag{8}$$

where

$$\text{BTHG} = \frac{\text{THG}}{P + \sqrt{\left(H_X(\text{THG})\right)^2 + \left(H_Y(\text{THG})\right)^2 + \left(\text{THG}\right)^2}} \tag{9}$$

In Equation (9), H_X and H_Y are the directional Hilbert transforms of THG, and the p value is a parameter that controls the balance effect of different amplitudes induced by different sources with different properties [27,28]. Following the suggestion of [27], $p = 1$ within this study. Similar to previously introduced representative edge detectors, the maximal amplitude values are located over the source edges, which can be utilized as an indicator describing the geometric distribution features of various causative sources.

The introduced TA [19], TA-THG [20], TAHG [24], ILF [25], THVH [26], and TBHG [27] filters are equalizing methods; in other words, the basic motivation for constructing these filters is to balance the amplitudes derived from shallow and deep buried structures with various characteristics. However, the adaptiveness of these filters is guaranteed by the premise that the processed potential field anomalies contain independently situated causative sources only, or they will produce false artifacts (erroneous edge information) with low resolutions when dealing with spatially superimposed and imposed sources (more common in field applications).

In order to address these issues further, ref. [18] proposed a fast sigmoid high-resolution filter (FS) and verified its effectiveness in producing clear and precise edge information related to causative sources regardless of the source spatial correlations from potential field data [18]. The FS filter is calculated using Equation (10):

$$\text{FS} = \frac{\left(\frac{\frac{\partial \text{THG}}{\partial z}}{\sqrt{\left(\frac{\partial \text{THG}}{\partial x} \right)^2 + \left(\frac{\partial \text{THG}}{\partial y} \right)^2}} \right) - 1}{1 + \left| \frac{\frac{\partial \text{THG}}{\partial z}}{\sqrt{\left(\frac{\partial \text{THG}}{\partial x} \right)^2 + \left(\frac{\partial \text{THG}}{\partial y} \right)^2}} \right|} \tag{10}$$

The FS strategy provides maximum amplitudes over the horizontal boundaries of the causative sources. The peaks of the FS filter can be implemented to balance the horizontal boundaries of buried sources with different geometric features [18].

2. Hyperbolic Domain (Gudermannian Function) Filters

This work proposes two novel filters based on the hyperbolic domain (also termed the Gudermannian function) to extract high-resolution horizontal boundaries of different anomalous sources located at different depths. Mathematically, the Gudermannian function is an odd function [29–31] which obtains an almost identical shape to the arctangent function commonly used for edge enhancement of buried structures in the literature [9]. The first filter is a combination of the hyperbolic domain (Gudermannian function), the second-order derivative of the field, and the corresponding directional derivatives. The second approach is comprised of the hyperbolic domain (Gudermannian function) and the improved total horizontal derivative obtained by the directional Hilbert transforms. This hyperbolic domain edge detection (GD_T) filter is given by

$$GD_T = 2atan\left[\tanh\left(2 \times \left(-\lambda + \frac{\frac{\partial T}{\partial z}}{\sqrt{\left(\frac{\partial T}{\partial x}\right)^2 + \left(\frac{\partial T}{\partial y}\right)^2}}\right)\right)\right] \tag{11}$$

where

$$T = \left|\left(\frac{\partial^2 M}{\partial x \partial z}\right)^2 + \left(\frac{\partial^2 M}{\partial y \partial z}\right)^2\right| \tag{12}$$

In Equation (12), $\frac{\partial^2 M}{\partial x \partial z}$ and $\frac{\partial^2 M}{\partial y \partial z}$ are the field horizontal gradients in the z direction.

Here, in order to improve the effectiveness of the TBHG filter in increasing the resolution of edge detection and removing spurious edges, we introduce the second filter built on the Gudermannian function. This hyperbolic domain filter (GD_H) is given by

$$GD_H = 2atan\left[\tanh\left(2 \times \left(-\lambda + \frac{\frac{\partial HD}{\partial z}}{\sqrt{\left(\frac{\partial HD}{\partial x}\right)^2 + \left(\frac{\partial HD}{\partial y}\right)^2}}\right)\right)\right] \tag{13}$$

where

$$HD = \frac{ITH^2}{2 + \sqrt{(H_X(ITH))^2 + (H_Y(ITH))^2 + (ITH)^2}} \tag{14}$$

where $(H_x[\cdot], H_y[\cdot])$ are the 2-D directional Hilbert transforms [28] and $ITH = \sqrt{\left(\frac{\partial^2 M}{\partial x \partial z}\right)^2 + \left(\frac{\partial^2 M}{\partial y \partial z}\right)^2}$.

λ is a positive number that should be specified by the interpreter before the implementation of GD_T and GD_H regarding the measured potential field anomalies. It is worth stating that the distinction between the two filters is essentially that the GD_T filter uses the second-order derivative of the field; however, the GD_H filter combines the directional Hilbert transforms and the second-order derivative of the field instead. The amplitude maxima of the two Gudermannian-function-based filters can be utilized to identify the edges of the buried sources.

To validate the effectiveness of balancing amplitudes from sources at different depths and mitigating superimposed or imposed source effects of the proposed new filters (GD_T and GD_H), we perform the Gudermannian-function-based filters on four synthetic data sets, including two profile anomalies and two plane gravity and magnetic responses derived from extremely imposed and superimposed sources with and without noise corruption, a real-world aeromagnetic data set acquired from the Central Puget Lowland (C.P.L) of the United States, and a real-world gravity data set acquired from the Jalal Abad mine, Iran. Moreover, detailed comparative studies are presented with previously described traditional and popular edge detectors (e.g., THG, AS, TA, TA-THG, TAHG, ILF, THVH, TBHG, and FS). The obtained results verified that the proposed filters can address the aforementioned issues with the best performance in terms of producing output images with good resolution, balanced amplitudes, and avoiding drawing false edges.

3. Evaluation of Parameter λ

A 2-D geomagnetic fault model is considered to investigate the influence of λ of the newly presented GD_T and GD_H filters and its contribution to enhancing the edges and controlling the resolution. The basic characteristic parameters of the buried fault structure are the strength of the geomagnetic field (47,000 nT), the fault dip = $90°$, the strike = $0°$, the inclination = $90°$, and the susceptibility contrast = 0.02 SI. The profile anomaly has a data interval of 50 m along the *x* direction. The geomagnetic anomaly is shown in Figure 1a. Subsequently, the GD_T and GD_H edge enhancement filters are applied to the calculated geomagnetic data with increasing values of λ from 0 to 9 (Figure 1b–k). The filtered results show that the maximum amplitudes of the GD_T and GD_H filters in all cases are equal and perfectly match the lateral boundary of the fault plane. Furthermore, both of the novel filters produce sharp edges when the λ = 0.5 and maintain its sharpness until the λ reaches 8. The extracted edges are unreliable if λ is larger than 8. Therefore, the value of λ can be selected from 0.5 to 8 in order to maintain the efficiency of GD_T and GD_H and obtain reliable edge information. The GD_T and GD_H methods' maximum amplitudes indicate the edges of buried sources in radians. The amplitude changes between $-\pi/2$ and $\pi/2$.

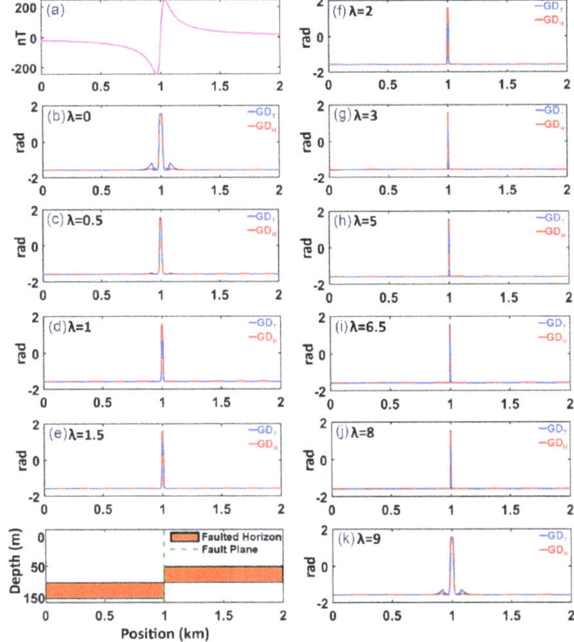

Figure 1. Evaluating the performance of the filters with the parameter varying from 0 to 9 using a simulated geomagnetic anomaly (**a**) the geomagnetic response (nT) over the fault model. The processed results of the GD_T and GD_H filters at (**b**) λ = 0, (**c**) λ = 0.5, (**d**) λ = 1, (**e**) λ = 1.5, (**f**) λ = 2, (**g**) λ = 3, (**h**) λ = 5, (**i**) λ = 6.5, (**j**) λ = 8, and (**k**) λ = 9. Notably, the schematic representation of the 2-D synthetic fault model is placed at the bottom left of this figure.

4. Application to Simulated Profile Data

A 2-D synthetic gravity model consisting of a vertical dyke-like model situated at a depth of 300 m with a density contrast of 1000 kg/m^3 is created. The gravity anomaly generated over this two-dimensional model with a data interval of 50 m along the *x* direction is shown in Figure 2a. Figure 2b–l shows the application of different filters, including THG, AS, TA, TA-THG, TAHG, ILF, THVH, TBHG, FS, GD_T, and GD_H. The maximum values of the processed results of THG (Figure 2b) and AS (Figure 2c) indicate the edges and the center of the buried gravity model, respectively. Hence, both THG and AS can be utilized as indicators for describing the geometric features of the buried structure in terms of the boundaries and geometric center, even though the provided information suffers from low resolution. Figure 2d shows that TA is able to generate a sharp response over the geometric center of the gravity source, making TA a more useful tool than the described THG and AS methods. However, if the source shifts, the edges are visible [13]. Figure 2e–i gives the results of the TA-THG, TAHG, ILF, THVH, and TBHG filters. These techniques are generally effective in outlining the edge information of the vertical dyke-like structure, but the extracted boundaries are diffusive and lack simplicity. Comparatively, FS obtains a relative high-resolution result (Figure 2j). The processed result is sharper, which makes the subsequent interpretation process easier. Moreover, as the results presented in Figure 2k–l show, the newly proposed GD_T and GD_H filters yield the merit of delineating the horizontal boundaries of the subsurface gravity anomalous body with maximal clarity without compromising the effectiveness of avoiding the generation of annoying artifacts or erroneous edges simply by finding the maximal amplitude values within the responses calculated.

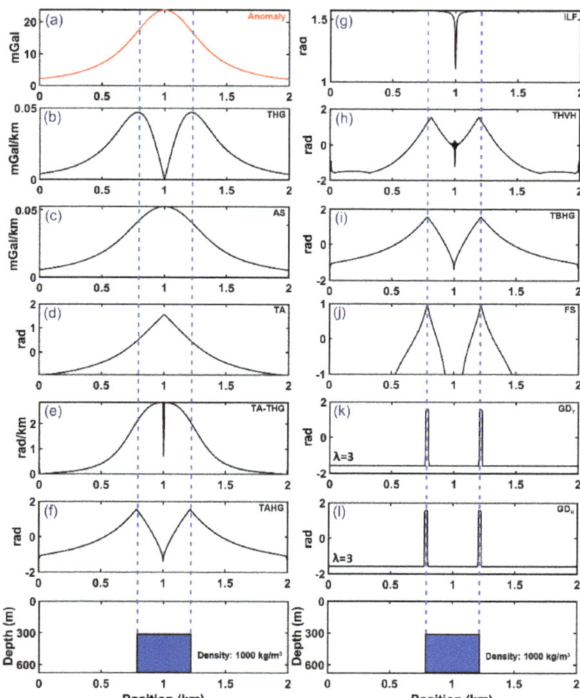

Figure 2. Produced results of various edge detection techniques over the synthetically constructed 2-D gravity dyke model: (**a**) the calculated gravity anomaly response over the dyke body, (**b**) THG, (**c**) AS, (**d**) TA, (**e**) TA-THG, (**f**) TAHG, (**g**) ILF, (**h**) THVH, (**i**) TBHG, (**j**) FS, (**k**) GD_T, and (**l**) GD_H. Likewise, the cross section of the synthetic dyke-like model is illustrated at the bottom of this figure.

5. Application to Synthetic Data

This section utilizes several 3-D synthetic gravity and magnetic models with and without the consideration of noise corruption to examine the effectiveness of the GD_T and GD_H edge determination filters further. It should be noted that, in this study, $\lambda = 0.5$ in the mathematical equations of the GD_T and GD_H filters assigned for the rest of the experiments, including processing simulated 3-D potential field anomalies and the real-world aeromagnetic and gravity data application.

5.1. The Gravity Model

Figure 3a,b shows the side and bird's-eye views of the designed gravity model, which contains five buried dyke-like prismatic sources (G1, G2, G3, G4, and G5) with different properties, including different geometric parameters and positive and negative density contrasts. Amongst the causative sources, three prisms share the same size but different density contrasts and depths (G1, G2, and G3), and the remaining two prisms yield different sizes, depths, and density contrasts (G4 and G5). The characteristic parameters of the synthetic 3-D gravity model are given in Table 1. The 2-D gravity anomaly of the synthetic model was produced on 12 km × 12 km grid nodes with a node interval of 0.05 km along the x and y directions.

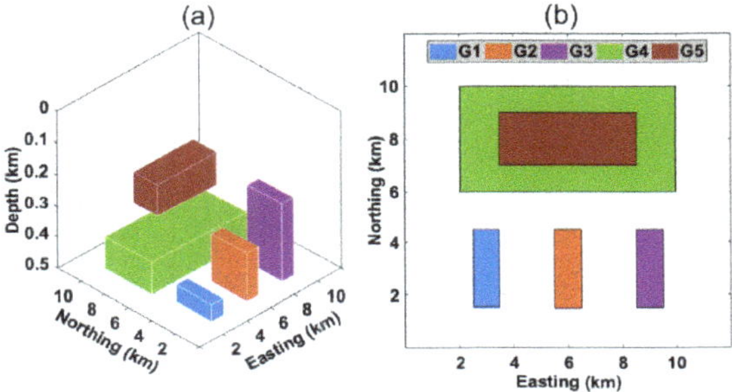

Figure 3. (**a**) 3-D view of the synthetic gravity model; (**b**) top view of the complex 3-D model with five buried sources.

Table 1. Density contrasts and geometric parameters of the constructed gravity model displayed in Figure 3.

Source/Label	G1	G2	G3	G4	G5
x-Coordinate of the Geometric Center (m)	3000	6000	9000	6000	6000
y-Coordinate of the Geometric Center (m)	3000	3000	3000	8000	8000
Prism Width (m)	1000	1000	1000	8000	5000
Prism Length (m)	3000	3000	3000	4000	2000
Depth of the Top (m)	450	350	250	400	200
Depth of the Bottom (m)	950	850	750	700	300
Density Contrast (kg/m^3)	3000	−2000	2000	−2500	2000

Figure 4a shows the calculated gravity anomaly from the constructed 3-D model in Figure 3. As shown in Figure 4b,c, the THG and AS maps are dominated by the strong amplitudes generated by the buried shallow sources, while the signal intensities from sources at deeper positions are degraded much more intensively, so that the edge information of deeper sources simply cannot be recognized satisfactorily from the processed results of THG and AS due to the deficiency of the two filters to balance the amplitudes produced by structures at different depths. The TA extracts the edges of bodies G1, G3, and G4 successfully; however, it is less effective for the interpretation of bodies G2 and G5 (Figure 4d). Although the horizontal boundaries can be outlined by the zero contours of TA, it produces false edges between sources, obtaining positive and negative density contrasts [5]. Additionally, Figure 4e displays the result of TA-THG, indicating that TA-THG is capable of reflecting all the boundaries; however, the displayed edges are very blurry. The TAHG approach can equalize the low and high amplitudes derived from different sources simultaneously and can also avoid generating erroneous edges in the produced results. Nevertheless, the TAHG method provides the boundary information in a low-resolution manner. Figure 4g depicts the processed result of ILF with respect to the gravity anomaly in Figure 4a. It can be observed that the maximum amplitudes of ILF cannot reflect the boundaries of G1, G2, G3, and G4, and ILF also creates a false contour inside the source G5. Figure 4h–i displays the filtered results of the THVH and TBHG methods, respectively. Unlike the filters described above, THVH and TBHG are suitable for balancing the weak and strong amplitudes produced by these sources at different depths, but both of the THVH and TBHG filters produce low-resolution outputs, and further, THVH also produces unwanted artifacts. Finally, Figure 4j–l stores the results of the FS, GD_T, and GD_H strategies. The last three methods can perfectly mitigate the amplitude effect without generating annoying false edges, that is, they can enhance the boundaries of all sources with different properties and only reflect the true boundaries. Furthermore, compared with the FS filter, the recommended GD_T and GD_H filters based on the Gudermannian function obtain superior resolution, namely, the produced maps containing the boundary information of the subsurface gravity structures are clearer.

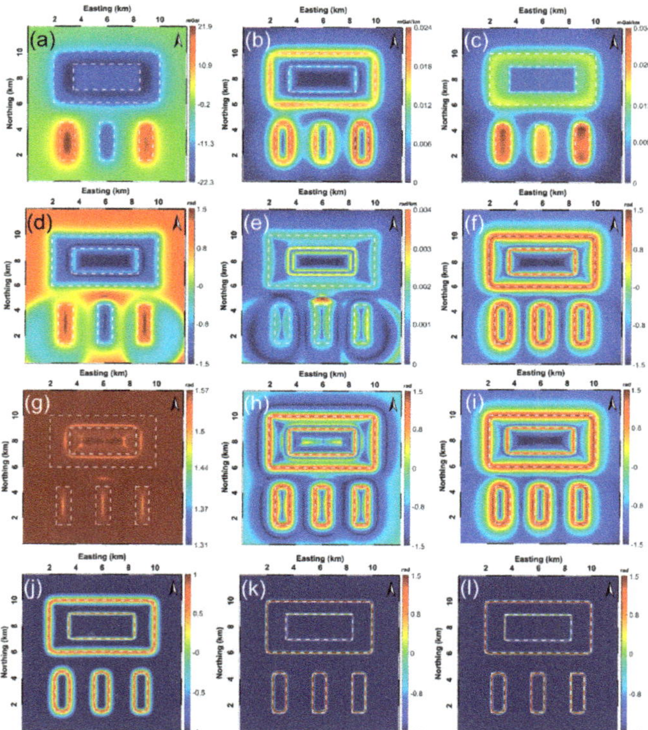

Figure 4. Detailed comparison of the effectiveness of the aforementioned edge enhancement strategies via the synthetic complex 3-D gravity model displayed in Figure 3 (dashed white lines represent the true boundaries of the sources buried): (**a**) synthetically produced 2-D gravity anomaly with respect to the five-prism model with positive and negative density contrasts, correspondingly processed results of (**b**) THG, (**c**) AS, (**d**) TA, (**e**) TA-THG, (**f**) TAHG, (**g**) ILF, (**h**) THVH, (**i**) TBHG, (**j**) FS, (**k**) GD_T, and (**l**) GD_H.

5.2. The Gravity Model with Noise Contamination

In order to test the robustness of the involved filtering strategies for resisting noise corruption, 3% of the amplitude of the anomaly displayed in Figure 4a was considered to generate the noisy gravity anomaly (Figure 5a). Before applying the aforementioned edge enhancement filters (THG, AS, TA, TA-THG, TAHG, ILF, THVH, TBHG, FS, GD_T, and GD_H), the constantly utilized upward continuation technique regarding the noise-contaminated gravity data was performed to relieve the noise effect (the upward continued height is 0.15 km).

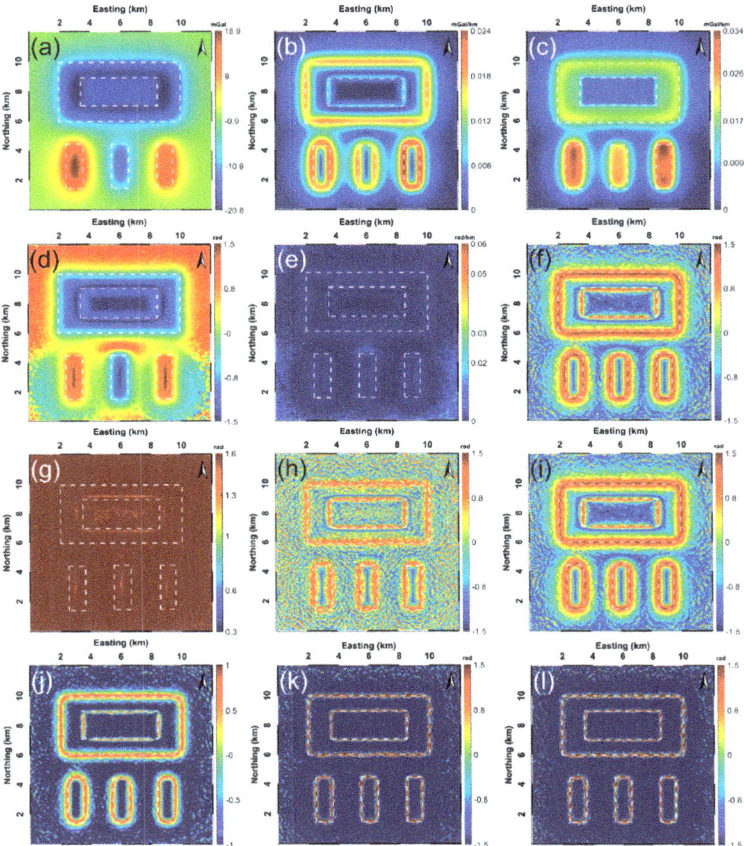

Figure 5. Performing the filtering strategies within this study on the noise-corrupted gravity data (likewise, dashed white lines represent the true boundaries of the sources buried): (**a**) the generated noisy gravity anomaly (the noise degree = 3%), processed results of (**b**) THG, (**c**) AS, (**d**) TA, (**e**) TA-THG, (**f**) TAHG, (**g**) ILF, (**h**) THVH, (**i**) TBHG, (**j**) FS, (**k**) GD_T, and (**l**) GD_H.

Figure 5b,c displays the processed results of the THG and AS techniques under consideration of the noise effect, respectively. The outcomes are still dominated by large amplitudes due to the shallow sources situated; however, the small amplitudes controlled by deeply buried sources are also ambiguously represented. The TA method still yields the ability to balance the weak and strong signals caused by different sources, but it sacrifices resolution in terms of detecting the edges of the G1, G2, G3, and G4 sources much wider than the true ones (Figure 5d). Figure 5e illustrates the filtered result of TA-THG associated with the noisy gravity response. It can be clearly seen that, in this situation, the maximum amplitudes of TA-THG cannot coherently correlate with the lateral boundaries of the imposed gravity model. In Figure 5f, the TAHG filtering method is able to detect the edges of all the buried sources regardless of the depth at which they are located, but this edge information is diffusively exhibited and easily affected by the noise component remaining. Figure 5g displays the result of the ILF filter. The ILF produces an unsatisfactory result considering the imposed gravity structure [26]. The filtered maps of THVH and TBHG are displayed in Figure 5h–i, respectively. Both of the two methods are able to detect all the edges. Nevertheless, THVH obtains lower anti-noise ability compared with the TBHG technique, and TBHG encounters the same problem as TAHG, that is, the retrieved edge information has a low-resolution. The yielded maps at the bottom row of Figure 5 keep the filtered results of the FS, GD_T, and GD_H methods, which demonstrate that not only can the three strategies delineate the edges of all structures buried at different depths with comparatively high resolution, but their performances can be mildly eroded by the noise perturbation. Therefore, the recommended GD_T and GD_H filters share the same capability as the FS method to robustly attenuate the noise influence so that the boundaries are more visible and clearer. Hitherto, we can conclude that the proposed GD_T and GD_H methods are useful tools for producing high-resolution and balanced amplitude edge detection results.

5.3. The Geomagnetic Model

To further investigate the capability of the GD_T and GD_H filters, a superimposed synthetic geomagnetic model (the inclination and declination angles are set to $90°$ and $0°$, respectively), consisting of ten dyke-like prisms named M1-M10 at different depths and extents, is simulated (Figure 6). The detailed model parameters of the established complex model are listed in Table 2. The total magnetic intensity (TMI) is estimated on a 12 km × 12 km grid with a grid spacing of 0.05 km along the x and y directions. Figure 7a shows the simulated RTP geomagnetic anomaly.

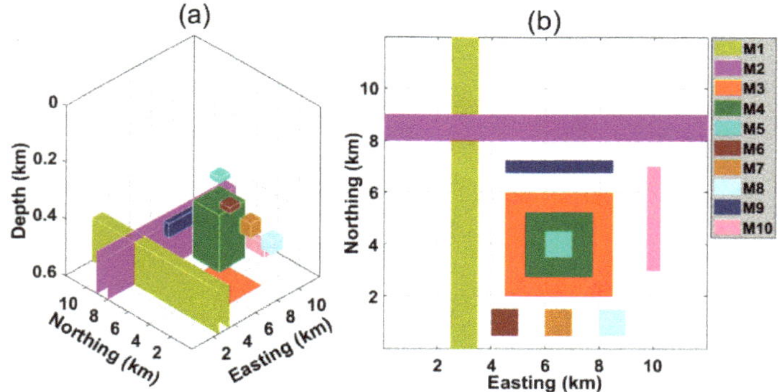

Figure 6. (**a**) 3-D view of the theoretical geomagnetic model; (**b**) top view of the second model with ten buried dyke-like geomagnetic structures.

Table 2. Geomagnetic susceptibilities and geometric parameters of the superimposed geomagnetic model.

Source/Label	M1	M2	M3	M4	M5	M6	M7	M8	M9	M10
x-Coordinate of the Center (m)	3000	5000	6500	6500	6500	4500	6500	8500	6500	10,000
y-Coordinate of the Center (m)	5000	8500	4000	4000	4000	1000	1000	1000	7000	5000
Prism Width (m)	1000	1000	4000	2500	1000	1000	1000	1000	500	500
Prism Length (m)	20,000	20,000	4000	2500	1000	1000	1000	1000	4000	4000
Depth of the Top (m)	450	450	600	300	200	200	300	400	400	500
Depth of the Bottom (m)	950	950	1100	500	220	220	350	450	450	550
Strike Azimuth (°)	0	90	0	0	0	0	0	0	90	0
Magnetic Susceptibility (SI)	0.021	0.023	0.019	0.025	0.027	0.03	0.032	−0.024	0.025	−0.026

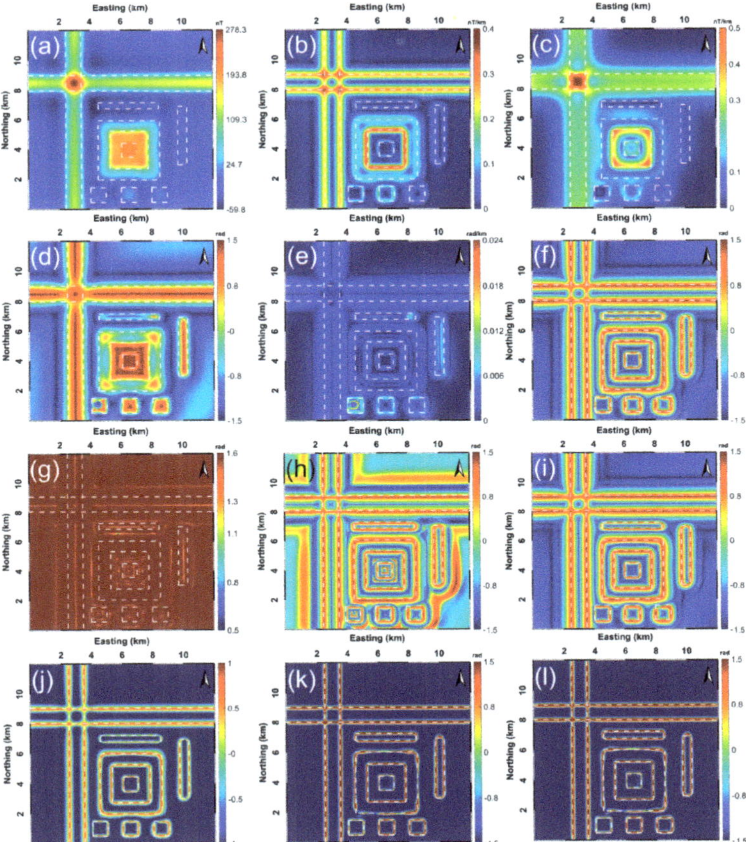

Figure 7. Detailed comparison of the effectiveness of the aforementioned edge detectors via the synthetic complex 3-D geomagnetic model displayed in Figure 6 (dashed white lines represent the true boundaries of the sources buried): (**a**) synthetic RTP geomagnetic response derived from the superimposed model with positive and negative magnetic susceptibilities, performed results of (**b**) THG, (**c**) AS, (**d**) TA, (**e**) TA-THG, (**f**) TAHG, (**g**) ILF, (**h**) THVH, (**i**) TBHG, (**j**) FS, (**k**) GD_T, and (**l**) GD_H.

Following the same processing procedure, we performed the 11 edge detectors on the RTP geomagnetic anomaly generated. Correspondingly filtered results are presented in Figure 7b–l, with the white dashed lines representing the true boundaries of all the causative bodies buried. Figure 7b–c displays the results of the THG and AS filters, showing coherent conclusions generated in the 3-D gravity data experiment containing an imposed effect. The THG and AS filters cannot balance the amplitudes caused by sources buried at different depths; this feature makes them inapplicable for dealing with superimposed structures. Figure 7d shows the extracted edges using the TA method. Although the edges of some sources can be distinguished successfully by reading the zero contours from the processed map of TA, there are chances to generate false contours between buried sources, which make the subsequent interpretation process inconvenient to continue. Also, the edges of M5 and M9 are faintly detected. Figure 7e shows the result created by applying the TA-THG method. It can be seen that, in this case, the maximum amplitudes of TA-THG cannot delineate the horizontal boundaries of M1-M10. Figure 7f presents the result of the TAHG method. Obviously, TAHG detects all the edge information via the maximal amplitudes in the filtered map in a more effective way without causing false edges than the TA method. The only problem with TAHG is that the coherent edges obtain low resolutions. Figure 7g shows the recuperated edges of the synthetic 3-D complex geomagnetic model in Figure 6 using the ILF method. The edges are faint and drawn very diffusely, making this filter not suitable for outlining the edges of the imposed and superimposed structures. Figure 7h–i displays the edges detected from the THVH and TBHG methods, respectively. As depicted in the two maps, the THVH and TBHG methods provide low-resolution edge information related to the causative sources. Although the THVH method can locate all the true boundaries, it inevitably produces fake positive contours around the sources M6, M7, M8, and M10. The amplitude result of the FS filter is shown in Figure 7j. Compared to the results mentioned above, the FS clearly outlines the horizontal boundaries of the ten superimposed sources and simultaneously enhances the visibility and sharpness of the buried structures. The final edge detection results of the GD_T and GD_H filters are displayed in Figure 7k–l, validating that the newly designed GD_T and GD_H methods are distinguished in producing balanced and coherent horizontal boundaries of the superimposed geomagnetic sources without generating disturbing artifacts compared with the other methods discussed. The FS, GD_T, and GD_H filters' maximum values lie directly over the source edge and enhance the horizontal boundaries to be more visible and sharper, compared to the THG, AS, TA, TA-THG, TAHG, ILF, THVH, and TBHG methods.

5.4. The Geomagnetic Anomaly with Noise Corruption

This section discusses the performance of the involved filters in dealing with the noise-corrupted RTP geomagnetic anomaly presented in Figure 8a. The random noise component has an amplitude of 3% of the noise-free data. For the purpose of maintaining the coherence of testing, the GD_T and GD_H filters are implemented on the noisy geomagnetic data with the comparison of the selected nine filters (THG, AS, TA, TA-THG, TAHG, ILF, THVH, TBHG, and FS).

Notably, the upward continuation method is utilized again regarding the noisy 2-D geomagnetic anomaly at a height of 0.15 km prior to the filtering process of the involved edge detectors in order to better simulate real applications.

Retrieved edge information with the noise effect using the THG and AS methods is displayed in Figure 8b,c, which are still incapable of balancing the amplitudes induced by sources situated at different depths. Instead, both of the filters are robust enough to resist noise contamination. Figure 8d depicts the result of the TA technique, creating many spurious edges that cannot reflect the true ones and showing less robustness in suppressing the noise effect. Moreover, the TA-THG method is still unable to draw horizontal borders via the maximum amplitudes in the noisy case. Relatively, Figure 8f depicts the edges (maximal amplitudes) that can be extracted coherently but are partially affected by the noise effect and low resolution by applying the TAHG filter. Especially, the geomagnetic body M9 failed to be recovered. According to the filtered output shown in Figure 8g, as expected, the ILF filter fails to delineate all the edges regardless of the thin, deep, and shallow structures. Figure 8h–j shows the final results of implementing the THVH, TBHG, and FS filters. All three filters yield corresponding results describing the true edges in Figure 6. However, FS yields the most clear and compact signals, namely, FS generates the highest resolution result with the most effective anti-noise ability amongst the three filters, due to its unique mathematical structure. Finally, results from the implementation of the GD_T and GD_H filters are displayed in Figure 8k,l, respectively, showing that the GD_T and GD_H filters provide successful estimations of all the boundaries of the buried sources with different characteristics, even with the involvement of 3% degrees of noise content. In other words, the GD_T and GD_H approaches are very useful for generating high-resolution, balanced amplitude results and are less sensitive to noise edge detection results.

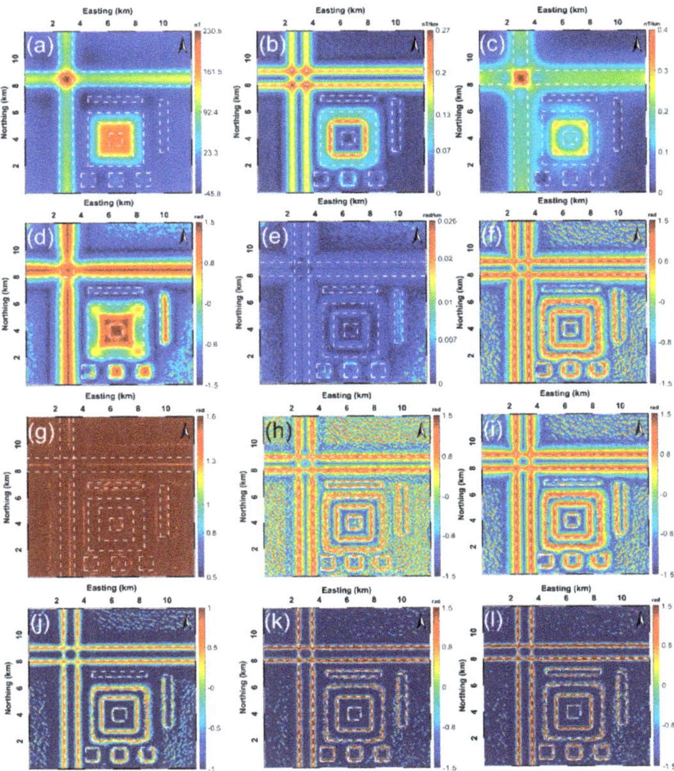

Figure 8. Performing the filtering strategies within this study on the noise-corrupted geomagnetic data (dashed white lines display the true edges of the buried sources): (**a**) the geomagnetic data of the synthetic model with random noise with an amplitude of 3% of the anomaly amplitude, produced results using (**b**) THG, (**c**) AS, (**d**) TA, (**e**) TA-THG, (**f**) TAHG, (**g**) ILF, (**h**) THVH, (**i**) TBHG, (**j**) FS, (**k**) GD_T, and (**l**) GD_H.

6. Application to Field Aeromagnetic Data

This section emphasizes the test of the practicability of the aforementioned traditional nine filters (THG, AS, TA, TA-THG, TAHG, ILF, THVH, TBHG, and FS) and the proposed GD_T and GD_H filters via a high-resolution field geomagnetic response belonging to the Central Puget Lowland (C.P.L), in the northwestern United States, published in 1997 by the U.S. Geological Survey (USGS). Figure 9 displays the geological map of the C.P.L study area and the adjacent area at a scale of $1 : 250,000$ (adapted from [32]). The study area is located within the geological complex above the Juan de Fuca plate that is actively subducting beneath the western margin of North America [5,33,34]. The high-resolution aeromagnetic data of the study area are displayed in Figure 10a. The flight height was 300 m, and the flight lines were in the E–W direction with a grid interval of 400 m [35,36]. The acquired aeromagnetic data were reduced to the pole (RTP) using an inclination angle of 69.4° and a declination angle of 19°, to reposition the anomalies from their casual sources, as depicted in Figure 10b [5]. To attenuate the short-wavelength noise effects, the filter using a 0.3 km upward continuation distance of the RTP aeromagnetic data was implemented to stabilize the latter filtering process (Figure 11a). Figure 11b–l displays the results of the THG, AS, TA, TA-THG, TAHG, ILF, THVH, TBHG, FS, GD_T, and GD_H edge determination filters, respectively.

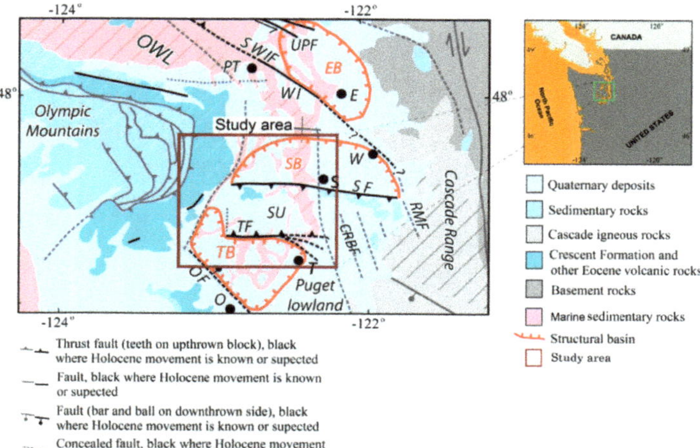

Figure 9. The location and the simplified geologic map of the Central Puget Lowland area (adjacent regions are shown; adapted from [5,32]). EB Everett basin, SB Seattle basin, TB Tacoma basin, SU Seattle uplift, PT Port Townsend, S Seattle, T Tacoma, WI Whidbey Island, CRBF Coast Range boundary fault, OF Olympia fault, RMF Rattlesnake Mountain fault, SWIF Southern Whidbey Island fault, SF Seattle fault, TF Tacoma fault, UPF Utsalady Point fault.

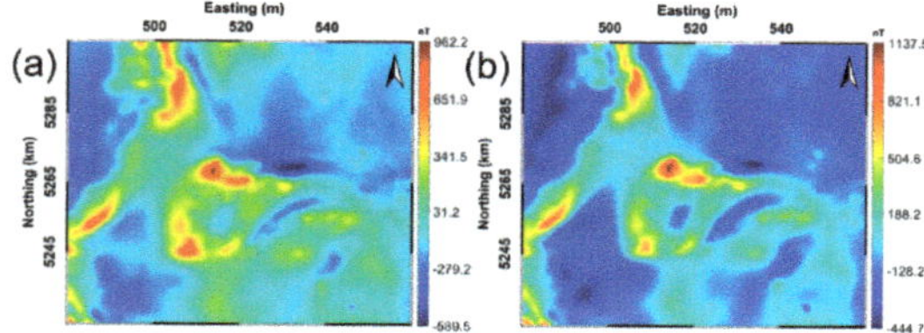

Figure 10. (**a**) The acquired field aeromagnetic anomaly (nT); (**b**) the corresponding RTP result.

Figure 11b,c exhibits the edge information delineated by performing the THG and AS filters on the upward continued RTP aeromagnetic data, respectively. As can be observed, the THG and AS methods are dominated by the large amplitudes produced by shallow structures buried, so the generated signals are not balanced effectively [5]. Figure 11d illustrates the map obtained by the TA method. Although this technique can balance the shallow and deep anomalies simultaneously, it cannot clearly determine the geological edges within the study area. Figure 11e shows the filtered result of the TA-THG filter. TA-THG fails to effectively delineate the horizontal boundaries via the maximum amplitudes; this phenomenon shows high consistency with the synthetic tests. The filtered signals using the TAHG filter are presented in Figure 11f, confirming that TAHG is an effective method for highlighting the geometric distribution of the buried geological structures within the study area. Figure 11g displays the output of the ILF filter. Considering the test results generated synthetically and practically, ILF is not suitable for edge determination and should not be considered as an edge detection tool in this study. Figure 11h–i keeps the lateral boundaries estimated by the THVH and TBHG methods. Although the lateral boundaries can be recognized clearly by the maximum contours of THVH and TBHG, the THVH filter produces additional spurious contours, degrading the efficiency of subsequent interpretations. Finally, Figure 11j–l shows the edges delineated by the FS, GD_T, and GD_H edge determination filters. The FS filter is validated again to be a prominent filter that is capable of equalizing the amplitudes caused by geological structures with different properties. Comparatively, the recommended Gudermannian-based GD_T, GD_H, and FS approaches provide even better results than the other filters and display more details than the filters of THG, AS, TA, TA-THG, TAHG, ILF, THVH, and TBHG in terms of balanced amplitude results and high-resolution edges without suspicious artifacts. These features make the GD_T, GD_H, and FS methods valuable tools for quantitatively extracting edge information from potential field data.

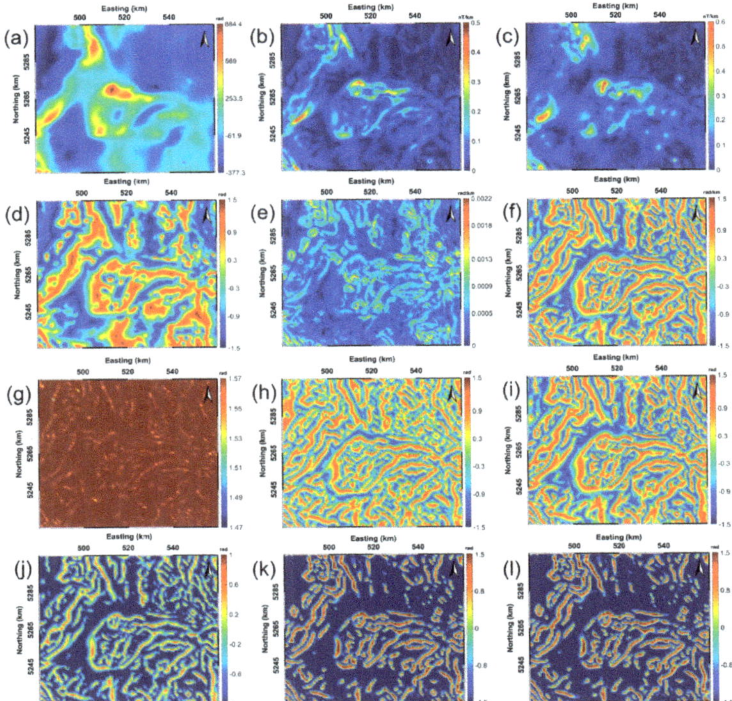

Figure 11. Processed results of the RTP aeromagnetic data using the aforementioned 11 edge detectors: (**a**) reduction-to-the-pole magnetic anomaly (the upward continuation height = 0.3 km), retrieved results of (**b**) THG, (**c**) AS, (**d**) TA, (**e**) TA-THG, (**f**) TAHG, (**g**) ILF, (**h**) THVH, (**i**) TBHG, (**j**) FS, (**k**) GD_T, and (**l**) GD_H.

In addition to the detection of horizontal boundaries, the depth estimation of causative bodies is also important for the interpretation of magnetic data [37]. In this work, to estimate the top depth of the structures retrieved, we used the popular tilt-depth (TD) technique [38]. The significant advantages of the tilt-depth method are that it does not depend on the structural index (SI) or the window size (WS) and can automatically yield depth information. Figure 12 shows the result of applying the tilt-depth method to the study area. The histogram of the estimated depths is shown in Figure 12c, which describes that 50% of these structures obtained exist at 0.3–1.5 km depth. The depth of the majority of the lateral boundaries varies between 0.2–5 km, being disturbed randomly all over the Central Puget Lowland area. For detailed comparisons, the lateral boundaries in Figure 11k,l are superimposed on the tilt-depth map. Clearly, the positions of the buried source points are mostly correlated with the lateral boundaries determined by the GD_T and GD_H filters. Also, we can see that many edges match the geological features. The determined boundaries illustrate a good correlation with the trending faults and other anomalies in the western and southeastern parts of the area. Therefore, with the help of the superimposed map from the processed results of the Gudermannian function filters and the tilt-depth method, it is possible to make reliable interpretations of the study area qualitatively and quantitatively.

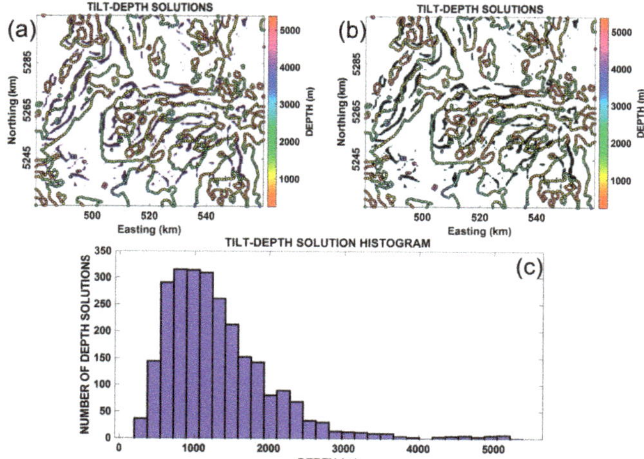

Figure 12. (**a**) The top depths derived from the tilt-depth technique (retrieved geological edges from Figure 11k are superimposed on the tilt-depth map); (**b**) the top depths derived from the TD technique (retrieved geological edges from Figure 11l are superimposed on the TD map); (**c**) histogram of the top depth estimated from the TD method regarding the first field application.

7. Application to Field Gravity Data

This section emphasizes the test of the practicability of the aforementioned traditional nine filters (THG, AS, TA, TA-THG, TAHG, ILF, THVH, TBHG, and FS) and the proposed GD_T and GD_H filters via a real gravity anomaly belonging to the hematite ore body in the Jalal Abad area, Iran. The study area is located in the southeast of Iran, Zarand city, Kerman province, to the north of the Jalal Abad iron mine. Figure 13 shows the geological map of the study area [39]. The host rock of this area is the Rizzo series of alluvial and volcanic rocks [39]. Small masses of igneous rocks such as microgabbro and some dikes and sills, together with diorite and diabase outcrops, can be seen in the area. The younger dikes and sills intruded into alluvial and volcanoclastic rocks. It was found that dikes and sills are older or occur simultaneously with iron mineralization in the area [40]. Iron mineralization is deep in the Jalal Abad ore deposit, and very few outcrops of iron can be seen in the entire area of the site. The general form of deposits around the mine site is similar to the stretched lens along the northwest–southeast, which is located in a folded structure. Hematite has been created mainly from magnetite through the secondary oxidation process and is obvious in areas where abundant fractures are present. Pyrite and chalcopyrite are sulfide minerals found in the Jalal Abad deposit [39]. Figure 14a shows the gravity anomaly map of the area after the corrections and the required pre-processing tasks. The gravity survey was conducted at a spacing of 20×40 m. In this map, a low-intensity gravity anomaly can be seen, but due to the presence of iron outcrops, it is more likely to be attributed to iron anomalies. This area was selected to show the efficiency of the proposed methods for the edge detection of iron and hematite deposits. To attenuate the short-wavelength noise effects, the filter using a 50 m upward continuation distance of the gravity data was implemented to stabilize the latter filtering process (Figure 14b).

Figure 15a–k displays the results of the THG, AS, TA, TA-THG, TAHG, ILF, THVH, TBHG, FS, GD_T, and GD_H edge determination filters, respectively.

As can be seen, THG and AS are dominated by anomalies in the Jalal Abad area, possibly caused by the shallow structures. The maps of AS and THG are blurred and unreliable for finding clear boundaries for subsurface sources. Figure 15c shows the images produced with the TA method. The TA is relatively insensitive to the depth of a buried source and resolves shallow and deep source anomalies equally well, allowing the anomalies to be identified in the study area. The TA method was used to determine the horizontal location of shallow and deep sources in the Jalal Abad mine. Again, TA-THG cannot delineate the boundaries of the buried sources (Figure 15d). In Figure 15e, the maximum amplitude of the TAHG filter works well for shallow and deep sources. TAHG can define the horizontal boundaries of the buried sources. However, one of the disadvantages of this method is the generation of low-resolution boundaries. Again, ILF cannot delineate the boundaries of the buried sources (Figure 15f). Figure 15g shows the results of applying the THVH filter. This filter produces some false boundaries that are inconsistent with the lateral boundaries. Figure 15h,i shows the results created with the TBHG and FS techniques, respectively. The results obtained with these two filters are of low resolution. Figure 15j,k shows the results obtained with the GD_T and GD_H edge determination filters. It is clear that the proposed filters were much more successful than the others in improving the source boundaries.

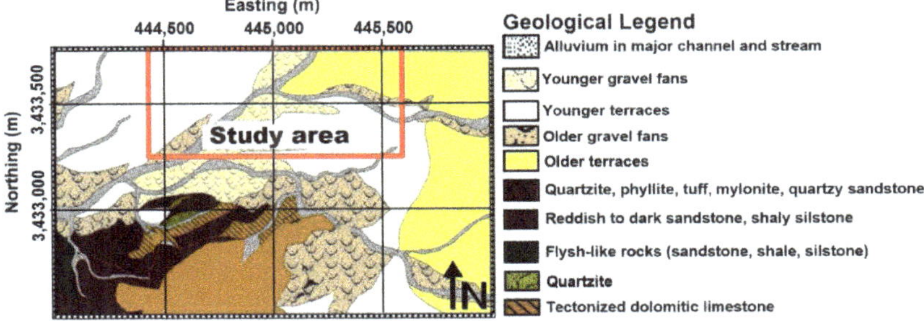

Figure 13. The simplified geologic map of the Jalal Abad mine area [39,40].

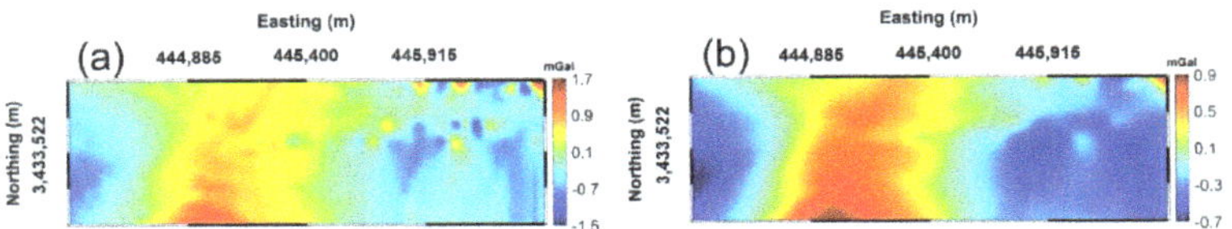

Figure 14. (**a**) The acquired field gravity anomaly (mGal); (**b**) the upward continued result at 50 m.

In addition to the detection of horizontal boundaries, the depth estimation of causative bodies is also important for the interpretation of gravity data [37]. In this paper, to estimate the top depth of the structures retrieved, we applied the popular tilt-depth (TD) technique [21,41]. Figure 16 shows the result of applying the tilt-depth method to the study area. The histogram of the estimated depths is shown in Figure 16c, which shows that 50% of these structures obtained exist at 10–40 m depth. The top depth of the majority of the lateral boundaries varies between 5 and 120 m, being disturbed randomly all over the Jalal Abad area. For detailed comparisons, the lateral boundaries in Figure 15i,k are superimposed on the tilt-depth map. Clearly, the positions of the buried source points are mostly correlated with the lateral boundaries determined by the GD_T and GD_H filters. Also, we can see that many edges match the geological features. The determined boundaries illustrate a good correlation with the trending structures and other anomalies within the study area. Therefore, with the help of the superimposed map from the processed results of the Gudermannian function filters and the tilt-depth method, it is possible to make reliable interpretations of the study area both qualitatively and quantitatively. TD results are in good agreement with the available drilling data and other works [39].

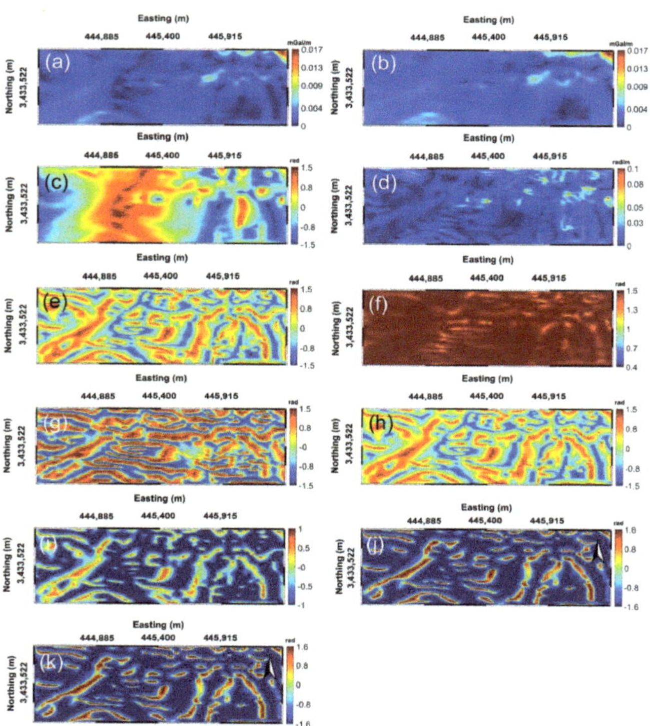

Figure 15. Results of edge detection in the Jalal Abad mine: (**a**) THG, (**b**) AS, (**c**) TA, (**d**) TA-THG, (**e**) TAHG, (**f**) ILF, (**g**) THVH, (**h**) TBHG, (**i**) FS, (**j**) GD_T, and (**k**) GD_H.

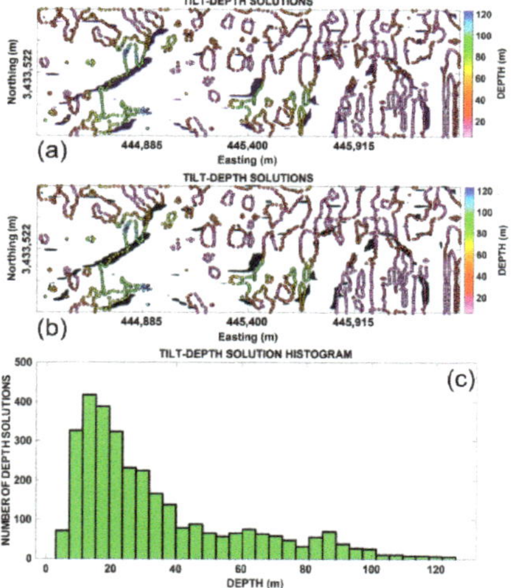

Figure 16. (**a**) The top depths derived from the TD technique (retrieved geological edges from Figure 15j are superimposed on the tilt-depth map); (**b**) the top depths derived from the TD technique (retrieved geological edges from Figure 15k are superimposed on the TD map); (**c**) histogram of the top depth estimated from the TD depth estimation method regarding the second field application.

8. Conclusions

We presented two effective edge enhancement methods, termed GD_T and GD_H, based on the Gudermannian function and the second-order derivative of the field. The efficiency of the two newly presented methods was testified using both synthetic potential data sets with various assumptions and two field anomalies including a real aeromagnetic response from the United States and ground-based gravity data from Iran. The processed results from the application of the two filters were compared further with those of other existing representative strategies. Synthetic experiments designed for different scenarios (2-D and 3-D; imposed and superimposed; noise-free and noise-contaminated) have verified that the proposed filters can effectively balance the large and small amplitudes due to causative sources situated at different depths, without compromising the resolution and accuracy of subtle details. Furthermore, both GD_T and GD_H yield the merit of avoiding erroneous artifacts when the anomalous bodies were assigned opposite-sign geophysical properties. The retrieved results from the application of the two filters to the acquired aeromagnetic and gravity data are in good agreement with the main geological structures within the study areas and the determined lateral information via the popular tilt-depth method. Hitherto, combining the conclusions generated from the theoretical experiments and field applications, the GD_T and GD_H filters are valuable tools for producing reliable, accurate, and robust edge detection results from processing potential field data sets. Both filters are recommended for locating potential ore and mineral deposits.

Author Contributions: Conceptualization, A.A. and H.A.; methodology, A.A., K.S. and H.A.; software, A.A. and H.A.; validation, A.A., K.S. and H.A.; formal analysis, K.S.; investigation, A.A., K.S., H.A. and V.E.A.; data curation, A.A., K.S. and H.A.; writing—original draft preparation, A.A., K.S., H.A. and V.E.A.; writing—review and editing, A.A., K.S., H.A., V.E.A. and C.L.; visualization, A.A., K.S., H.A., V.E.A. and C.L.; supervision, A.A., K.S. and H.A.; project administration: A.A., K.S., H.A., V.E.A. and C.L. All authors have read and agreed to the published version of the manuscript.

Funding: This research did not receive any specific grant from funding agencies in the public, commercial, or not-for-profit sectors.

Data Availability Statement: Data may be available from the corresponding authors upon reasonable request.

Acknowledgments: We thank the anonymous reviewers for their guidance and constructive comments which have greatly improved the quality of the paper. We also thank the United States Geological Survey (accessible at https://earthexplorer.usgs.gov (accessed on 18 August 2023)) for permission to use the geologic map and aeromagnetic data in Figures 9 and 10a. We are grateful for the help of Luan Thanh Pham (Vietnam National University, Vietnam) for preparing Figures 9 and 10a and Madankav Engineering Co., Ltd., for preparing Figures 13 and 14a.

Conflicts of Interest: The authors declare no conflict of interest.

References

1. Cooper, G.R.J.; Cowan, D.R. Enhancing potential field data using filters based on the local phase. *Comput. Geosci.* **2006**, *32*, 1585–1591. [CrossRef]
2. Sun, Y.; Yang, W.; Zeng, X.; Zhang, Z. Edge enhancement of potential field data using spectral moments. *Geophysics* **2016**, *81*, G1–G11. [CrossRef]
3. Dwivedi, D.; Chamoli, A. Source edge detection of potential field data using wavelet decomposition. *Pure Appl. Geophys.* **2021**, *178*, 919–938. [CrossRef]
4. Pham, L.T.; Oksum, E.; Do, T.D. Edge enhancement of potential field data using the logistic function and the total horizontal gradient. *Acta Geod. Geophys.* **2019**, *54*, 143–155. [CrossRef]
5. Pham, L.T.; Van Vu, T.; Le Thi, S.; Thi Trinh, P. Enhancement of Potential Field Source Boundaries Using an Improved Logistic Filter. *Pure Appl. Geophys.* **2020**, *177*, 5237–5249. [CrossRef]
6. Pham, L.T.; Oksum, E.; Le, D.V.; Ferreira, F.J.F.; Le, S.T. Edge detection of potential field sources using the soft sign function. *Geocarto Int.* **2021**, *37*, 4255–4268. [CrossRef]
7. Prasad, K.N.D.; Pham, L.T.; Singh, A.P. Structural mapping of potential field sources using BHG filter. *Geocarto Int.* **2022**, *37*, 11253–11280. [CrossRef]

8. Alvandi, A.; Toktay, H.D.; Pham, L.T. Capability of improved Logistics filter in determining lateral boundaries and edges of gravity and magnetic anomalies Tuzgolu, Area Turkey. *Iran. J. Min. Eng.* **2022**, *17*, 57–72. (In Persian) [CrossRef]
9. Ibraheem, I.M.; Tezkan, B.; Ghazala, H.; Othman, A.A. A New Edge Enhancement Filter for the Interpretation of Magnetic Field Data. *Pure Appl. Geophys.* **2023**, *180*, 2223–2240. [CrossRef]
10. Alvandi, A.; Ardestani, V.E. Edge detection of potential field anomalies using the Gompertz function as a high-resolution edge enhancement filter. *Bull. Geophys. Oceanogr.* **2023**, *64*, 279–300. [CrossRef]
11. Chen, A.G.; Zhou, T.F.; Liu, D.J.; Zhang, S. Application of an enhanced theta-based filter for potential field edge detection: A case study of the luzong ore district. *Chin. J. Geophys.* **2017**, *60*, 203–218.
12. Weihermann, J.D.; Ferreira, F.J.F.; Oliveira, S.P.; Cury, L.F.; de Souza, J. Magnetic interpretation of the Paranaguá Terrane, southern Brazil by signum transform. *J. Appl. Geophys.* **2018**, *154*, 116–127. [CrossRef]
13. Nasuti, Y.; Nasuti, A. NTilt as an improved enhanced tilt derivative filter for edge detection of potential field anomalies. *Geophys. J. Int.* **2018**, *214*, 36–45. [CrossRef]
14. Nasuti, Y.; Nasuti, A.; Moghadas, D. STA: A novel approach for enhancing and edge detection of potential field data. *Pure Appl. Geophys.* **2019**, *176*, 827–841. [CrossRef]
15. Eldosouky, A.M. Aeromagnetic data for mapping geologic contacts at Samr El-Qaa area, North Eastern Desert. *Egypt. Arab J. Geosci.* **2019**, *12*, 2. [CrossRef]
16. Cordell, L.; Grauch, V.J.S. Mapping Basement Magnetization Zones from Aeromagnetic Data in the San Juan Basin. In *The Utility of Regional Gravity and Magnetic Anomaly Maps*; SEG Publication: Houston, TX, USA, 1985; pp. 181–197. [CrossRef]
17. Roest, W.R.J.; Verhoef, J.; Pilkington, M. Magnetic interpretation using the 3-D analytic signal. *Geophysics* **1992**, *57*, 116–125. [CrossRef]
18. Oksum, E.; Le, D.V.; Vu, M.D.; Nguyen, T.H.T.; Pham, L.T. A novel approach based on the fast-sigmoid function for interpretation of potential field data. *Bull. Geophys. Oceanogr.* **2021**, *62*, 543–556.
19. Miller, H.G.; Singh, V. Potential field tilt—A new concept for location of potential field sources. *J. Appl. Geophys.* **1994**, *32*, 213–217. [CrossRef]
20. Verduzco, B.; Fairhead, J.D.; Green, C.M.; MacKenzie, C. New insights into magnetic derivatives for structural mapping. *Lead. Edge* **2004**, *23*, 116–119. [CrossRef]
21. Alvandi, A.; Ghanati, R. Using magnetic data for estimating the location of lateral boundaries and the depth of the shallow salt dome of Aji-Chai, East Azerbaijan Province, Iran. *Int. J. Min. Geo-Eng.* **2023**, *57*, 251–258. [CrossRef]
22. Alvandi, A.; Toktay, H.D.; Ardestani, V.E. Edge detection of geological structures based on a logistic function: A case study for gravity data of the Western Carpathians. *Int. J. Min. Geo-Eng.* **2023**, *57*, 267–274. [CrossRef]
23. Othman, A.A.; Ibraheem, I.M. Origin of El-Maghara Anticlines, North Sinai Peninsula, Egypt: Insights from Gravity Data Interpretation Using Edge Detection Filters. *Arab. J. Sci. Eng.* **2023**. [CrossRef]
24. Ferreira, F.J.F.; de Souza, J.; Bongiolo, A.B.S.; de Castro, L.G. Enhancement of the total horizontal gradient of magnetic anomalies using the tilt angle. *Geophysics* **2013**, *78*, J33–J41. [CrossRef]
25. Ma, G. Edge detection of potential field data using improved local phase filter. *Explor. Geophys.* **2013**, *44*, 36–41. [CrossRef]
26. Zhang, X.; Yu, P.; Tang, R.; Xiang, Y.; Zhao, C.J. Edge enhancement of potential field data using an enhanced tilt angle. *Explor. Geophys.* **2014**, *46*, 276–283. [CrossRef]
27. Eshaghzadeh, A.; Dehghanpour, A.; Kalantari, R.A. Application of the tilt angle of the balanced total horizontal derivative filter for the interpretation of potential field data. *Boll. Geofis. Teor. Appl.* **2018**, *59*, 161–178.
28. Cooper, G.R.J. Balancing images of potential field data. *Geophysics* **2009**, *74*, L17–L20. [CrossRef]
29. Gudermann, C. *Grundriss der Analytischen Sphärik*; DuMont-Schauberg: Köln, Germany, 1830. (In German)
30. Zwillinger, D. *CRC Standard Mathematical Tables and Formulae*; Chapman and Hall/CRC: Boca Raton, FL, USA, 2002. [CrossRef]
31. Romakina, L.N. The inverse Gudermannian in the hyperbolic geometry. *Integral Transform. Spec. Funct.* **2018**, *29*, 384–401. [CrossRef]
32. Sherrod, B.L.; Blakely, R.J.; Weaver, C.S.; Kelsey, H.M.; Barnett, E.; Liberty, L.; Meagher, K.L.; Pape, K. Finding concealed active faults: Extending the southern Whidbey Island fault across the Puget Lowland, Washington. *J. Geophys. Res.* **2008**, *113*, B05313. [CrossRef]
33. Johnson, S.Y.; Potter, C.J.; Miller, J.J.; Armentrout, J.M.; Finn, C.; Weaver, C.S. The southern Whidbey Island fault: An active structure in the Puget Lowland, Washington. *Geol. Soc. Am. Bull.* **1996**, *108*, 334–354. [CrossRef]
34. Saltus, R.W.; Blakely, R.J.; Haeussler, P.J.; Wells, R.E. Utility of aeromagnetic studies for mapping of potentially active faults in two forearc basins: Puget Sound, Washington, and Cook Inlet, Alaska. *Earth Planets Space* **2005**, *57*, 781–793. [CrossRef]
35. Blakely, R.J.; Wells, R.E.; Weaver, C.S. Puget Sound Aeromagnetic Maps and Data. U.S. Geological Survey Open-File Report. 1999; pp. 99–514. Available online: https://pubs.usgs.gov/of/1999/of99-514, (accessed on 15 March 2020).
36. Blakely, R.J.; Sherrod, B.L.; Hughes, J.F.; Anderson, M.L.; Wells, R.E.; Weaver, C.S. Saddle Mountain fault deformation zone, Olympic Peninsula, Washington: Western boundary of the Seattle uplift. *Geosphere* **2009**, *5*, 105–125. [CrossRef]
37. Alvandi, A.; Toktay, H.D.; Nasri, S. Application of direct source parameter imaging (direct local wave number) technique to the 2-D gravity anomalies for depth determination of some geological structures, for depth determination of some geological structures. *Acta Geophys.* **2022**, *70*, 659–667. [CrossRef]

38. Salem, A.; Williams, S.; Fairhead, J.D.; Ravat, D.J.; Smith, R. Tilt-depth method: A simple depth estimation method using first-order magnetic derivatives. *Lead. Edge* **2007**, *26*, 1502–1505. [CrossRef]
39. Joulidehsar, F.; Moradzadeh, A.; Doulati Ardejani, F. An Improved 3D Joint Inversion Method of Potential Field Data Using Cross-Gradient Constraint and LSQR Method. *Pure Appl. Geophys.* **2018**, *175*, 4389–4409. [CrossRef]
40. Madankav. Report of Geophysical Investigation of Jalal Abad Iron Mine. Madankav Company: Tehran, Iran, 2013. (In Persian)
41. Oruç, B. Edge detection and depth estimation using a tilt angle map from gravity gradient data of the Kozaklı-Central Anatolia Region, Turkey. *Pure Appl. Geophys.* **2010**, *168*, 1769–1780. [CrossRef]

Article

Mapping of the Structural Lineaments and Sedimentary Basement Relief Using Gravity Data to Guide Mineral Exploration in the Denizli Basin

Fatma Figen Altinoğlu 

Department of Geological Engineering, Pamukkale University, Denizli 20160, Turkey; faltinoglu@pau.edu.tr

Abstract: The Aegean Graben System is a complex tectonic structure in Western Anatolia and the Denizli Graben is a member of this system that hosts many geothermal springs, ore deposits, and travertine areas. In this study, the gravity data were analyzed to determine the subsurface geological structures and the depth model of the basin. The Bouguer gravity anomaly has a NW–SE pattern that is consistent with the general trend of the Denizli Basin. The pre-Neogene basement depths range from 0.1 km to 2.3 km. The Denizli Basin is composed of the Çürüksu Basin and the Laodikia sub-basin. The basins have undulated structures with many depressions; the deepest depression region is in the northern part of the Çürüksu Basin, which is close to the Pamukkale Fault Zone. In addition, the new gravity lineament map was obtained by using new-generation edge detection techniques: the tilt angle of the horizontal gradient amplitude (TAHG), and fast sigmoid-edge detection (FSED) of gravity data. The new proposed lineament map shows that the Denizli Basin has complex structures consisting of NW–SE, E–W, and NE–SE trending lineaments, and the major NW–SE trending faults and NE–SW trending lineaments control the main structural configuration. The uplifts and depressions in the basin deposit and the intersection area of lineaments are promising prospective areas for mineral deposits and have energy resource potential.

Keywords: gravity; edge detection; Denizli Basin; structural mapping

Citation: Altinoğlu, F.F. Mapping of the Structural Lineaments and Sedimentary Basement Relief Using Gravity Data to Guide Mineral Exploration in the Denizli Basin. *Minerals* **2023**, *13*, 1276. https://doi.org/10.3390/min13101276

Academic Editors: Luan Thanh Pham, Saulo Pomponet Oliveira, Le Van Anh Cuong and Michael S. Zhdanov

Received: 15 August 2023
Revised: 20 September 2023
Accepted: 25 September 2023
Published: 29 September 2023

1. Introduction

Researching mineral deposits, hydrocarbon and geothermal potential, hot springs, and travertine occurrences requires knowledge of tectonic structures such as faults, fractures, their interactions, intersection zones, and the progression of these structures at the subsurface. The potential field methods in geophysics are the widespread methods to identify boundaries of subsurface structures and various techniques have been used to define the edges of density and magnetization structures [1–5]. On the other hand, identifying the presence and location of smooth depressions and uplifts in a sedimentary basin's basement relief may assist in locating stratigraphic and structural oil traps [6,7]. In such research, potential field data on larger-scale surveys [8–10] and seismic data in more local target areas are used very effectively [11–13].

Western Anatolia is one of the major continental extension regions in the world [14,15], characterized by active extensional tectonics [16–19] and a thin continental crust [20,21]. Due to the compression and extension regime in Western Anatolia, many horst-graben structures extending in various directions called the Aegean Graben System, have been formed in the region [22] (Figure 1a). On the other hand, the Aegean Graben System is also known as a remarkable region in Western Anatolia that hosts many geothermal fields, mineral deposits, oil, gas, and hydrocarbons. Therefore, the tectonic evolution and geological structure of the region have attracted the attention of many researchers [14–18,20–33]. In this topic, there are many studies in the literature investigating hydrocarbon resources [34–37], geothermal potential [38–47], and travertine formation [48–53] in the Western Anatolian grabens.

The research area of this study includes the NW–SE trending Denizli Graben as part of the Aegean Graben System, which was formed as an extensional basin during the Late Miocene-Quaternary neotectonic evolution of SW Turkey (Figure 1b). The area is of great economic significance since it comprises geothermal areas, numerous hot springs, travertines, and marble quarries. It has been subjected to a number of geological studies due to its tectonic position [19,32,54,55]. However, geophysical studies focusing on determining the depth structure of the basin and its subsurface geologic structural elements have been few in the literature. A review of the results of previous research on this topic is as follows: Sarı and Şalk [56] investigated the sedimentary thickness of Aegean grabens using 2D and 3D analysis of Bouguer gravity anomalies and reported a sediment thickness of over 2 km in the Denizli Graben. Altinoğlu et al. [57] provided the 3D topography of the upper/lower crustal boundary of the Denizli Basin by inversion of gravity data and used conventional edge detection methods such as the horizontal gradient, analytic signal, and tilt angle to detect lineaments around the basin for the first time. Altinoğlu [58] determined the topography of the sedimentary basement in the southeastern part of the Denizli region and reported a thickness of 2–2.2 km in the Honaz region. Ekinci et al. [59] determined the sediment topography along Aegean grabens by evaluating profile gravity data, one of which is located close to the northeast boundary of the Denizli Basin. They concluded that the maximum sediment thickness in the Aegean Graben System is approximately 1.5 km. They have also used the geothermal wells drilled by MTA. The lithological log of the geothermal well, which is drilled in the Denizli Basin [59,60], is given in Figure 1c. Apart from the studies on the sedimentary depth structure of the study area, some other studies investigated the deeper crust and thermal structure of the region. In the study by Bilim [61], the tectonic and thermal structure of the eastern part of Western Anatolia was examined using aeromagnetic and Bouguer gravity anomalies. In their study, the northeast part of Denizli city has been determined with shallow CPD. Kaypak and Gökkaya [62] investigated the upper crustal velocity structure in the Denizli basin with 3D local earthquake tomography and reported that the main geothermal fields in the Denizli region are mostly located along graben systems and main fault systems. Irmak [63] determined the focal mechanisms of the small earthquakes in the Denizli Basin. Erbek-Kiran et al. [64] studied the upper crustal structure of the Denizli Graben using Bouguer gravity data and seismic data. The Bouguer anomalies in their study were evaluated to determine the basement topography of the upper-lower crust boundary by inversion of gravity data. Based on the interpretation of two seismic reflection datasets, they concluded that sedimentary sequences in the Denizli Basin can be considered oil traps.

Knowledge of the sediment base topography and interconnected lineaments may yield significant gains in the course of the investigation of a region's mineral deposits and energy resources potential. Most of the previous geophysical studies in the Denizli Graben were focused on deeper crustal architecture, and the studies regarding shallow sediments and basement characteristics mainly address a narrow part of the basin area. This research aims to determine the three-dimensional (3D) depth variation of the Denizli sedimentary basin by inverting its gravity anomalies and delineating subsurface linear structural elements such as faults or contacts using recent enhanced techniques of gravity data interpretation with the expectation of providing key information for further mineral and energy resource studies in the region. For this purpose, firstly, the 3D sedimentary basement topography of the basin area was determined by an iterative inversion process of the gravity data utilized in the frequency and space domain. Next, the gravity lineaments were obtained using two different edge detector filters that are capable of balancing anomalies of shallow and deep sources. Finally, some of the featured findings obtained from the study, which improve previous geological information, are discussed in the context of potential sites for future hydrocarbon, geothermal, and mineral explorations.

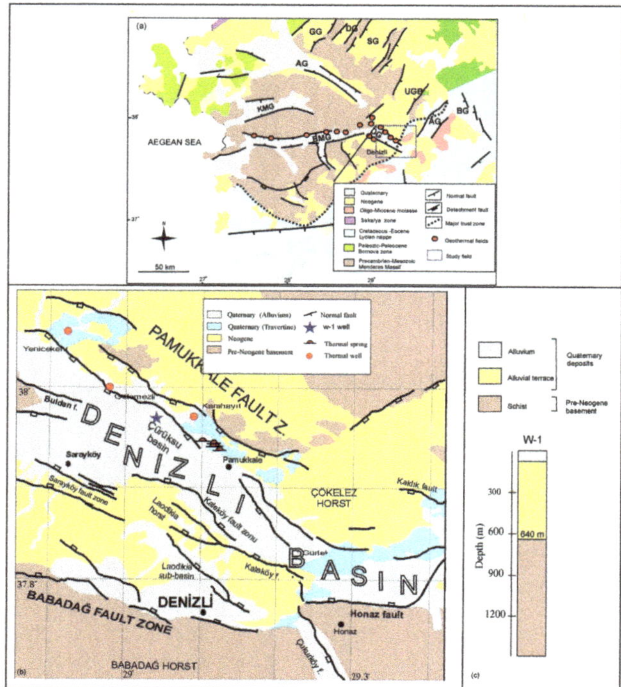

Figure 1. (**a**) The simplified geological map of Western Anatolia, modified from [30,31]. The dashed box shows the area coverage of the study area. GG: Gördes Graben, DG: Demirci Graben, SG: Selendi Graben, AG: Alaşehir Graben, KMG: Küçük Menderes Graben, BMG: Büyük Menderes Graben, DG: Denizli Graben, UGB: Uşak-Güre Basin, AG: Acıgöl Graben, and BG: Baklan Graben. (**b**) The geological map of the study area, modified from [53,65]. (**c**) Lithological log of the geothermal well drilled by General Directory of Mineral Exploration and Research Company of Turkey (MTA) in Denizli Basin (modified from [59,60]).

2. Tectonic Settings of Denizli Basin

The Denizli Basin area is a transition region of Aegean grabens with various orientation and extension directions: E–W trending Büyük Menderes Graben, Küçük Menderes Graben, NW–SE-trending Alaşehir Graben, and NE-SW trending Baklan Graben, Acıgöl Graben and Burdur Graben (Figure 1a). The NW–SE trending Denizli Basin is approximately 50 km in length and 7–28 km in width. The formation of the basin started around 14 million years ago, with an estimated average slip rate of 0.14–0.15 mm/year [15,19,66]. The basin was developed on pre-Oligocene metamorphic rocks found in the Menderes Massif, Lycian Nappes, and an Oligocene-Lower Miocene molassic sequence [66] (Figure 1a). The sedimentary fill of the Denizli Basin comprises Early Miocene to Quaternary units, including alluvial, fluvial, and lacustrine deposits, superimposed on the metamorphic bedrock. The geological substrate of the region consists of Paleozoic schist-quartzite, gneiss, and marbles, which form the pre-Neogene basement of the Western Anatolian extensional province.

The Denizli Basin is bordered by the Çökelez Horst and Pamukkale Fault Zone to the north, the Babadağ Horst and Babadağ Fault Zone to the SW, and Honazdağ Horst and Honaz Fault to the south. The Kaleköy Fault Zone extending in the central part of the basin divides the basin into two Quaternary basins, namely the Çürüksu Basin in the north and the Laodikia sub-basin in the south [54] (Figure 1b).

3. Data and Methods

The Bouguer gravity data of the study region were obtained from the General Directory of Mineral Exploration and Research Company of Turkey (MTA) and Turkish Petroleum Corporation (TPAO). No detailed information is available on instruments. The gravity values were tied to MTA and General Command of Mapping base stations related to the Potsdam 981260.00 mGal absolute gravity value accepted by the International Union of Geodesy and Geophysics in 1971. The latitude correction was carried out by using the Gravity Formula of 1967 and the terrain reduction was carried out with the aid of the Hammer tables out to Zone J. The Bouguer density was assumed to be 2.67 gr cm^{-3}. Then, the data were gridded over 1 km^2 areas. The Bouguer gravity anomaly map of the Denizli Graben with 10 mGal intervals is given in Figure 2.

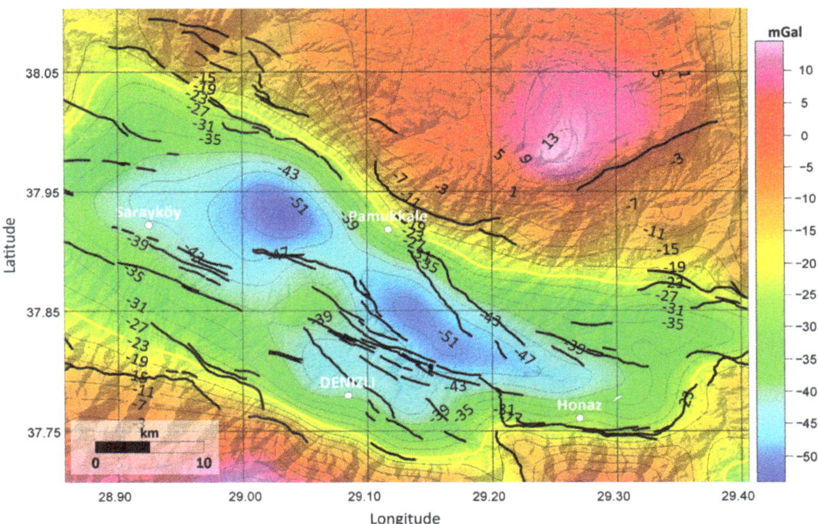

Figure 2. The Bouguer gravity anomaly map of the study area. The black lines indicate the active faults in the region [67].

The Bouguer anomaly map indicates the entire impact of shallow and deep sources' gravity responses due to the density heterogeneity of the subsurface. Therefore, regional-residual separation for the relevant anomalies in determining the depth model of sedimentary basins is the first step of the processing of Bouguer gravity data. Many techniques are used in the literature to separate anomalies, such as subtracting a fixed value referenced to a geological unit from the Bouguer anomaly, trend surface analysis, polynomial fitting, and a variety of filtering schemes in both the time and Fourier domains. Due to the study area being small and the basin bordered by pre-Neogene basements in the north and south, the anomaly value passing through the basin boundary was determined. In the Bouguer gravity anomaly map, −23 mGal contours pass through the basin boundary as shown with the yellow contour line in Figure 2. To increase the impact of shallow geological features, residual anomalies were created by subtracting these values from gravity anomalies. The residual gravity anomaly map is given in Figure 3.

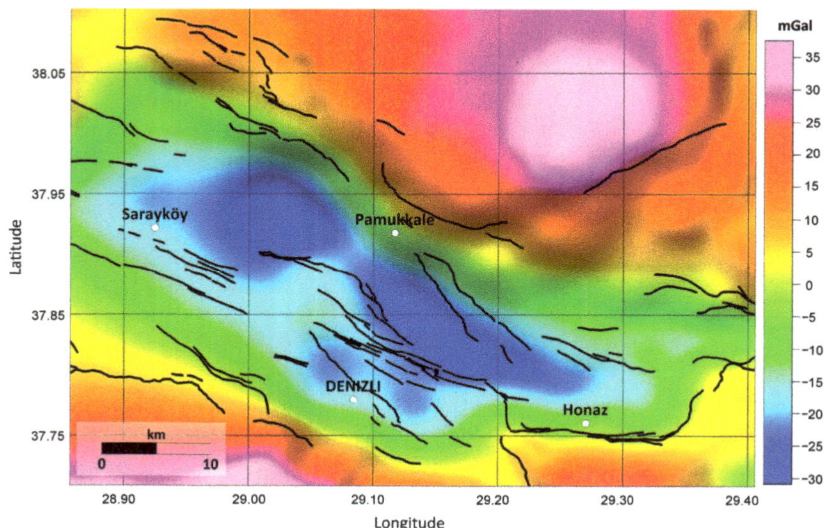

Figure 3. The residual gravity anomaly map of the study area. Black lines indicate the active faults in the region [67].

In the inversion of gravity data to recover the sediment-basement relief, a constant value of density contrast is often used for shallow basins, while varying density contrasts with depth are considered for deep basins. As the Aegean grabens are not expected to be a deep basin, a constant density contrast value of 0.5 g cm^{-3} was used in the inversion process, based on the information obtained from the boreholes drilled for geothermal exploration by MTA [56,59]. Due to the low density of sediments in comparison to a surrounding rock density, a sedimentary cover is generally associated with negative gravity anomalies [59]. Therefore, only the negative anomaly corresponding to the basin fill is taken into consideration in the inversion scheme, whereas the positive anomalies in the observed gravity anomaly are ignored.

3.1. Basin Depth Model

The gravity anomalies of the study area are inverted by using an iterative procedure proposed in [68] which combines the advantages of both the frequency domain and space domain techniques applied for the forward calculation of the gravity anomalies and the modification of calculated depth values after each iteration. The algorithm consists of three main steps to invert the gravity anomalies for the basement depth h.

The infinite horizontal slab equation is first used to generate an initial approximation of the depth to the basement interface at each observation [69]:

$$h^1_{(i,j)} = -\frac{1}{\lambda} ln\left(1 - \frac{\lambda \Delta g_{obs(i,j)}}{2\pi \Delta \rho_0}\right) \tag{1}$$

where $\Delta g_{obs(i,j)}$ is the observed field at a mesh point (i, j), $\Delta \rho = \Delta \rho_0 e^{-\lambda z}$ is the density model where λ is the decay constant of the decrease in density with depth z, and $\Delta \rho_0$ is the density contrast at the surface. The FFT-based algorithm in [70] is used in the next step to calculate the gravity response of the density interface of the initial model:

$$\Delta g = \frac{2\pi \gamma \Delta \rho_0}{\lambda}\left(1 - e^{-\lambda z_0}\right) + 2\pi \gamma \Delta \rho_0 e^{-\lambda z_0} \times F^{-1}\left[\frac{e^{-|k|z_0}}{|k| + \lambda}\left(F\left[1 - e^{-\lambda \Delta h}\right] - \sum_{n=1}^{\infty} \frac{(-|k|)^n}{n!} F\left[e^{-\lambda \Delta h} \Delta h^n\right]\right)\right] \tag{2}$$

where k is the wavenumber, γ is the gravitational constant, $F[\ ]$ and $F^{-1}[\ ]$ represent the Fourier and inverse Fourier transforms, respectively, and z_0 is the mean depth of the basement interface as a reference level in the sense that $h = z_0 + \Delta h$ describes the depth of the interface. The depth model is modified using the space domain relation in the final stage found in [71]:

$$h_{(i,j)}^{(t+1)} = \frac{\Delta g_{obs(i,j)}}{\Delta g_{calc(i,j)}} h_{(i,j)}^{(t)} \tag{3}$$

where t denotes the number of iterations. When the desired fit between the calculated and observed gravity fields, as measured by the root mean square error (RMS) between them, occurs between the second and third steps, the gravity response of the modified depths is calculated once more, and the process is repeated until the inverted model is satisfied.

3.2. Detection of Lineaments

To identify the lineaments from potential field data, edge detection techniques play an important role, and they are generally based on vertical and horizontal derivatives of the field. The total horizontal gradient amplitude (THG) [72] is a popularly used method to highlight the boundaries of potential field sources. The authors of [73] showed that the use of the maximum values of the total horizontal gradient can outline the source edges. The THG filter is defined as:

$$\text{THG} = \sqrt{\left(\frac{\partial F}{\partial x}\right)^2 + \left(\frac{\partial F}{\partial y}\right)^2}, \tag{4}$$

where $\partial F/\partial x$ and $\partial F/\partial y$ are the gradients of the field F in horizontal x and y directions, respectively.

Some authors used higher-order derivatives to increase the resolution of the edge detection results [74–76]. The main disadvantage of these methods based on the amplitude of gravity gradients is that they cannot equalize the amplitudes of anomalies caused by sources located at different depths.

Miller and Singh [76] showed that the phase-based methods that can produce balanced results using the arctangent function of the ratio of the vertical gradient to the horizontal gradient amplitude, called the tilt angle (TA), can equalize large and small amplitudes at the same time. The TA filter is given as:

$$\varnothing = \tan^{-1}\left[\frac{\frac{\partial F}{\partial z}}{\sqrt{\left[\left(\frac{\partial F}{\partial x}\right)^2 + \left(\frac{\partial F}{\partial y}\right)^2\right]}}\right] \tag{5}$$

Ferreira et al. [77] proposed the use of the tilt angle of the horizontal gradient amplitude (TAHG) method for balancing signals from shallow and deep structures more effectively. This method represents an improvement over the tilt angle method. The TAHG filter is defined as:

$$\text{TAHG} = \tan^{-1}\frac{\frac{\partial THG}{\partial z}}{\sqrt{\left(\frac{\partial THG}{\partial x}\right)^2 + \left(\frac{\partial THG}{\partial y}\right)^2}} \tag{6}$$

where $\partial THG/\partial x$, $\partial THG/\partial y$, and $\partial THG/\partial z$ are the horizontal and vertical derivatives of the total horizontal gradient of F. The technique offers the advantage of equalizing the total horizontal gradient (THG) while being less dependent on depth considerations. This method showcases superior resolution in detecting body limits compared to conventional approaches.

The other method called fast sigmoid-edge detection (FSED) by the authors of [78] employs a modified fast sigmoid function of the ratio of the derivatives of the THG in order to improve the resolution and accuracy of estimated edges. The source edges are identified using its maximum values. The formulation of the FSED is given as:

$$\text{FSED} = \frac{R-1}{1+|R|},$$ (7)

where

$$R = \frac{\frac{\partial THG}{\partial z}}{\sqrt{\left(\frac{\partial THG}{\partial x}\right)^2 + \left(\frac{\partial THG}{\partial y}\right)^2}}$$ (8)

The FSED filter can balance the large and small amplitude anomalies due to sources of different depths and properties, as well as the edges of the sources being determined with higher resolution and more accurately. As another advantage, this filter also does not produce false edges when the anomalous sources contain opposite sign density contrasts simultaneously [78].

4. Results

The Bouguer gravity anomalies range from -50 to 10 mGal, with a maximum amplitude variation of 60 mGal (Figure 2). High positive anomalies are seen on the northern and southern parts of the map, which is explained by the presence of metamorphic basement units with a high density. The residual gravity anomaly map of the study area is given in Figure 3, the NW–SE trending anomalies are consistent with trend of the Denizli Basin. The anomaly contours at the northern and southern edges of the basin display a dense trend corresponding to the basin's extension. This type of contour arrangement in gravity maps may indicate the presence of linear structures such faults or contacts. Here, the positive residual anomaly pattern observed in the northern and southern boundaries along the graben is consistent with the regional geology of pre-Neogene basement units.

4.1. Sedimentary Basin Depth Model

Mapping the sediment- pre-Neogene basement topography is an important aspect of the geophysical investigation of oil, gas, and mineral resources. In the inversion of gravity data, the iteration ends when the RMS between the calculated and observed gravity data at any point of the iteration increases relative to the preceding step. In this case, the divergence criterion in the RMS error is used as the algorithm's termination model. A window length of the three data points from the edges of the data was also not taken into consideration throughout the calculations in order to prevent edge effects. The termination of the iterative process has been achieved at the nineteenth iteration (Figure 4). The RMS error between the observed and calculated gravity anomalies of the initial depth approximation is 1.22 mGal and decreases to 0.32 mGal for the optimum solution. The obtained RMS error of 0.32 mGal corresponds to approximately 1% of the maximum amplitude of the input data. In gravity inversion depth modeling, this fit is a sufficient fit in terms of modeling.

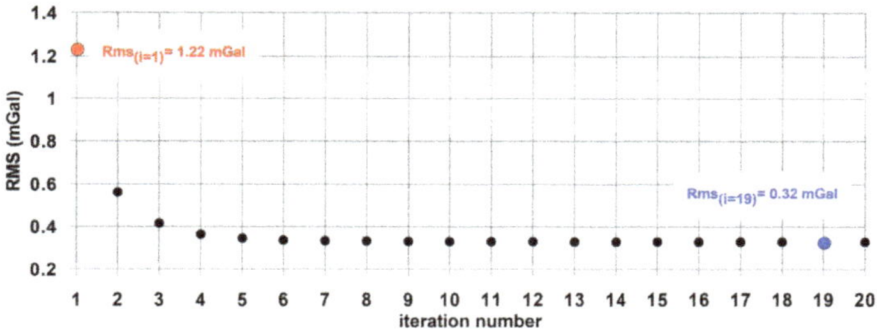

Figure 4. The graph of RMS error variation against the iteration number.

The 3D sediment-basement depth configuration map has been enhanced after the nineteenth iteration was given in Figure 5. The pre-Neogene basement depths range from 0.1 km to 2.3 km. The Denizli Basin consists of two basins [19,32]. In the Laodikia sub-basin, there are two depression areas, D1 and D2, with a maximum thickness of 1.7 km and a ridge between these depressions. The Çürüksu Basin has an undulation structure. Five depressions are observed within the basin. The deepest depression region, D3, is located in the northern part of the basin with the largest areal extension settlement. Pamukkale travertine, the largest and most famous travertine formation that gives the name to Pamukkale Fault Zone, and many other geothermal field occurrences are close to the D3 depression. Another notable feature is a ridge located between the depressions D3 and D4 in the sediment topography. This ridge has a relatively shallow burial depth (about 1.2–1.6 km) and may contain structural traps such as faulted anticlines and fault block traps. This undulated structure was also given by the interpretation of seismic reflection data in the study [64]. This area could be a promising prospective area for hydrocarbon and geothermal exploration. The Çürüksu Basin narrows towards the Honaz district and three other depressions (D5, D6, and D7) are observed within the basin. In previous studies, Sarı and Şalk [56] analyzed the gravity anomalies of Western Anatolia with variable density contrast by 2D and 3D inversion techniques and they reported a sediment thickness of 2 km in the Denizli Graben. Altinoğlu [58] studied in the SW part of the Denizli region and reported a sediment thickness of 2–2.2 km in the Honaz region. In the study of Ekinci et al. [59], the authors conducted a study using gravity data inversion with global optimization to map basement reliefs of Western Anatolia Grabens. For a profile passing through the Denizli Graben located close to the northern edge of the graben, they presented an undulated structure and reported a basement depth of 1.4 km in the Pamukkale region and 0.86 km in the west of the Honaz district. In this study [59], the lithological log of some boreholes drilled by MTA for geothermal exploration in the Western Anatolian Grabens is given. In the sediment-basement depth map, the basement depth value of 700 m at the location corresponding to the W-1 borehole drilled in the Denizli Basin is very close to the geological information obtained from the lithological log of this well ($h_{W-1} = 640$ m). The gravity and seismic data of the Denizli region were studied by [64]. They determined the sediment thickness as 2.64 km in the Çürüksu Graben by evaluation of two seismic profiles in the Denizli Graben. The results of this study are relatively consistent with the results of previous studies.

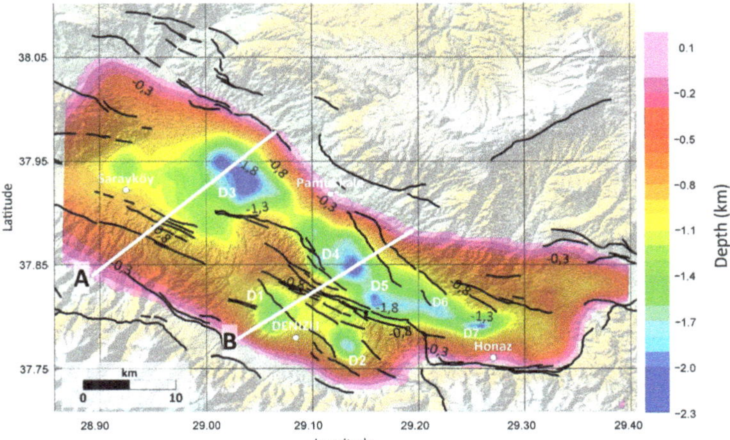

Figure 5. The sediment pre-Neogene basement relief map of Denizli Graben. The black lines show the active fault of the region given by MTA [67], and the white lines indicate the gravity profiles. The topographic elevation data were downloaded from [79].

The cross-section data from the inverted basin model, with the calculated and observed gravity anomalies along two profiles, are shown in Figure 6. Profile A starts from the north of Babadağ district and ends in the west of Pamukkale in the SW–NE direction. As demonstrated in Profile A, the sediment thickness is thin in the southwest and thickens towards the northeast. The NE boundary of the basin is steep. The observed and calculated anomalies are mostly in accordance and the maximum difference between the observed and calculated anomalies is 0.05 mGal in Profile A. Profile B starts from the west of Denizli province and first passes the Laodeika sub-basin, then the Karakova Horst, separating the Loadeika sub-basin and the Çürüksu Basin, and finally the Çürüksu Basin in the north. The pre-Neogene basement surface is at 1.5 km in the Laodeika sub-basin (D1 depression) and 2.2 km in the Çürüksu Basin (D4 depression). The pre-Neogene basement topography has been uplifted as seen in the central part of the basin model. The maximum difference between the observed and calculated anomalies is 0.2 mGal in Profile B.

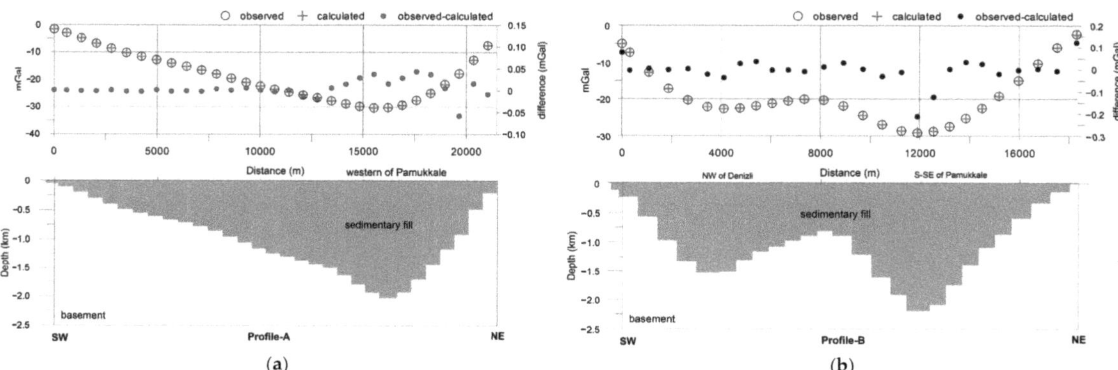

Figure 6. The inverted basin model across Profiles A and B with the observed and calculated gravity anomalies. (**a**) Profile A, and (**b**) Profile B (the location of profiles is given in Figure 5).

4.2. Lineament Detection

Lineaments are basically geological structures that represent faults, joints, lithological contacts, and shear zones which are important locations for gathering data in regional geological and tectonic investigations. Edge detection algorithms applied to gravity data are particularly effective for mapping linear features and identifying the boundaries of a geological structure [80–84]. The TAHG and FSED edge detection methods were applied to the residual gravity anomaly map of the Denizli region to determine lineaments. The maximum values of these maps are located at the boundaries of subsurface features and causative source bodies [78,80]. These new-generation filters have a higher resolution than traditional filters such as horizontal gradient, analytic signal, and tilt angle, which were previously applied to Western Anatolia gravity data [85]. These traditional filters based on the amplitude of the gravity gradients mainly have the disadvantage of being unable to equalize the amplitudes of anomalies caused by source bodies located at different depths. More detailed information about the filters can be found in [78].

The results indicate that the maximum positive anomalies of TAHG and FSED maps (Figure 7a,b) are consistent with the regional NW–SE structural trend and indicate many linear structures with different trends. On the other hand, the FSED map produced higher-resolution lineaments than the TAHG map.

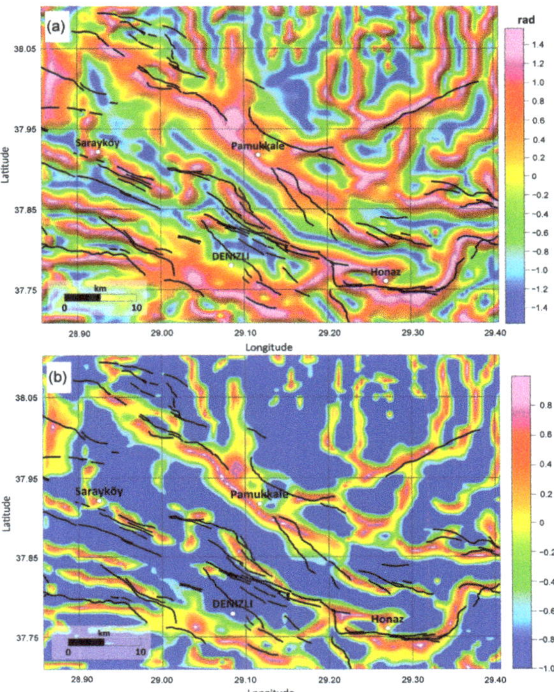

Figure 7. (**a**) Tilt angle of the horizontal gradient amplitude (TAHG), (**b**) FSED map of the study area. The black lines show the active fault of the region given by MTA [67].

The main boundary faults of the Denizli Graben, the Pamukkale Fault Zone in the north, the Babadağ Fault Zone in the southeast, and the Honaz fault in the south, are well delineated in the TAHG and FSED maps. Additionally, the Kaleköy Fault Zone, which separates the Çürüksu Basin and Laodeika sub-basin, can easily be seen on the maps. The newly identified linear structures will be referred to as "L" in the text. As a result of this study, many lineaments were identified, but buried structures and intersecting lineaments were emphasized. The new proposed lineament map of the study area is given in Figure 8.

Lineaments L1, L2, L3, and L4 are probably buried structures under the sedimentary fill (Figure 8). Lineament L1 extends from the Sarayköy fault to the north in an approximately N–S direction. The intersection of this new lineament with the Sarayköy fault is promising for hydrocarbon and geothermal exploration. This is because there are many geothermal fields and hot springs around this region (Kızıldere, Ortakçı, Umut thermal, Buldan, Yenice-Kamara). Lineament L2 is located between Sarayköy and Pamukkale and is extended in the N–S direction and borders the D3 depression in the sediment. The NW–SE trending L3 lineament is extended from the north of the Kaleköy Fault Zone and also borders the D3 depression from the south. The N–S trending lineament, L4, borders the Karakova Horst from the west. Lineament L5 may have been the extension of the Kaleköy fault towards the Honaz district, which was also suggested by [58]. This segment is probably buried in the sediment fill and so is undefined in the active fault map. Another NE–SW trending Lineament, L6, has been detected west of the Karateke village, in the western part of the Honaz region. The existence of travertine occurrences and springs around this region was mentioned in [44,86] when the authors declared that the travertine occurrences in the Denizli Basin are related to the fault system. In addition, the L7 lineament may have been the westward continuation of the Kaklık fault.

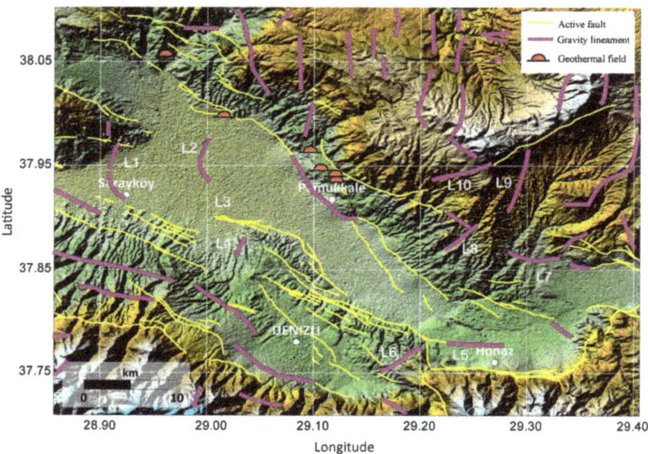

Figure 8. The compiled structural lineament map of the study area is superimposed on a topographic map [79]. The active faults of the region are yellow lines [67], and the new proposed lineaments are shown with purple lines.

The other NE–SW directional lineaments have been identified in the east-northeast part of the study area, south of the Çökelezdağ Horst. Some other newly detected NE–SW direction lineaments (L8, L9, and L10) are extended from south of the Çökelez Horst to the Pamukkale Fault Zone. These NE–SW trending lineaments may have developed under the influence of the NW–SE trending expansion phase, as suggested in [86]. The intersection areas of these divergent linear structures can be possible sites for mineral deposition investigations.

5. Conclusions

The Denizli Graben area is an important structure of the Western Anatolia Horst Graben system. The basin hosts many geothermal areas, hot springs, and travertine deposits. The Denizli Graben consists of Çürüksu Graben and Laodikia sub-graben, as mentioned in previous geological studies. The sediment–pre-Neogene basement interface topography of the whole Denizli Basin is established in this study. The deepest and the largest areal depression region of Çürüksu Graben is in the Pamukkale region with a 2.3 km depth. The basin narrows towards the south. The ridges between two depressions may be a tectonic uplift or basement ridge and could be the initial exploration area for hydrocarbon or geothermal exploration. The produced lineament map of the region includes many lineaments in different orientations which are unspecified in the active fault map of MTA. These lineaments and their intersection points with each other or known faults are suggested as the key structures to look for in geothermal, travertine, and mineral exploration studies.

Funding: This research received no external funding.

Data Availability Statement: The data presented in this study are not publicly available due to the fact that they must be taken from the General Directory of Mineral Exploration and Research Company of Turkey (MTA) and Turkish Petroleum Corporation (TPAO).

Conflicts of Interest: The authors declare no conflict of interest.

References

1. Satyakumar, A.V.; Pandey, A.K.; Singh, A.P.; Tiwari, V.M. Delineation of structural and tectonic features in the Mahanadi basin, eastern India: New insights from remote sensing and land gravity data. *J. Asian Earth Sci.* **2022**, *227*, 105116. [CrossRef]
2. Pham, L.T.; Oksum, E.; Kafadar, O.; Trinh, P.T.; Nguyen, D.V.; Vo, Q.T.; Le, S.T.; Do, T.D. Determination of subsurface lineaments in the Hoang Sa islands using enhanced methods of gravity total horizontal gradient. *Vietnam J. Earth Sci.* **2022**, *44*, 395–409.
3. Alrefaee, H.; Soliman, M.; Merghelani, T. Interpretation of the subsurface tectonic setting of the Natrun Basin, north Western Desert, Egypt using Satellite Bouguer gravity and magnetic data. *J. Afr. Earth Sci.* **2022**, *187*, 104450. [CrossRef]
4. Liu, J.; Li, S.; Jiang, S.; Wang, X.; Zhang, J. Tools for Edge Detection of Gravity Data: Comparison and Application to Tectonic Boundary Mapping in the Molucca Sea. *Surv. Geophys.* **2023**, 1–30. [CrossRef]
5. Kamto, P.G.; Oksum, E.; Pham, L.T.; Kamguia, J. Contribution of advanced edge detection filters for the structural mapping of the Douala Sedimentary Basin along the Gulf of Guinea. *Vietnam. J. EARTH Sci.* **2023**. [CrossRef]
6. Chakravarthi, V.; Shankar, G.B.K.; Muralidharan, D.; Harinarayana, T.; Sundararajan, N. An integrated geophysical approach for imaging subbasalt sedimentary basins: Case study of Jam River Basin, India. *Geophysics* **2007**, *72*, B141–B147. [CrossRef]
7. Silva, J.B.; Oliveira, A.S.; Barbosa, V.C. Gravity inversion of 2D basement relief using entropic regularization. *Geophysics* **2010**, *75*, I29–I35. [CrossRef]
8. Babu, H.R. Basement structure of the Cuddapah Basin from gravity anomalies. *Tectonophysics* **1993**, *223*, 411–422. [CrossRef]
9. Blakely, R.J.; Christiansen, R.L.; Guffanti, M.; Wells, R.E.; Donnelly-Nolan, J.M.; Muffler, L.J.P.; Clynne, M.A.; Smith, J.G. Gravity anomalies, Quaternary vents, and Quaternary faults in the southern Cascade Range, Oregon and California: Implications for arc and backarc evolution. *J. Geophys. Res. Solid Earth* **1997**, *102*, 22513–22527. [CrossRef]
10. Aydogan, D. Extraction of lineaments from gravity anomaly maps using the gradient calculation: Application to Central Anatolia. *Earth Planets Space* **2011**, *63*, 903–913. [CrossRef]
11. Cherepanova, Y.; Artemieva, I.M.; Thybo, H.; Chemia, Z. Crustal structure of the Siberian craton and the West Siberian basin: An appraisal of existing seismic data. *Tectonophysics* **2013**, *609*, 154–183. [CrossRef]
12. Anukwu, G.C.; Khalil, A.E.; Nawawi, M.; Younis, A.M. Delineation of shallow structures in the vicinity of Ulu Slim hot spring using seismic refraction and MASW techniques. *NRIAG J. Astron. Geophys.* **2020**, *9*, 7–15. [CrossRef]
13. Salaun, N.; Toubiana, H.; Mitschler, J.-B.; Gigou, G.; Carriere, X.; Maurer, V.; Richard, A. High-resolution 3D seismic imaging and refined velocity model building improve the image of a deep geothermal reservoir in the Upper Rhine Graben. *Geophysics* **2020**, *39*, 857–863. [CrossRef]
14. Jackson, J.A.; Mckenzie, D.P. The relationship between plate motions and seismic moment tensors and rates of active de-formation in the Mediterranean and Middle East. *Geophys. J. Int.* **1988**, *93*, 45–73. [CrossRef]
15. Westaway, R. Neogene evolution of the Denizli region of western Turkey. *J. Struct. Geol.* **1993**, *15*, 37–53. [CrossRef]
16. McKenzie, D. Active tectonics of the Alpine—Himalayan belt: The Aegean Sea and surrounding regions. *Geophys. J. Int.* **1978**, *55*, 217–254. [CrossRef]
17. Şengör, A.M.C.; Yilmaz, Y. Tethyan evolution of Turkey: A plate tectonic approach. *Tectonophysics* **1981**, *75*, 181–241. [CrossRef]
18. Bozkurt, E. Origin of NE-trending basins in western Turkey. *Geodin. Acta* **2003**, *16*, 61–81. [CrossRef]
19. Kaymakçı, N. Kinematic development and paleostress analysis of the Denizli Basin (Western Turkish): Implications of spatial variation of relative paleostress magnitudes and orientations. *J. Asian Earth Sci.* **2006**, *27*, 207–222. [CrossRef]
20. Jolivet, L.; Faccenna, C.; Huet, B.; Labrousse, L.; Le Pourhiet, L.; Lacombe, O.; Lecomte, E.; Burov, E.; Denèle, Y.; Brun, J.-P.; et al. Aegean tectonics: Strain localisation, slab tearing and trench retreat. *Tectonophysics* **2012**, *597-598*, 1–33. [CrossRef]
21. Faccenna, C.; Becker, T.W.; Auer, L.; Billi, A.; Boschi, L.; Brun, J.P.; Capitanio, F.A.; Funiciello, F.; Horvàth, F.; Jolivet, L.; et al. Mantle dynamics in the Mediterranean. *Rev. Geophys.* **2014**, *52*, 283–332. [CrossRef]
22. Dewey, J.F.; Şengör, A.M.C. Aegean and surrounding regions: Complex multiplate and continuum tectonics in a convergent zone. *Geol. Soc. Am. Bull.* **1979**, *90*, 84–92. [CrossRef]
23. Bozkurt, E.; Mittwede, S.K. Introduction: Evolution of continental extensional tectonics of western Turkey. *Geodin. Acta* **2005**, *18*, 153–165. [CrossRef]
24. Şengör, A.M.C.; Görür, N.; Şaroglu, F. Strike-slip faulting and related basin formation in zones of tectonic escape: Turkey as a case study. In *Strike-slip Deformation, Basin Formation and Sedimentation*; Biddle, K.T., Christie-Blick, N., Eds.; SEPM Society for Sedimentary Geology: Tulsa, OK, USA, 1985; Volume 37, pp. 227–264.
25. Le Pichon, X.; Angelier, J. The hellenic arc and trench system: A key to the neotectonic evolution of the eastern mediterranean area. *Tectonophysics* **1979**, *60*, 1–42. [CrossRef]
26. Le Pichon, X.; Angelier, J. The Aegean sea. *Philos. Trans. R. Soc. Lond.* **1981**, *72*, 300–357.
27. Dewey, J.F. Extensional collapse of orogens. *Tectonics* **1988**, *7*, 1123–1139. [CrossRef]
28. Seyitoglu, G.; Scott, B.C. Age of the Alasehir graben (west Turkey) and its tectonic implications. *Geol. J.* **1996**, *31*, 1–11. [CrossRef]
29. Dilek, Y.; Whitney, D.L. Cenozoic crustal evolution in central Anatolia: Extension, magmatism, and landscape development. In *Proceedings, Third International Conference on the Geology of the Eastern Mediterranean*; Panayides, I., Xenophontos, C., Malpas, J., Eds.; Geological Survey Department: Nicosia, Cyprus, 2000; pp. 183–192.
30. Koçyiğit, A.; Yusufoğlu, H.; Bozkurt, E. Evidence from the Gediz graben for episodic two-stage extension in western Turkey. *J. Geol. Soc.* **1999**, *156*, 605–616. [CrossRef]

31. Sözbilir, H. Extensional tectonics and the geometry of related macroscopic structures: Field evidence from the Gediz de-tachment, western Turkey. *Turk. J. Earth Sci.* **2001**, *10*, 51–67.

32. Koçyiğit, A. The Denizli graben-horst system and the eastern limit of western Anatolian continental extension: Basin fill, structure, deformational mode, throw amount and episodic evolutionary history, SW Turkey. *Geodin. Acta* **2005**, *18*, 167–208. [CrossRef]

33. Gessner, K.; Gallardo, L.A.; Markwitz, V.; Ring, U.; Thomson, S.N. What caused the denudation of the Menderes Massif: Review of crustal evolution, lithosphere structure, and dynamic topography in southwest Turkey. *Gondwana Res.* **2013**, *24*, 243–274. [CrossRef]

34. Yilmaz, M.; Gelisli, K. Stratigraphic–structural interpretation and hydrocarbon potential of the Alaşehir Graben, western Turkey. *Pet. Geosci.* **2003**, *9*, 277–282. [CrossRef]

35. Çiftçi, N.B.; Temel, R.; Iztan, Y.H. Hydrocarbon occurrences in the western Anatolian (Aegean) grabens, Turkey: Is there a working petroleum system? *AAPG Bull.* **2010**, *94*, 1827–1857. [CrossRef]

36. Altan, Z.; Ocakoğlu, N.; Böhm, G.; Sarıkavak, K.T. Seismic Events in The Upper Miocene–Pliocene Sed-Imentary Succession In The Gulf Of Izmir (Western Anatolia): Implications for Hydrocarbon Prospectivity. *J. Pet. Geol.* **2020**, *43*, 209–224. [CrossRef]

37. Demircioğlu, D.; Ecevitoğlu, B.; Seyitoğlu, G. Evidence of a rolling hinge mechanism in the seismic records of the hydrocarbon-bearing Alaşehir graben, western Turkey. *Pet. Geosci.* **2010**, *16*, 155–160. [CrossRef]

38. Tüfekçi, N.; Süzen, M.L.; Güleç, N. GIS based geothermal potential assessment: A case study from Western Anatolia, Turkey. *Energy* **2010**, *35*, 246–261. [CrossRef]

39. Erkan, K. Geothermal investigations in western Anatolia using equilibrium temperatures from shallow boreholes. *Solid Earth* **2015**, *6*, 103–113. [CrossRef]

40. Baba, A.; Sözbilir, H. Source of arsenic based on geological and hydrogeochemical properties of geothermal systems in Western Turkey. *Chem. Geol.* **2012**, *334*, 364–377. [CrossRef]

41. Şimşek, Ş. Geothermal model of Denizli, Sarayköy-Buldan area. *Geothermics* **1985**, *14*, 393–417. [CrossRef]

42. Özgür, N. Geochemical signature of the Kizildere geothermal field, western Anatolia, Turkey. *Int. Geol. Rev.* **2002**, *44*, 153–163. [CrossRef]

43. De Filippis, L.; Faccenna, C.; Billi, A.; Anzalone, E.; Brilli, M.; Soligo, M.; Tuccimei, P. Plateau versus fissure ridge travertines from Quaternary geothermal springs of Italy and Turkey: Interactions and feedbacks between fluid discharge, paleoclimate, and tectonics. *Earth Sci. Rev.* **2013**, *123*, 35–52. [CrossRef]

44. Brogi, A.; Alçiçek, M.C.; Yalçıner, C.; Capezzuoli, E.; Liotta, D.; Meccheri, M.; Rimondi, V.; Ruggieri, G.; Gandin, A.; Boschi, C.; et al. Hydrothermal fluids circulation and travertine deposition in an active tectonic setting: Insights from the Kamara geothermal area (western Anatolia, Turkey). *Tectonophysics* **2016**, *680*, 211–232. [CrossRef]

45. Brogi, A.; Alçiçek, M.C.; Liotta, D.; Capezzuoli, E.; Zucchi, M.; Matera, P.F. Step-over fault zones controlling geothermal fluid-flow and travertine formation (Denizli Basin, Turkey). *Geothermics* **2020**, *89*, 101941. [CrossRef]

46. Tarcan, G.; Özen, T.; Gemici, Ü.; Çolak, M.; Karamanderesi, İ.H. Geochemical assessment of mineral scaling in Kızıldere geothermal field, Turkey. *Environ. Earth Sci.* **2016**, *75*, 1317. [CrossRef]

47. Karamanderesi, İ.H.; Ölçenoğlu, K. Geology of the Denizli Sarayköy (Gerali) Geothermal Field, Western Anatolia, Turkey. In Proceedings of the World Geothermal Congress, Antalya, Turkey, 24–29 April 2005.

48. Mesci, B.L.; Tatar, O.; Piper, J.D.A.; Gürsoy, H.; Altunel, E.; Crowley, S. The efficacy of travertine as a palaeoenvironmental indicator: Palaeomagnetic study of neotectonic examples from Denizli, Turkey. *Turk. J. Earth Sci.* **2013**, *22*, 191–203. [CrossRef]

49. Altunel, E.; Karabacak, V. Determination of horizontal extension from fissure-ridge travertines: A case study from the Denizli Basin, southwestern Turkey. *Geodin. Acta* **2005**, *18*, 333–342. [CrossRef]

50. Çakır, Z. Along-strike discontinuity of active normal faults and its influence on Quaternary travertine deposition; examples from western Turkey. *Turk. J. Earth Sci.* **1999**, *8*, 67–80.

51. Erees, F.S.; Aytas, S.; Sac, M.M.; Yener, G.; Salk, M. Radon concentrations in thermalwaters related to seismic events along faults in the Denizli Basin, Western Turkey. *Radiat. Meas.* **2007**, *42*, 80–86. [CrossRef]

52. Van Noten, K.; Topal, S.; Baykara, M.O.; Özkul, M.; Claes, H.; Aratman, C.; Swennen, R. Pleistocene-Holocene tectonic reconstruction of the Ballık travertine (Denizli Graben, SW Turkey): (De) formation of large travertine geobodies at inter-secting grabens. *J. Struct. Geol.* **2019**, *118*, 114–134. [CrossRef]

53. Ozkul, M.; Varol, B.; Alcicek, M.C. Depositional environments and petrography of Denizli travertines. *Bull. Miner. Res. Explor.* **2002**, *125*, 13–29.

54. Hançer, M. Study of the structural evolution of the Babadağ-Honaz and Pamukkale fault zones and the related earthquake risk potential of the Buldan region in SW Anatolia, east of the Mediterranean. *J. Earth Sci.* **2013**, *24*, 397–409. [CrossRef]

55. Seyïtoğlu, G.; Işik, V. Late Cenozoic Extensional Tectonics in Western Anatolia: Exhumation of the Menderes Core Complex and Formation of Related Basins. *Bull. Miner. Res. Explor.* **2015**, *151*, 47–106. [CrossRef]

56. Sarı, C.; Şalk, M. Sediment thickness of the Western Anatolia graben structures determined by 2D and 3D analysis using graviy data. *J. Asian Earth Sci.* **2006**, *26*, 39–48. [CrossRef]

57. Altinoğlu, F.F.; Sari, M.; Aydin, A. Detection of Lineaments in Denizli Basin of Western Anatolia Region Using Bouguer Gravity Data. *Pure Appl. Geophys.* **2015**, *172*, 415–425. [CrossRef]

58. Altinoğlu, F.F. Structural interpretation of SW part of Denizli, Turkey, based on gravity data analysis. *Arab. J. Geosci.* **2020**, *13*, 1–16. [CrossRef]
59. Ekinci, Y.L.; Balkaya, Ç.; Göktürkler, G.; Özyalın, Ş. Gravity data inversion for the basement relief delineation through global optimization: A case study from the Aegean Graben System, western Anatolia, Turkey. *Geophys. J. Int.* **2020**, *224*, 923–944. [CrossRef]
60. Akkuş, I.; Akıllı, H.; Ceyhan, S.; Dilemre, A.; Tekin, Z. *Inventory of Turkey Geothermal Sources, Inventory Series-201*; MTA General Directorate of Mineral Research and Exploration Publications: Ankara, Turkey, 2005.
61. Bilim, F. Investigation into the tectonic lineaments and thermal structure of Kutahya-Denizli region western Anatolia, from using aeromagnetic, gravity and seismological data. *Phys. Earth Planet. Inter.* **2007**, *165*, 135–146. [CrossRef]
62. Kaypak, B.; Gökkaya, G. 3D imaging of the upper crust beneath the Denizli geothermal region by local earthquake tomog-raphy, western Turkey. *J. Volcanol. Geotherm. Res.* **2012**, *211–212*, 47–60. [CrossRef]
63. Irmak, T.S. Focal mechanisms of small-moderate earthquakes in Denizli Graben (SW Turkey). *Earth Planets Space* **2013**, *65*, 943–955. [CrossRef]
64. Erbek-Kiran, E.; Ates, A.; Dolmaz, M.N. Upper Crustal Structure of Denizli Graben (Western Turkey) From Bouguer Gravity Data and Seismic Reflection Sections. *Surv. Geophys.* **2022**, *43*, 1947–1966. [CrossRef]
65. Sun, S. *Denizli Uşak Arasının Jeolojisi ve Linyit Olanakları. [Geology and Lignite Occurrences of Denizli and Uşak Region]*; MTA Report, No: 9985; MTA: Ankara, Turkey, 1990; (in Turkish with English abstract).
66. Sözbilir, H. Geometry and origin of folding in the Neogene sediments of the Gediz Graben, western Anatolia, Turkey. *Geo-Din. Acta* **2002**, *15*, 277–288. [CrossRef]
67. Emre, Ö.; Duman, T.Y.; Özalp, S.; Elmacı, H. *1:250,000 Scale Active Fault Map Series of Turkey, Denizli (NJ 35-12) Quad-Rangle*; Serial Number: 12; General Directorate of Mineral Research and Exploration: Ankara, Turkey, 2011.
68. Pham, L.T.; Oksum, E.; Do, T.D. GCH_gravinv: A MATLAB-based program for inverting gravity anomalies over sedimentary basins. *Comput. Geosci.* **2018**, *120*, 40–47. [CrossRef]
69. Cordell, L. Gravity analysis using an exponential density-depth function; San Jacinto Graben, California. *Geophysics* **1973**, *38*, 684–690. [CrossRef]
70. Granser, H. Three-dimensional interpretation of gravity data from sedimentary basins using an exponential density-depth function. *Geophys. Prospect.* **1987**, *35*, 1030–1041. [CrossRef]
71. Cordell, L.; Henderson, R.G. Iterative three-dimensional solution of gravity anomaly data using a digital computer. *Geophysics* **1968**, *33*, 596–601. [CrossRef]
72. Cordell, L.; Grauch, V.J.S. Mapping basement magnetization zones from aeromagnetic data in the San Juan Basin, New Mexico. In *The Utility of Regional Gravity and Magnetic Map*, 1st ed.; Hinze, W.J., Ed.; Society of Exploration Geophysicists: Tulsa, OK, USA, 1985; pp. 181–197.
73. Roest, W.R.; Verhoef, J.; Pilkington, M. Magnetic interpretation using the 3-D analytic signal. *Geophysics* **1992**, *57*, 116–125. [CrossRef]
74. Verduzco, B.; Fairhead, J.D.; Green, C.M.; MacKenzie, C. New insights into magnetic derivatives for structural mapping. *Geophysics* **2004**, *23*, 116–119. [CrossRef]
75. Wijns, C.; Perez, C.; Kowalczyk, P. Theta map: Edge detection in magnetic data. *Geophysics* **2005**, *70*, L39–L43. [CrossRef]
76. Miller, H.G.; Singh, V. Potential field tilt—A new concept for location of potential field sources. *J. Appl. Geophys.* **1994**, *32*, 213–217. [CrossRef]
77. Ferreira, F.J.; de Souza, J.; Bongiolo, d.B.e.S.A.; de Castro, L.G. Enhancement of the total horizontal gradient of magnetic anomalies using the tilt angle. *Geophysics* **2013**, *78*, J33–J41. [CrossRef]
78. Oksum, E.; Le, D.V.; Vu, M.D.; Nguyen, T.H.T.; Pham, L.T. A novel approach based on the fast sigmoid function for inter-pretation of potential field data. *Bull. Geophys. Oceanogr.* **2021**, *62*, 543–556.
79. USGS. Available online: https://earthexplorer.usgs.gov (accessed on 8 September 2023).
80. Sahoo, S.D.; Pal, S.K. Mapping of Structural Lineaments and Fracture Zones around the Central Indian Ridge (10°S–21°S) using EIGEN 6C4 Bouguer Gravity Data. *J. Geol. Soc. India* **2019**, *94*, 359–366. [CrossRef]
81. Essa, K.S.; Nady, A.G.; Mostafa, M.S.; Elhussein, M. Implementation of potential field data to depict the structural lineaments of the Sinai Peninsula, Egypt. *J. Afr. Earth Sci.* **2018**, *147*, 43–53. [CrossRef]
82. Bencharef, M.H.; Eldosouky, A.M.; Zamzam, S.; Boubaya, D. Polymetallic mineralization prospectivity modelling using multi-geospatial data in logistic regression: The Diapiric Zone, Northeastern Algeria. *Geocarto Int.* **2022**, *37*, 15392–15427. [CrossRef]
83. Pham, L.T.; Prasad, K.N.D. Analysis of gravity data for extracting structural features of the northern region of the Central Indian Ridge. *Vietnam J. Earth Sci.* **2023**, *45*, 147–163.
84. Jorge, V.T.; Oliveira, S.P.; Thanh, L.P.; Duong, V.H. A balanced edge detector for aeromagnetic data. *Vietnam. J. Earth Sci.* **2023**. [CrossRef]

85. Altınoğlu, F.F. Investigation of Tectonics of Western Anatolia by Geophysical Methods. Ph.D. Thesis, University of Pamukkale, Denizli, Turkey, 2012; p. 225.
86. Hançer, M. Structural Evolution of the Northeast—Southwest Trending Tectonic Lineament and a Model for Graben Formation in the Denizli Region of Western Anatolian (West of the Zagros Fold-and-Thrust Belt). *Dev. Struct. Geol. Tecton.* **2019**, *3*, 83–100.

Article

Mineral Prospectivity Mapping for Orogenic Gold Mineralization in the Rainy River Area, Wabigoon Subprovince

Pouran Behnia *, Jeff Harris, Ross Sherlock, Mostafa Naghizadeh and Rajesh Vayavur

Mineral Exploration Research Centre, Harquail School of Earth Sciences, Laurentian University, 935 Ramsey Lake Road, Sudbury, ON P3E 2C6, Canada; 603jharris@gmail.com (J.H.); rsherlock@laurentian.ca (R.S.); mnaghizadeh@laurentian.ca (M.N.); rvayavur@laurentian.ca (R.V.)
* Correspondence: pbehnia@laurentian.ca

Abstract: Random Forest classification was applied to create mineral prospectivity maps (MPM) for orogenic gold in the Rainy River area of Ontario, Canada. Geological and geophysical data were used to create 36 predictive maps as RF algorithm input. Eighty-three (83) orogenic gold prospects/occurrences were used to train the classifier, and 33 occurrences were used to validate the model. The non-Au (negative) points were randomly selected with or without spatial restriction. The prospectivity mapping results show high performance for the training and test data in area-frequency curves. The F1 accuracy is high and moderate when assessed with the training and test data, respectively. The mean decrease accuracy was applied to calculate the variable importance. Density, proximity to lithological contacts, mafic to intermediate volcanics, analytic signal, and proximity to the Cameron-Pipestone deformation zone exhibit the highest variable importance in both models. The main difference between the models is in the uncertainty maps, in which the high-potential areas show lower uncertainty in the maps created with spatial restriction when selecting the negative points.

Keywords: mineral prospectivity mapping (MPM); random forest; Rainy River; orogenic gold

Citation: Behnia, P.; Harris, J.; Sherlock, R.; Naghizadeh, M.; Vayavur, R. Mineral Prospectivity Mapping for Orogenic Gold Mineralization in the Rainy River Area, Wabigoon Subprovince. *Minerals* **2023**, *13*, 1267. https://doi.org/10.3390/min13101267

Academic Editors: Luan Thanh Pham, Saulo Pomponet Oliveira and Le Van Anh Cuong

Received: 12 August 2023
Revised: 22 September 2023
Accepted: 25 September 2023
Published: 28 September 2023

1. Introduction

Mineral prospectivity mapping (MPM) has seen many advancements over the past 30 years, beginning with the original work by Bonham-Carter [1,2]. MPM involves the use of geoscience data in the form of evidence or predictor maps representing vectors for various types of mineralization. These data are input into machine learning algorithms to produce maps that identify areas with higher probabilities for various types of mineral deposits. These maps should not only accurately predict the known deposits in a given study area but also reveal areas that are prospective for discovering new deposits. The final stage of the MPM process is evaluating how predictive the map is with respect to the training set of known deposits. This can be accomplished using cross-validation, the efficiency of classification/prediction graphs [3], and receiver operator curves (ROC) [4,5].

Machine learning algorithms used to model the geoscience data comprise two basic types: data-driven and knowledge-driven [1,2]. Data-driven techniques require a set of training data (e.g., mineral deposits or prospects) for modeling. Meanwhile, knowledge-driven techniques require the geoscientist to weigh each predictor map with a subjective weight; the predictor maps are then summed to produce an MPM map. Many types of data-driven algorithms exist, including weights of evidence [1,2,6–11], logistic regression [6,10,12,13], evidential belief modeling [14,15], support vector machine (SVM) [16,17], neural networks [18–25], and random forest, among others. Knowledge-driven techniques comprise algorithms such as Boolean, index overlay, and fuzzy logic [1,2,26–29].

Recently, numerous studies have described the random forest algorithm [30,31] as a highly powerful approach for producing MPM maps [32–46]. As such, this paper adopts the random forest data-driven machine learning algorithm to produce an MPM map for a portion of the Rainy River greenstone belt in Ontario, Canada, based on orogenic gold. The random nature of the random forest algorithm regarding bootstrapping and random permutation of predictor maps through each decision tree prevents overfitting of the model.

This research is being conducted under the Metal Earth Project at the Laurentian University in Sudbury, Ontario, Canada. The fundamental objective is to elucidate why some greenstone belts are fertile while others are essentially barren. This is accomplished by collecting new 3D geophysical, seismic, lithogeochemical, and geochronological data, as well as collecting and compiling existing geoscience data to produce mineral prospectivity maps for various mineral commodities.

Gold deposits in the Superior geologic province exhibit a wide range of mineralization as well as structural and geometric characteristics [47,48]. In the Abitibi geologic subprovince, a significant portion of the gold deposits are syn-volcanic and synmagmatic, formed during the volcanic construction of the Abitibi belt. However, over 60% of the gold endowment originates from quartz-carbonate veins formed during late-stage tectonic inversion of the extensional basins [49]. In the Wabigoon subprovince, many gold deposits and prospects are interpreted as orogenic gold, including the areas of mineralization within the Cameron Lake area hosted by the Cameron-Pipestone deformation zone [50,51]. This type of gold mineralization is the focus of the present study.

2. Study Area

Figure 1 is an overview of the Wabigoon tectonic province and Rainy River transect. Figure 2 presents the regional geology of the Rainy River transect and the study area in the north [52]. This area was selected based on the availability of geoscience data used in the modeling. Synvolcanic gold is present in the southern portion of the area. However, due to insufficient synvolcanic gold bodies in the area to apply a data-driven method, we applied the MPM—for the orogenic gold mineralization in the northern region.

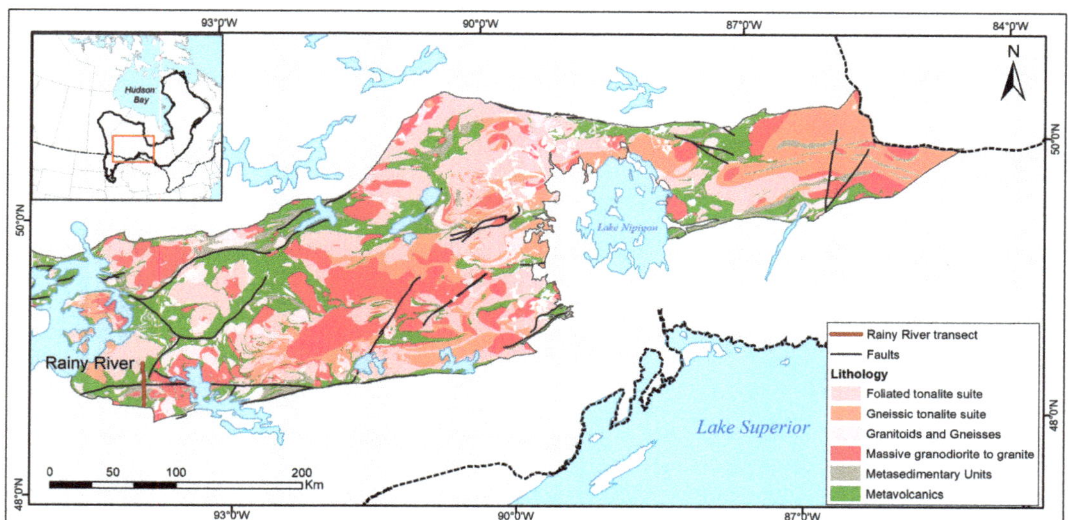

Figure 1. Simplified geology of the Wabigoon and Rainy River transects.

The area comprises volcanic lithologies organized as anastomosing belts surrounding large granitoid batholiths [52,53]. These are divided into four main sequences [54]: (1) a lower mafic tholeiitic unit; (2) intermediate to felsic volcanic rocks comprising volcaniclastic rocks, intrusions, and calc-alkalic flows; (3) an upper unit comprising mafic tholeiitic flows; and (4) an upper sequence of sedimentary rocks comprising turbiditic and alluvial/fluvial sedimentary rocks. All lithologies were affected by regional greenschist-grade metamorphism and amphibolite-grade metamorphism related to the emplacement of the granitoid batholiths. There are also several major regional structures that -could act as potential conduits for ore-bearing fluids [55].

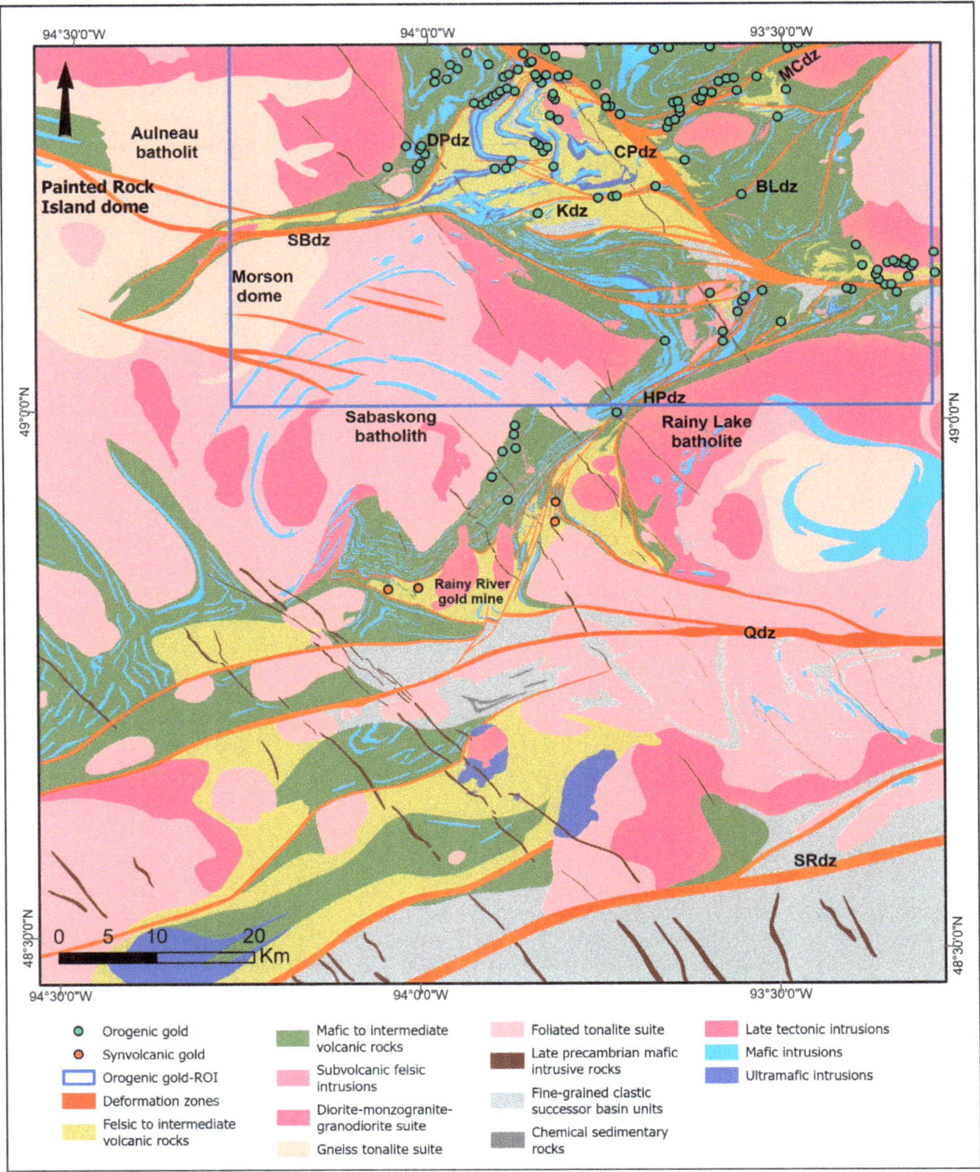

Figure 2. Geologic map of the Rainy River area [52]. Deformation zones: BLdz = Brooks Lake, CPdz = Cameron-Pipestone, DPdz = DogPaw, HPdz = Helena-Pipestone, Kdz = Kakagi, MCdz = Monte Cristo, PHdz = Phinney, Qdz = Quetico, SBdz = Sabaskong Bay, SRdz = Seine River. A blue rectangle shows the study area selected for MPM.

The volcanic units are bound in the north by the Sasbaskonk batholith and in the south by the Quetico Fault (Figure 2). The volcanic units around the transect are intruded by the Flemming-Kingsford and Jackfish Lake plutons [55]. The southern part of the Rainy River area is bounded by the Quetico subprovince, which comprises metasedimentary rocks of turbiditic origin and rare conglomerates. The east–west striking dextral and sub-vertical Quetico shear zone forms the boundary of the Quetico subprovince and the Rainy River greenstone belt [54,56,57].

The largest gold deposit in the area is the Rainy River deposit, operated by NewGold Inc. This deposit is hosted by a subaqueous felsic volcanic complex with gold strongly correlated with a sericite alteration assemblage [51,58,59]. Work by Pelletier et al. [58,59] suggests that mineralization is early, pre-deformation, and likely synvolcanic in origin. The mafic volcanic rocks host most of the orogenic gold mineralization [51,58,59], particularly in the northern portion of the map area (Figure 2), which tends to have undeveloped mineral occurrences. Launay et al. [55] provide additional details on the geology and mineralization of the study area.

3. Materials and Methods

3.1. Random Forest (RF)

A random forest is an ensemble of individual decision trees [30,31] that can be used for classification and regression. A decision tree is composed of several leaves (nodes) and branches (edges). In each node, a decision is made, and a split is performed. The splitting process is continued until the node is pure. To train the classifier, a bagging procedure called bootstrapping is employed [60]. In this method, a subset of the training data is randomly selected by each tree to train the classifier; the remaining data, also called out-of-bag (OOB) data, are used for the validation [60]. Each tree uses a random subset of all variables (predictor maps in this study) that helps reduce the correlation between the trees. Combining the trees provides the majority vote. The final outputs are a classification map (in the case of MPM a two-class map) and a probability map for each class. The probability (prospectivity) is calculated by the number of trees that predicted gold divided by the total number of trees at each data point (pixel); this yields a number between 0 and 1. For example, if there are 10 trees at location A and no trees predicted the presence of gold, the probability would be 0; if 10 trees predicted gold, the probability would be 1. Additional details on the RF algorithm and how it can be applied to MPM have been previously described [30,31,35,37,46].

We used EnMap-BOX 1.4 [60] to apply the RF classification. The number of features was set to the square root of all features (i.e., 6), and the impurity function was set to the Gini coefficient. We tested 300–1000 trees to assess the learning curves for each forest. In most cases, the OOB was stabilized after 500 trees with an average accuracy of 72%–78%.

3.2. Predictor Maps

Based on the characteristics of orogenic gold mineralization in the area, the following lithological, structural, and alteration criteria were considered as vectors to mineralization. We also included the geophysical data to evaluate their potential predictive power for the mineralization of interest.

3.2.1. Lithology

The main host rocks for mineralization in the area are mafic to intermediate and felsic to intermediate volcanics (Figure 2). These units were extracted from the geology map [55] and used as the lithologic predictor maps. As some of the gold occurrences were associated with diorite-monzogranite granitoids, these rocks were also included as another predictor map. Figure 3a–c shows the binary maps of the three lithologies used to create the MPM. In addition, contacts between these rock units were selected, and proximity zones up to 500 m were created around them with an interval of 100 m (Figure 3d).

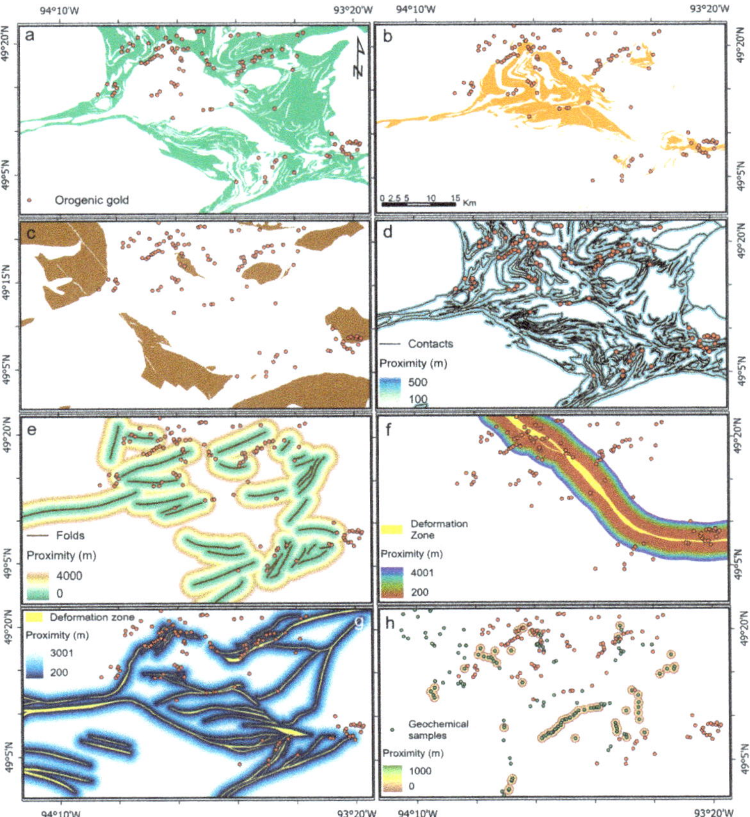

Figure 3. Lithological, structural, and alteration predictor maps: (**a**) Mafic to intermediate volcanics; (**b**) felsic to intermediate volcanics; (**c**) diorite-monzogranite granitoids; (**d**) proximity to lithological contacts; (**e**) proximity to folds; (**f**) proximity to the Cameron-Pipestone deformation zone; (**g**) proximity to deformation zones; (**h**) proximity to chlorite-carbonate-pyrite alteration. Green points represent geochemical sample locations.

3.2.2. Structure

A structural compilation was obtained from the Ontario Geological Survey (OGS) digital database. Several regional-scale structures exist in the Rainy River area [55]. Features such as deformation zones (fault and shear zones) and fold hinges can act as potential conduits and traps for the migration of mineralized fluids and the deposition of orogenic gold, respectively. Hence, these features are considered important vectors to mineralization. To create the structural predictive maps, proximity zones up to 4 km with an interval of 200 m were created around the folds based on the field geologist's expert knowledge (Figure 3e). The same proximity zones of up to 3 km were created around the deformation zones. Since the Cameron-Pipestone deformation zone appears to be more important than the other zones (many gold prospects occur close to this deformation zone), we applied it as an individual predictor map and extended the proximity zones up to 4 km (Figure 3f,g).

3.2.3. Alteration

The geochemical data for gold and its pathfinder elements were sparse and, thus, excluded from the integration model. However, as more analyses were available on the major elements, these data were extracted from the Metal Earth database and applied to calculate the chlorite-carbonate-pyrite alteration index (CCPI = 100 (MgO + FeO)/(MgO + FeO + Na_2O + K_2O)) using the ioGAS software, version 8.1. A centered log-ratio (clr) transformation [61] was applied to the data to minimize the closure effects.

As the samples were not sufficiently dense to warrant interpolation, we employed a zone of influence approach. Anomalous samples were selected using Q-Q plots, and proximity zones up to 1 km with a 250 m interval were created around the samples (Figure 3h).

3.2.4. Airborne Magnetic Data

The airborne magnetic data used for this study was extracted from the Ontario Geological Survey (OGS) Master Grid compilation [62]. The downloaded magnetic grid—originates from a compilation of many surveys that were leveled to a common datum of 300 m with a cell size of 200 m. In our MPM study, we used the total field reduced to the pole and the analytical signal images (Figure 4a,b).

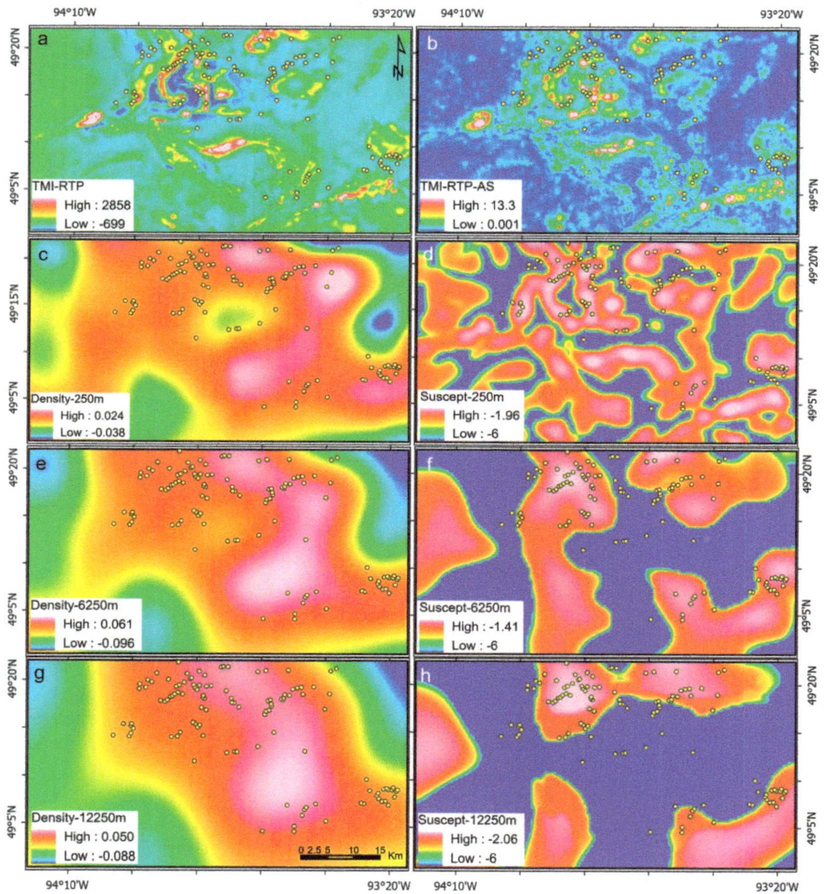

Figure 4. Airborne magnetic data and a subset of susceptibility and density maps created by the inversion of gravity and total magnetic anomalies. (**a**) Total magnetic intensity reduced to pole; (**b**) Analytical signal; (**c,e,g**) density slices at elevations of −250 m, −6250 m, and −12,250 m, respectively. (**d,f,h**) susceptibility at elevations of −250 m, −6250 m, and −12,250 m, respectively. Orogenic gold is shown as yellow points.

3.2.5. Susceptibility and Density Data

The analytical processing of geophysical data was carried out on 3D density and susceptibility models. These models were inverted from the GSC gravity Bouguer anomaly and the total magnetic anomaly [63] on a 1×1 km horizontal grid, respectively. To prevent spatial aliasing artifacts, the total magnetic anomaly and digital elevation maps of the study area were smoothed by a Gaussian window before sampling on a 1×1 km horizontal grid. The vertical grids of the 3D density and susceptibility models were exponentially increased toward the deeper parts of the model. Both gravity and magnetic inversions were carried out by imposing model smoothness constraints. The inverted density and susceptibility 3D models were re-sampled/interpolated into a regular $500 \times 500 \times 500$ m grid to generate co-localized density and susceptibility volumes suitable for further data analysis. Figure 4c,e,g depicts the density slices at elevations of -250, -6250, and $-12{,}250$ m, respectively. Figure 4d,f,h depicts the susceptibility slices at elevations of -250, -6250, and $-12{,}250$ m, respectively.

All predictor maps (totaling 36) were resampled to 50 m to ensure the retention of details from the lithology maps when converting to the raster format. Table 1 provides a summary of the predictor maps used in the RF modeling procedure.

Table 1. Summary of predictor maps.

Data	Variable	Source
Lithology (4 maps)	Mafic-intermediate volcanics Felsic-intermediate volcanics Diorite-monzogranite granitoids Proximity to lithological contacts	Launay, 2021, Metal Earth [52]
Structure (3 maps)	Cameron-Pipestone DZ * Deformation zones (other) Folds	Launay, 2021, Metal Earth [52]
Alteration (1 map)	Chlorite-carbonate-pyrite index (CCPI)	Geochemical data, Metal Earth
Airborne Magnetic (2 maps)	Total magnetic intensity (TMI) reduced to pole, Analytic signal	Ontario Geological Survey
3D geophysical data (26 maps)	Density: -250 m to $-12{,}250$ m with a 1 km interval Susceptibility: -250 m to $-12{,}250$ m with a 1 km interval	Geological Survey of Canada

* Deformation zone.

3.2.6. Training (Au) Data

Mineral occurrence data were obtained from the Ontario Mineral Inventory database (OMI). Our target variable was orogenic gold bodies, comprising 116 orogenic gold prospects and occurrences available in the study area. We selected 83 gold prospects and occurrences as the training data and retained 33 gold occurrences as the test data to validate the MPM. Although random forest applies internal cross-validation, it is useful to validate the model with test data that were not included in the model.

In addition to the location of gold deposits, random forest requires additional negative or non-gold locations. However, selecting these points is relatively challenging. Rahimi et al. [64] used a clustering algorithm and selected random points from unfavorable clusters. Behnia [13] created random points from the low favorability zones of an MPM created using the weights of evidence method. Meanwhile, Carranza and Laborte [34] and Sun et al. [65] selected the negative points following a point pattern analysis. In this study, we selected the negative points using two strategies. In set-1, 83 random points were created some distance away from the known orogenic gold bodies, following a point pattern analysis. Figure 5 presents the results of the point-pattern analysis for orogenic gold prospects/occurrences in the area. The distance between each deposit and its nearest neighbor and the corresponding probability of finding a gold deposit within that distance were calculated and plotted. There was a 100% probability that another deposit occurred within 7 km of a deposit (Figure 5).

However, using this distance to create random points away from the existing deposits causes the negative points to localize in a small area. Thus, instead, we selected a 2 km distance, corresponding to an 84% probability of finding a neighboring deposit next to any given deposit, to create 83 random points. For set-2, 83 random points were created with no restriction. This was repeated five times for each set for a total of ten training data sets, in which the gold prospects/occurrences were the same while the non-gold points differed. Hence, each time the model was trained, it received 83 gold and 83 non-gold points. Figure 6 shows how the two datasets were created. Locations of the known gold prospects/occurrences (train and test) are also shown.

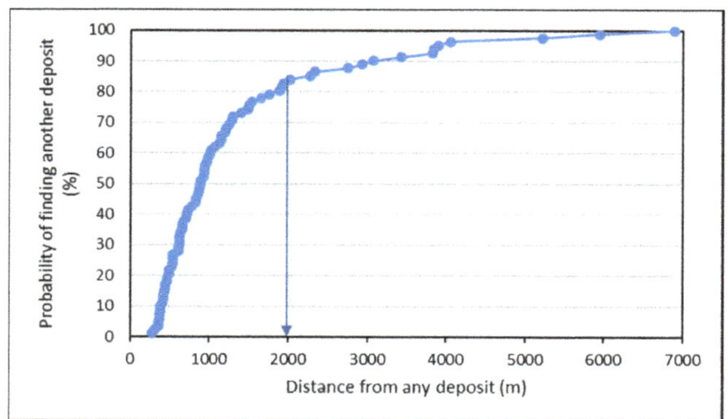

Figure 5. Point pattern analysis on orogenic gold bodies. A distance of 2 km corresponds to an 84% probability of finding a neighboring deposit to any given deposit.

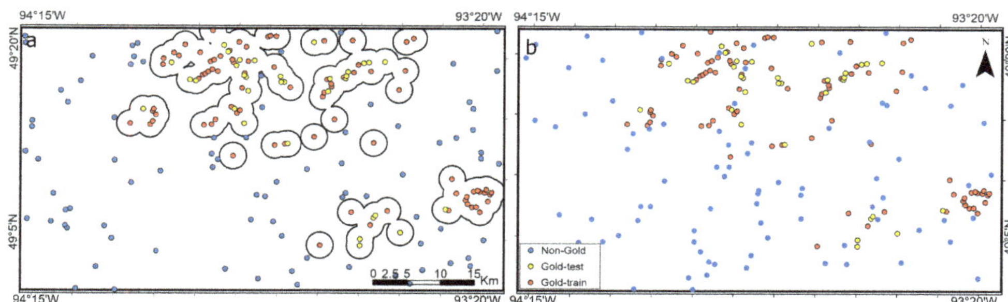

Figure 6. (**a**) Example of non-gold points randomly generated 2 km away from the known gold prospects/occurrences (set-1). (**b**) Non-gold points randomly generated with no restriction (set-2). Gold occurrences used as test data are also shown.

4. Results

4.1. Probability Maps

A 10-fold repetition of the model, each using a different suite of randomly selected non-gold points, resulted in ten probability maps. To better compare the two models, the mean probability value for each set was generated, creating two MPMs (Figure 7a,b). Although the two maps look similar, they also have certain differences. For example, some areas, such as the northern regions and area (A) in Figure 7a, exhibit a higher probability in set-1, while set-2 has a higher probability in the east–south area (B) in Figure 7b. Moreover, uncertainty due to the use of different sets of training data was mapped based on the standard deviation of each of the two sets comprising five MPMs (Figure 8a,b). The uncertainty range is similar for the two sets; however, it is slightly lower in set-1 (0–0.274 vs. 0–0.287). As shown in Figure 8, the uncertainty is low in areas of gold mineralization and in areas with low probability/prospectivity in the eastern and western regions. Set-1 has lower uncertainty in high-probability areas where several gold prospects/occurrences exist, particularly in the north–central (black circle, Figure 8) and north–eastern regions. Areas of high uncertainty primarily correspond with areas lacking gold mineralization.

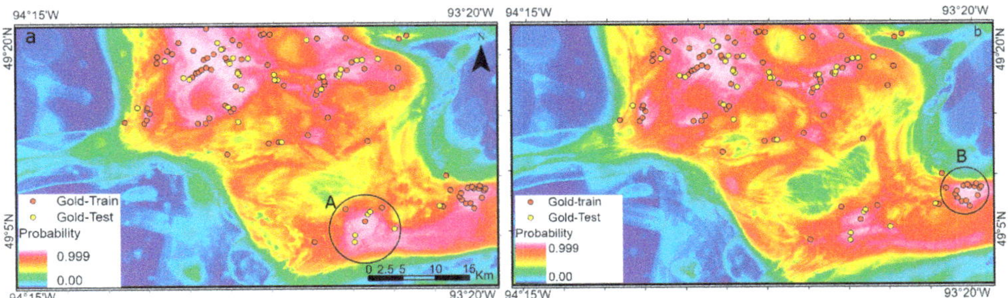

Figure 7. Mean probability maps of set-1 (**a**) and set-2 (**b**). Note the higher probabilities in the northern parts and area (A) in set-1 and area (B) in Set-2.

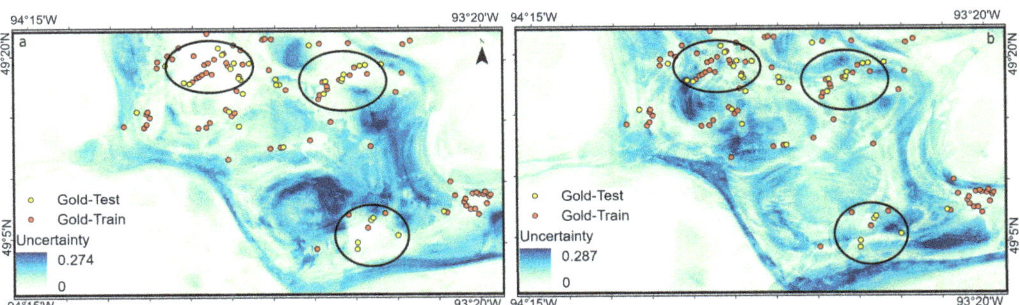

Figure 8. Uncertainty map based on the standard deviation for set-1 (**a**) and set-2 (**b**). Note the lower uncertainty in set-1 in areas shown in black circles.

To assess the correlation between the MPMs, Pearson correlation coefficients were calculated. The correlations between the maps in set-1 and set-2 range from 0.881 to 0.955 and 0.867 to 0.923, respectively, showing a slightly higher correlation for set-1. The correlations between all ten maps range from 0.835 to 0.955.

The variable importance can be measured using different methods, including the mean decrease impurity and the mean decrease accuracy. Impurity-based feature importance measures can be biased in favor of variables with high cardinality (i.e., variables with many unique values). Hence, we used mean decrease accuracy or permutation feature importance, which do not have such a bias [66]. In this method, accuracy is measured on the OOB samples for each tree. The OOB values for a single variable are then randomly permuted, and the accuracy is recalculated. The average of the differences in the original and permuted accuracies for a variable represents the raw importance of that variable [60,66]. The normalized variable importance is the raw importance value divided by the standard deviation of the variable.

The variable importance in ten experiments (set-1 and 2 random points) was relatively different. To better compare the results, we applied the average results from each set (Figure 9). The density (depth: 5250–12,250 m) exhibits the highest predictive power in both models. Proximity to lithological contacts and mafic to intermediate volcanics show high predictive power, particularly in set-2, whereas felsic to intermediate volcanics and diorite-monzogranite granitoids exhibit low importance. Proximity to the Cameron-Pipestone zone and to other deformation zones has medium to high importance. Moreover, the analytical signal has high (set-1) to medium (set-2) predictive power, whereas the total magnetic field shows lower importance. Susceptibility has medium to high predictive power, while proximity to folds and alteration zones have the lowest prediction power. This was expected as the alteration map lacked sufficient data throughout the study area; as such, the values were missing for most of the area. The main difference between the two sets is that in set-1, the analytical signal shows higher importance than the lithological contacts and mafic to intermediate volcanics, while in set-2, the opposite occurs. Furthermore, susceptibility shows higher variability in set-1, ranging from high to low, while in set-2, excluding an elevation of -250 m, medium to low predictive power is observed.

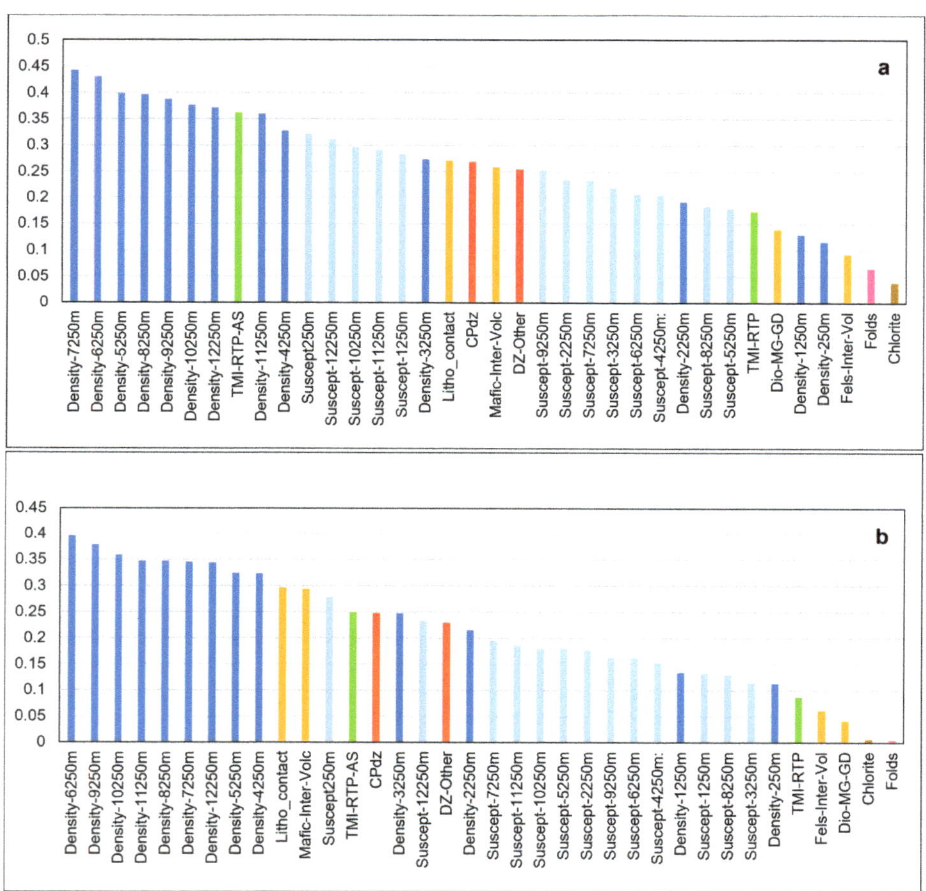

Figure 9. Mean variable importance for set-1 (**a**) and set-2 (**b**). Density has the highest importance in both models, followed by proximity to lithological contacts, susceptibility, mafic-intermediate volcanics, analytic signal, and Cameron-Pipestone deformation zone.

4.2. Validation

Area–frequency curves were used to assess the MPM performance. In these graphs, the cumulative frequency percent of the training/test data is plotted against the cumulative area percent of the probability map from high to low values. Figure 10 shows the efficiency of the classification and prediction curves for the RF mineral prospectivity maps. More specifically, Figure 10a presents plots for mean set-1 and mean set-2, whereas 10b shows a plot for the mean of all ten prospectivity maps. Figure 10b shows that the area under the curve (AUC) is high for classification defined by the original training points used for classification (0.958) and prediction (0.904) defined by an independent test set not used for classification indicating that the models do not suffer from overfitting. Approximately 80% of the gold occurrences are predicted within 6.4% of the most prospective areas on the probability map; this falls to 15% when examining the prediction rate.

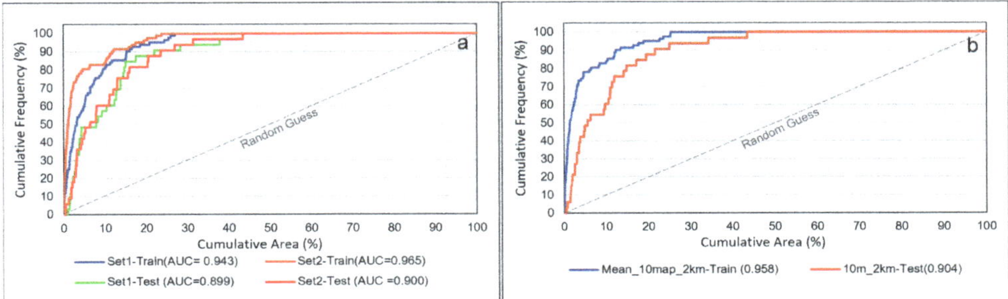

Figure 10. (**a**) Efficiency of classification and prediction curves for the mean prospectivity maps of set-1 and set-2. Set-2 performs slightly better for classification and prediction. (**b**) Efficiency of classification and prediction for the mean of ten prospectivity maps.

The classification accuracy was assessed using cross-validation. The average F1 accuracy is high, ranging from 0.952 to 0.982 when assessed by the training data and 0.814 to 0.893 when assessed by the test data. Although the accuracy changes slightly in each experiment, significant differences do not exist between the two training sets.

5. Discussion

The probability maps created using different non-deposit sites (set-1 and 2) are similar in appearance, with correlation coefficients ranging from 0.835 to 0.955. Based on the efficiency of classification curves (Figure 10), the probability maps created using the two sets of training data (5 maps for each set) have similar performances. Moreover, based on cross-validation, considerable differences do not exist in the F1 accuracy of the ten classification maps. The F1 accuracy for all the maps is very high when assessed by the training data used in the model (95.2%–98.2%) and moderate when assessed by the test data (81.4%–89.33%).

The variable importance for the two sets of training data is similar, with certain differences. Density (depth: 5250–12,250 m) has the highest predictive power in both models, followed by the analytic signal, susceptibility, proximity to lithological contacts, and Cameron-Pipestone deformation zone in set-1 and proximity to lithological contacts and mafic to intermediate volcanics in set-2. The high correlation with density is likely related to the presence of mafic-ultramafic intrusions in the Kakagi Lake area that localize the deformation corridors and, thus, the mineralization.

One main difference between the two sets is in the uncertainty maps. That is, the uncertainty in set-1 compared to set-2 is lower in areas of high probability where several orogenic mineralizations occur. This is likely due to restricting the creation of random points to 2 km from the known gold mineralization in set-1.

Figure 11a,b exhibits the most prospective areas (top 5%) based on the MPM produced from the average of set-1 and set-2 data points, respectively, overlaid on the lithology map. The prospective areas (A to H) are common in both models, except that areas G, H, and small patches in the south–east do not show up on the map created using set-1 training data. These areas have fewer or no gold prospects/occurrences in the vicinity, suggesting that selecting non-gold points a certain distance away from the mineralized bodies results in higher probabilities in the areas surrounding these bodies compared to other areas. Table 2 presents a geologic description of each prospective zone (A–H). Mafic to intermediate volcanics are present in all areas except C, in which mafic and ultramafic intrusions and felsic to intermediate volcanics predominate. All areas are associated with deformation zones. Although density does not exhibit a clear pattern, it is approximately medium to high and corresponds with higher values with increased depth. Moreover, the uncertainty is very low to low in all areas, and all but area H and small patches in the south–east have associated gold prospects/occurrences.

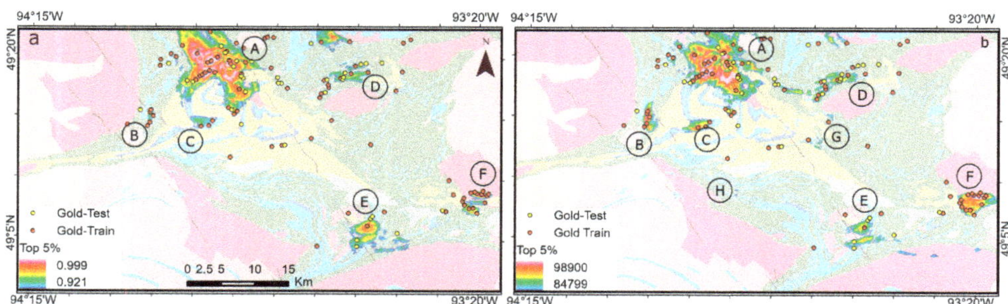

Figure 11. The most prospective areas show the top 5% of the MPM for (**a**) set-1 and (**b**) set-2. Most of the known orogenic gold prospects/occurrences lie in these areas.

Table 2. Geologic description of the prospective areas in Figure 11.

Area	Geology	Deformation Zone	Density	Uncertainty	Gold Mineralization
A	Mafic-intermediate volcanics, felsic-intermediate volcanics, mafic and ultramafic intrusions, diorite-monzogranite granitoids	Cameron-Pipestone, DogPaw	Medium near surface, higher in deeper parts	Very low	Yes
B	Mainly mafic-intermediate volcanics, and mafic intrusions	DogPaw	Medium to high in medium depths	Very low to low	Yes
C	Mainly mafic and ultramafic intrusions, as well as felsic-intermediate volcanics	Kakagi	Medium to high	Very low to low	Yes
D	Mainly mafic-intermediate volcanics, felsic-intermediate volcanics, mafic intrusions, and diorite-monzogranite granitoids	Monte Cristo	High	Very low to low	Yes
E	Mainly mafic intrusions, felsic-intermediate volcanics, mafic-intermediate volcanics, and felsic intrusions	Helena-Pipestone, Phinney	Medium to high, higher in deeper parts	Very low to low	Yes
F	Felsic-intermediate volcanics, diorite-monzogranite granitoids, and mafic-intermediate volcanics	Cameron-Pipestone	Variable	Very low to low	Yes
G	Mafic-intermediate volcanics, felsic-intermediate volcanics, and mafic intrusions	Cameron-Pipestone	Very high (in deeper parts) to medium	Very low	Yes
H	Mafic-intermediate volcanics	Phinney	Medium	Very low	No

6. Conclusions

Random forest classification was applied to orogenic gold mineralization using 36 predictor maps and two sets of training data. The AUC of the area-frequency curves for classification and prediction is high, indicating that the model is not overfitted. The two models, despite visual similarity and very close accuracies, have differences regarding uncertainty, feature importance, and top prospective areas, indicating that the way in which the negative points are selected is important and affects the results. Further studies are required to determine how to select these points. Selecting mineralized bodies other than the mineralization of interest with different geological, geochemical, and geophysical characteristics may be a better option for selecting the negative points.

Based on the variable importance values, the presence of mafic to intermediate volcanics and lithological contacts are important vectors to mineralization. Moreover, proximity to deformation zones, particularly the Cameron-Pipestone, shows high predictive power.

Most of the orogenic gold bodies are within the top 5% of high prospectivity areas (A–F in Figure 11). However, some high-probability areas have fewer (G) or no gold mineralization (H) and can, thus, serve as targets for further exploration.

Author Contributions: Conceptualization, P.B. and J.H.; methodology, P.B.; formal analysis, P.B., M.N. and R.V.; writing—original draft preparation, J.H., P.B., R.S. and M.N.; writing—review and editing, P.B., R.S. and J.H.; visualization, P.B.; supervision, R.S. All authors have read and agreed to the published version of the manuscript.

Funding: This research was funded by the Canada First Research Excellence Fund (CFREF-2015-00005) and is the Metal Earth publication MERC-ME-2023-28.

Data Availability Statement: The data presented in this study are available on request from the corresponding author.

Acknowledgments: We thank Gaëtan Launay for providing valuable advice on conceptual mineral deposit models. We also thank four anonymous reviewers for their constructive comments.

Conflicts of Interest: The authors declare no conflict of interest.

References

1. Bonham-Carter, G.F. *Geographic Information Systems for Geoscientists: Modeling with GIS. Pergamon*; Elsevier Science Inc.: New York, NY, USA, 1994.
2. Bonham-Carter, G.F.; Agterberg, F.P.; Wright, D.F. Integration of geological datasets for gold exploration in Nova Scotia. *PE&RS* **1988**, *54*, 1585–1592.
3. Chung, C.-J.F.; Fabbri, A.G. Validation of Spatial Prediction Models for Landslide Hazard Mapping. *Nat. Hazards* **2003**, *30*, 451–472. [CrossRef]
4. Chen, Y.; Wu, W. A prospecting cost-benefit strategy for mineral potential mapping based on ROC curve analysis. *Ore Geol. Rev.* **2016**, *74*, 26–38. [CrossRef]
5. Nykänen, V.; Lahti, I.; Niiranen, T.; Korhonen, K. Receiver operating characteristics (ROC) as validation tool for prospectivity models—A magmatic Ni–Cu case study from the Central Lapland Greenstone Belt, Northern Finland. *Ore Geol. Rev.* **2015**, *71*, 853–860. [CrossRef]
6. Raines, G.L.; Mihalasky, M.J. A Reconnaissance Method for Delineation of Tracts for Regional-Scale Mineral-Resource Assessment Based on Geologic-Map Data. *Nat. Resour. Res.* **2002**, *11*, 241–248. [CrossRef]
7. Carranza, E.J.M. Weights of Evidence Modeling of Mineral Potential: A Case Study Using Small Number of Prospects, Abra, Philippines. *Nat. Resour. Res.* **2004**, *13*, 173–187. [CrossRef]
8. Harris, D.; Zurcher, L.; Stanley, M.; Marlow, J.; Pan, G. A Comparative Analysis of Favorability Mappings by Weights of Evidence, Probabilistic Neural Networks, Discriminant Analysis, and Logistic Regression. *Nat. Resour. Res.* **2003**, *12*, 241–255. [CrossRef]
9. Harris, J.R.; Sanborn-Barrie, M.; Panagapko, D.A.; Skulski, T.; Parker, J.R. Gold prospectivity maps of the Red Lake greenstone belt: Application of GIS technology. *Can. J. Earth Sci.* **2006**, *43*, 865–893. [CrossRef]
10. Harris, J.R.; Wilkinson, L.; Heather, K.; Fumerton, S.; Bernier, M.A.; Ayer, J.; Dahn, R. Application of GIS Processing Techniques for Producing Mineral Prospectivity Maps—A Case Study: Mesothermal Au in the Swayze Greenstone Belt, Ontario, Canada. *Nat. Resour. Res.* **2001**, *10*, 91–124. [CrossRef]
11. Zuo, R.; Zhang, Z.; Zhang, D.; Carranza, E.J.M.; Wang, H. Evaluation of uncertainty in mineral prospectivity mapping due to missing evidence: A case study with skarn-type Fe deposits in Southwestern Fujian Province, China. *Ore Geol. Rev.* **2015**, *71*, 502–515. [CrossRef]
12. Chung, C.F.; Agterberg, F.P. Regression models for estimating mineral resources from geological map data. *Math. Geosci.* **1980**, *12*, 473–488. [CrossRef]
13. Carranza, E.J.M.; Hale, M. Logistic regression for geologically-constrained mapping of gold mineralization potential, Baguio district, Philippines. *Explor. Min. Geol.* **2001**, *10*, 165–175. [CrossRef]

14. Carranza, E.J.M.; Woldai, T.; Chikambwe, E.M. Application of Data-Driven Evidential Belief Functions to Prospectivity Mapping for Aquamarine-Bearing Pegmatites, Lundazi District, Zambia. *Nat. Resour. Res.* **2005**, *14*, 47–63. [CrossRef]
15. Carranza, E.J.M. Data-Driven Evidential Belief Modeling of Mineral Potential Using Few Prospects and Evidence with Missing Values. *Nat. Resour. Res.* **2015**, *24*, 291–304. [CrossRef]
16. Zuo, R.; Carranza, E.J.M. Support vector machine: A tool for mapping mineral prospectivity. *Comput. Geosci.* **2011**, *37*, 1967–1975. [CrossRef]
17. Abedi, M.; Norouzi, G.-H.; Bahroudi, A. Support vector machine for multi-classification of mineral prospectivity areas. *Comput. Geosci.* **2012**, *46*, 272–283. [CrossRef]
18. Singer, D.A.; Kouda, R. Application of a feedforward neural network in the search for Kuroko deposits in the Hokuroku district, Japan. *Math. Geosci.* **1996**, *28*, 1017–1023. [CrossRef]
19. Harris, D.; Pan, G. Mineral Favorability Mapping: A Comparison of Artificial Neural Networks, Logistic Regression, and Discriminant Analysis. *Nat. Resour. Res.* **1999**, *8*, 93–109. [CrossRef]
20. Brown, W.M.; Gedeon, T.D.; Groves, D.I.; Barnes, R.G. Artificial neural networks: A new method for mineral prospectivity mapping. *Aust. J. Earth Sci.* **2000**, *47*, 757–770. [CrossRef]
21. Porwal, A.; Carranza, E.J.M.; Hale, M. Artificial neural networks for mineral potential mapping. *Nat. Resour. Res.* **2003**, *12*, 155–171. [CrossRef]
22. Porwal, A.; Carranza, E.J.M.; Hale, M. A Hybrid Neuro-Fuzzy Model for Mineral Potential Mapping. *Math. Geosci.* **2004**, *36*, 803–826. [CrossRef]
23. Behnia, P. Application of Radial Basis Functional Link Networks to Exploration for Proterozoic Mineral Deposits in Central Iran. *Nat. Resour. Res.* **2007**, *16*, 147–155. [CrossRef]
24. Nykänen, V. Radial Basis Functional Link Nets Used as a Prospectivity Mapping Tool for Orogenic Gold Deposits Within the Central Lapland Greenstone Belt, Northern Fennoscandian Shield. *Nat. Resour. Res.* **2008**, *17*, 29–48. [CrossRef]
25. Parsa, M.; Carranza, E.J.M.; Ahmadi, B. Deep GMDH Neural Networks for Predictive Mapping of Mineral Prospectivity in Terrains Hosting Few but Large Mineral Deposits. *Nat. Resour. Res.* **2022**, *31*, 37–50. [CrossRef]
26. An, P.; Moon, W.; Rencz, A.N. Application of fuzzy theory for integration of geological, geophysical, and remotely sensed data. *CJEG* **1991**, *27*, 1–11.
27. Lusty, P.A.J.; Scheib, C.; Gunn, A.G.; Walker, A.S.D. Reconnaissance-scale prospectivity analysis for gold mineralization for gold mineralization in the southern Uplands-Dwon-Longford Terrane, Northern Island. *Nat. Resour. Res.* **2012**, *21*, 359–382. [CrossRef]
28. Knox-Robinson, C.M. Vectorial fuzzy logic: A novel technique for enhanced mineral prospectivity mapping, with reference to the orogenic gold mineralization potential of the Kalgoorlie Terrane, Western Australia. *Aust. J. Earth Sci.* **2000**, *47*, 929–941. [CrossRef]
29. Harris, J.; Grunsky, E.; Behnia, P.; Corrigan, D. Data- and knowledge-driven mineral prospectivity maps for Canada's North. *Ore Geol. Rev.* **2015**, *71*, 788–803. [CrossRef]
30. Breiman, L. Bagging Predictors. *Mach. Learn.* **1996**, *24*, 123–140. [CrossRef]
31. Breiman, L. Random Forests. *Mach. Learn.* **2001**, *45*, 5–32. [CrossRef]
32. Lachaud, A.; Marcus, A.; Vučetić, S.; Mišković, I. Study of the Influence of Non-Deposit Locations in Data-Driven Mineral Prospectivity Mapping: A Case Study on the Iskut Project in Northwestern British Columbia, Canada. *Minerals* **2021**, *11*, 597. [CrossRef]
33. Cracknell, M.J.; Reading, A.M. Geological mapping using remote sensing data: A comparison of five machine learning algorithms, their response to variations in the spatial distribution of training data and the use of explicit spatial information. *Comput. Geosci.* **2014**, *63*, 22–33. [CrossRef]
34. Rodriguez-Galiano, V.F.; Chica-Olmo, M.; Chica-Rivas, M. Predictive modelling of gold potential with the integration of multisource information based on random forest: A case study on the Rodalquilar area, Southern Spain. *Int. J. Geogr. Inf. Sci.* **2014**, *28*, 1336–1354. [CrossRef]
35. Carranza, E.J.M.; Laborte, A.G. Data-driven predictive mapping of gold prospectivity, Baguio district, Philippines: Application of Random Forests algorithm. *Ore Geol. Rev.* **2015**, *71*, 777–787. [CrossRef]
36. Harris, J.R.; Naghizadeh, M.; Behnia, P.; Mathieu, L. Data-driven gold potential maps for the Chibougamau area, Abitibi greenstone belt, Canada. *Ore Geol. Rev.* **2022**, *150*, 105176. [CrossRef]
37. Carranza, E.J.M.; Laborte, A.G. Data-Driven Predictive Modeling of Mineral Prospectivity Using Random Forests: A Case Study in Catanduanes Island (Philippines). *Nat. Resour. Res.* **2016**, *25*, 35–50. [CrossRef]
38. Behnia, P.; Harris, J.; Liu, H.; Jørgensen, T.R.C.; Naghizadeh, M.; Roots, E.A. Random forest classification for volcanogenic massive sulfide mineralization in the Rouyn-Noranda Area, Quebec. *Ore Geol. Rev.* **2023**, *161*, 105612. [CrossRef]
39. McKay, G.; Harris, J.R. Comparison of the Data-Driven Random Forests Model and a Knowledge-Driven Method for Mineral Prospectivity Mapping: A Case Study for Gold Deposits Around the Huritz Group and Nueltin Suite, Nunavut, Canada. *Nat. Resour. Res.* **2015**, *25*, 125–143. [CrossRef]
40. Harris, J.; Grunsky, E.C. Predictive lithological mapping of Canada's North using Random Forest classification applied to geophysical and geochemical data. *Comput. Geosci.* **2015**, *80*, 9–25. [CrossRef]
41. Zhang, S.; Xiao, K.; Carranza, E.J.M.; Yang, F. Maximum Entropy and Random Forest Modeling of Mineral Potential: Analysis of Gold Prospectivity in the Hezuo–Meiwu District, West Qinling Orogen, China. *Nat. Resour. Res.* **2019**, *28*, 645–664. [CrossRef]

42. Zhang, S.; Carranza, E.J.M.; Xiao, K.; Wei, H.; Yang, F.; Chen, Z.; Li, N.; Xiang, J. Mineral Prospectivity Mapping based on Isolation Forest and Random Forest: Implication for the Existence of Spatial Signature of Mineralization in Outliers. *Nat. Resour. Res.* **2021**, *31*, 1981–1999. [CrossRef]

43. Wang, J.; Zuo, R.; Xiong, Y. Mapping Mineral Prospectivity via Semi-supervised Random Forest. *Nat. Resour. Res.* **2020**, *29*, 189–202. [CrossRef]

44. Parsa, M.; Maghsoudi, A. Assessing the effects of mineral systems-derived exploration targeting criteria for random Forests-based predictive mapping of mineral prospectivity in Ahar-Arasbaran area, Iran. *Ore Geol. Rev.* **2021**, *138*, 104399. [CrossRef]

45. Hong, S.; Zuo, R.; Huang, X.; Xiong, Y. Distinguishing IOCG and IOA deposits via random forest algorithm based on magnetite composition. *J. Geochem. Explor.* **2021**, *230*, 106859. [CrossRef]

46. Ford, A. Practical Implementation of Random Forest-Based Mineral Potential Mapping for Porphyry Cu–Au Mineralization in the Eastern Lachlan Orogen, NSW, Australia. *Nat. Resour. Res.* **2020**, *29*, 267–283. [CrossRef]

47. Colvine, A.C.; Fyon, J.A.; Heather, K.B.; Marmont, S.; Smith, P.M.; Troop, D.G. *Archean Lode Gold Deposits in Ontario*; Ontario Geological Survey: Sudbury, ON, Canada, 1988; Paper 139; 136p.

48. Hodgson, C.J. Mesothermal lode-gold deposits. In *Mineral Deposits Modeling*; Kirkham, R.V., Sinclair, W.D., Thorpe, R.I., Duke, J.M., Eds.; Geological Association of Canada: St. John's, NL, Canada, 1993; pp. 635–678.

49. Dubé, B.; Mercier-Langevin, P. Gold Deposits of the Archean Abitibi Greenstone Belt, Canada. In *Geology of the World's Major Gold Deposits and Provinces*; Sillitoe, R.H., Goldfarb, R.J., Robert, F., Simmons, S.F., Eds.; Society of Economic Geologists, Inc.: Littleton, CO, USA, 2020; pp. 669–708. [CrossRef]

50. Melling, D.R. The occurrence of gold in the Cameron–Rowan Lake area: Preliminary report on the Monte Cristo and Victor Island gold prospects in the Monte Cristo Shear Zone. In *Summary of Field Work and Other Activities*; Ontario Geological Survey: Sudbury, ON, Canada, 1986; pp. 252–260.

51. Pelletier, M. The Rainy River gold deposit, Wabigoon Subprovince, Western Ontario: Style, Geometry, Timing and Structural Controls on ore Distribution and Grades. Master's Thesis, Université du Québec, Institut national de la recherche scientifique, Quebec City, QC, Canada, 2016; 404p.

52. Launay, G.; McRae, M.L.; Sherlock, R.L. Stratigraphy, Metallogeny and Crustal Architecture of the Rainy River Greenstone Belt. Laurentian University. 2021. Available online: https://merc.laurentian.ca/sites/default/files/mepresentationrainyriver2021_seg_v2.pdf (accessed on 10 June 2023).

53. Percival, J.A.; Skulski, T.; Sanborn-Barrie, M.; Stott, G.M.; Leclair, A.D.; Corkery, M.T.; Boily, M. Chapter 6 Geology and tectonic evolution of the Superior province, Canada. In *Tectonic Styles in Canada: The Lithoprobe Perspective*; Percival, J.A., Cook, F.A., Clowes, R.M., Eds.; Geological Association of Canada: St. John's, NL, Canada, 2012; Special Paper 49; p. 321.

54. Blackburn, C.E.; Johns, G.W.; Ayer, J.A.; Davis, D.W. Wabigoon Subprovince. In *Geology of Ontario*; Ontario Geological Survey: Sudbury, ON, Canada, 1991; Volume 4, pp. 303–382.

55. Launay, G.; McRae, M.L.; Sherlock, R.L. Preliminary results from stratigraphic and structural study of the Rainy River green-stone belt and the Quetico Subprovince. In *Summary of Field Work and Other Activities*; Ontario Geological Survey: Sudbury, ON, Canada, 2019; pp. 44-1–44-10, Open File Report 6360.

56. Davis, D.W.; Poulsen, K.H.; Kamo, S.L. New Insights into Archean Crustal Development from Geochronology in the Rainy Lake Area, Superior Province, Canada. *J. Geol.* **1989**, *97*, 379–398. [CrossRef]

57. Fernández, C.; Czeck, D.M.; Díaz-Azpiroz, M. Testing the model of oblique transpression with oblique extrusion in two natural cases: Steps and consequences. *J. Struct. Geol.* **2013**, *54*, 85–102. [CrossRef]

58. Pelletier, M.; Mercier-Langevin, P.; Crick, D.; Tolman, J.; Beakhouse, G.P.; Dubé, B. Targeted Geoscience Initiative 4. Lode gold deposits in ancient and deformed terranes: Preliminary observations on the nature and distribution of the deformed and metamorphosed hydrothermal alteration associated with the Archean Rainy River gold deposit, northwestern Ontario. In *Summary of Field Work and Other Activities*; Ontario Geological Survey: Sudbury, ON, Canada, 2014; pp. 41-1–41-10, Open File Report 6300.

59. Pelletier, M.; Mercier-Langevin, P.; Dubé, B.; Crick, D.; Tolman, J.; McNicoll, V.J.; Jackson, S.E.; Beakhouse, G.P. The Rainy River—Atypical Archean Au deposit, western Wabigoon Subprovince, Ontario. In *Targeted Geoscience Initiative 4: Contributions to the Understanding of Precambrian Lode Gold Deposits and Implications for Exploration*; Geological Survey of Canada: Ottawa, ON, Canada, 2014; pp. 193–207, Open File 7852.

60. Jakimow, B.; Rabe, A.; van der Linden, S.; Wirth, F.; Hostert, P. EnMAP-Box Manual, Version 1.4, Humboldt-Universität zu Berlin, 2012, Germany. Available online: https://bitbucket.org/hu-geomatics/enmap-box-idl/wiki/imageRF%20-%20Manual%20for%20Application (accessed on 10 November 2021).

61. Aitchison, J. *The Statistical Analysis of Compositional Data*; Chapman and Hall: London, UK, 1986; 416p.

62. Ontario Geological Survey. *Single Master Gravity and Aeromagnetic Data for Ontario*; Ontario Geological Survey: Sudbury, ON, Canada, 1999.

63. Geological Survey of Canada. The Canadian Gravity and Magnetic Anomaly Database: GSC. Available online: http://gdr.agg.nrcan.gc.ca/gdrdap/dap/search-eng.php (accessed on 12 October 2022).

64. Rahimi, H.; Abedi, M.; Yousefi, M.; Bahroudi, A.; Elyasi, G.-R. Supervised mineral exploration targeting and the challenges with the selection of deposit and non-deposit sites thereof. *Appl. Geochem.* **2021**, *128*, 104940. [CrossRef]

65. Sun, T.; Chen, F.; Zhong, L.; Liu, W.; Wang, Y. GIS-based mineral prospectivity mapping using machine learning methods: A case study from Tongling ore district, eastern China. *Ore Geol. Rev.* **2019**, *109*, 26–49. [CrossRef]
66. Pedregosa, F.; Varoquaux, G.; Gramfort, A.; Michel, V.; Thirion, B.; Grisel, O.; Blondel, M.; Prettenhofer, P.; Weiss, R.; Dubourg, V.; et al. Scikit-learn: Machine Learning in Python. *J. Mach. Learn. Res.* **2011**, *12*, 2825–2830. Available online: https://scikit-learn.org/stable/modules/permutation_importance.html#permutation-importance (accessed on 14 March 2022).

Article

Particle Swarm Optimization (PSO) of High-Quality Magnetic Data of the Obudu Basement Complex, Nigeria

Stephen E. Ekwok [1], Ahmed M. Eldosouky [2,3,*], Khalid S. Essa [4], Anthony M. George [1], Kamal Abdelrahman [5], Mohammed S. Fnais [5], Peter Andráš [6], Emmanuel I. Akaerue [1] and Anthony E. Akpan [1]

1   Applied Geophysics Programme, Department of Physics, University of Calabar, Calabar PMB 1115, Nigeria
2   Department of Geology, Suez University, Suez 43518, Egypt
3   Academy of Scientific Research & Technology, Cairo 4262104, Egypt
4   Geophysics Department, Faculty of Science, Cairo University, Giza 12613, Egypt; khalid_sa_essa@yahoo.com
5   Department of Geology and Geophysics, College of Science, King Saud University, Riyadh 11451, Saudi Arabia
6   Faculty of Natural Sciences, Matej Bel University in Banska Bystrica, Tajovského 40, 974 01 Banska Bystrica, Slovakia
*   Correspondence: ahmed.eldosouky@sci.suezuni.edu.eg

Abstract: The particle swamp optimization procedure was applied to high-quality magnetic data acquired from the Precambrian Obudu basement complex in Nigeria with the object of estimating the distinctive body parameters (depth (z), index angle (θ), amplitude coefficient (K), shape factor (Sf), and location of the origin (x_0)) of magnetic models. The magnetic models were obtained from four profiles that ran perpendicular to the observed magnetic anomalies within the study area. Profile A–A' with a length of 2600 m is characterized by inverted model parameters of K = 315.67 nT, z = 425.34 m, θ = 43°, Sf = 1.15, and x_0 = 1554.86 m, while profile B–B' with a length of 5600 m is described by K = 257.71 nT, z = 543.75 m, θ = 54°, Sf = 0.96, and x_0 = 3645.42 m model parameters. Similarly, profile C–C' with a length of 3000 m is defined by K = 189.53 nT, z = 560.87 m, θ = 48, Sf = 1.2, and x_0 = 1950 m. Profile D–D', which is well-defined by a 2500 m length, started at the crest of the observed magnetic anomaly and displays inverted model parameters of 247.23 nT, 394.16 m, 39°, 1.26, and 165.41 m. Correlatively, the estimated shape factor of the four models (Sf = 1.15, 0.96, 1.2, and 1.26) shows that the magnetic models are linked to thin sheets. Furthermore, quantitative interpretations of the models show that the PSO operation is rapid and proficient.

Keywords: particle swarm optimization; magnetic; mineral exploration

Citation: Ekwok, S.E.; Eldosouky, A.M.; Essa, K.S.; George, A.M.; Abdelrahman, K.; Fnais, M.S.; Andráš, P.; Akaerue, E.I.; Akpan, A.E. Particle Swarm Optimization (PSO) of High-Quality Magnetic Data of the Obudu Basement Complex, Nigeria. *Minerals* **2023**, *13*, 1209. https://doi.org/10.3390/min13091209

Academic Editor: Yosoon Choi

Received: 15 June 2023
Revised: 7 September 2023
Accepted: 12 September 2023
Published: 14 September 2023

1. Introduction

High-resolution magnetic data analysis has been applied extensively to mineral and ore explorations worldwide [1–3]. Additionally, it can be used in the investigation of hydrocarbon [4], engineering surveys [5,6], detection of UXO [2], mapping of geothermal anomalies [7,8], archaeological investigations [9], and delineation of subsurface hydrological structures [3,5]. Moreover, almost all magnetic data analysis procedures are performed believing that the subsurface features are simple geometrical structures, such as thin sheets, horizontal cylinders, spheres, and faults (caused by diverse ore and mineral-bearing bodies, as well as structural hydrocarbon traps), which are buried at various depths [5,10].

A number of inversion procedures are used in general to evaluate parameters connected with simple geologic models, such as geologic contacts, cylinders, spheres, and thin sheets [11–13]. In several tasks, these models assume notable roles [13]. However, while it may be challenging to precisely define the geological origins of these models resulting from subsurface magnetic bodies, they still hold significant utility in magnetic analysis for the determination of body parameters [13,14]. Many calculable inversion procedures were created to evaluate magnetic data across diverse geologic features [12,13]. Previous procedures applied to analyze the magnetic data have included parametric curves [15], the Werner deconvolution [16], and Euler deconvolution [17]. Additionally, linear least-squares [14], layered model inversion [18], gradient [19], and fair-function minimization [20] methods have been used. Nonetheless, some of these conventional inversion methodologies produce a significant number of unacceptable results because of misinterpretations between magnetic sources, noise sensitivity, and window sizes [21]. Furthermore, some of these procedures require the use of initial model parameters derived from geological information, rely on a limited dataset along a profile, assume knowledge of the shape factor, and have a longer processing time [21].

In recent years, global optimization procedures have been applied as an alternative to geophysical inversion techniques. These techniques include PSO [22], genetic algorithms [23], simulated annealing [24], and differential evolution [25,26]. Other notable inversion procedures involve simulated annealing [27], social spider optimization [28], ant-colony optimization [29], hybrid genetic algorithm [30], and hybrid genetic algorithm [30]. Genetic algorithms (GAs), proposed by [31], mimic the process of natural evolution to search for optimal solutions. GAs utilize populations of individuals representing potential solutions, subjecting them to selection, crossover, and mutation operations to evolve fitter solutions over generations. Simulated annealing (SA), as introduced by [32] in optimizations performed by simulated annealing, draws inspiration from the annealing process in metallurgy. This method explores the solution space by allowing temperature-controlled transitions between solutions, gradually reducing the exploration intensity to converge towards optimal solutions. Differential evolution (DE), a simple and efficient heuristic for global optimization over continuous spaces [33], operates through the mutation and recombination of parameter vectors to generate new candidate solutions. DE employs a population-based approach that prioritizes the best-performing solutions while maintaining diversity. The PSO is an efficient optimization technique for precisely and dependably resolving challenging issues [28]. The PSO technique developed by [34] is a stochastic computation tool. It focuses mostly on simulating the natural behavior of fish, insects, and birds as they look for food. The PSO algorithm uses a population of particles that traverse the solution space while adapting their trajectories based on historical information and the best-performing solutions encountered [35]. Location vectors that signify the parameter value and a speed vector are both present in every particle. Each particle or person, for instance, will have a position in a five-dimensional space that serves as a solution for the optimization issue [36]. In the realm of optimization, researchers and practitioners continually seek innovative methods to efficiently tackle complex and high-dimensional optimization problems across various domains [35]. The PSO has gained considerable attention due to its simplicity, ease of implementation, and effectiveness in solving a wide range of optimization challenges. It enhances exploration and exploitation capabilities by incorporating environmental factors that simulate real-world conditions [35].

The PSO technique has been applied to the inversion and modeling of geophysical data [37]. It also has connections to a wide range of problems, including machine learning [38], electromagnetic optimizations [33,39], model building [40], inverse scattering [41], biomedical imageries [42], hydrological issues [43], etc. In this research, the PSO procedure is applied to the inversion of magnetic data over causative bodies noticeable in observed magnetic data obtained from Obudu Precambrian basement rocks. The model parameters approximated involve the depth (z), index angle (θ), amplitude coefficient (K), location of the origin (x_0), and shape factor (Sf), and they are determined to be reliable and resolvable.

2. Location and Geology of the Investigated Area

The studied location was situated within the Precambrian Obudu basement complex (Figure 1), and it was located between longitude values of 9°00′ E and 9°30′ E and latitude values of 6°30′ N and 7°00′ N. According to [44], a series of tectono-thermal events with about three or more stages of distortion are responsible for the formation of the Nigerian basement complex. Three lithologic assemblages—the migmatite–gneiss complex, schist belts, and older granite sets—generally characterize the basement complex. In structurally regulated basins, the Cretaceous-Recent sedimentary sequence has been successfully maintained. The main rocks in the Nigerian basement complex are believed to be migmatite–gneiss complexes [45]. Age values related to the Kibaran ranged from 900 to 450 Ma, indicating the influence of the Pan-African occurrence that produced gneisses, migmatite, older granites, and connected lithological components [45].

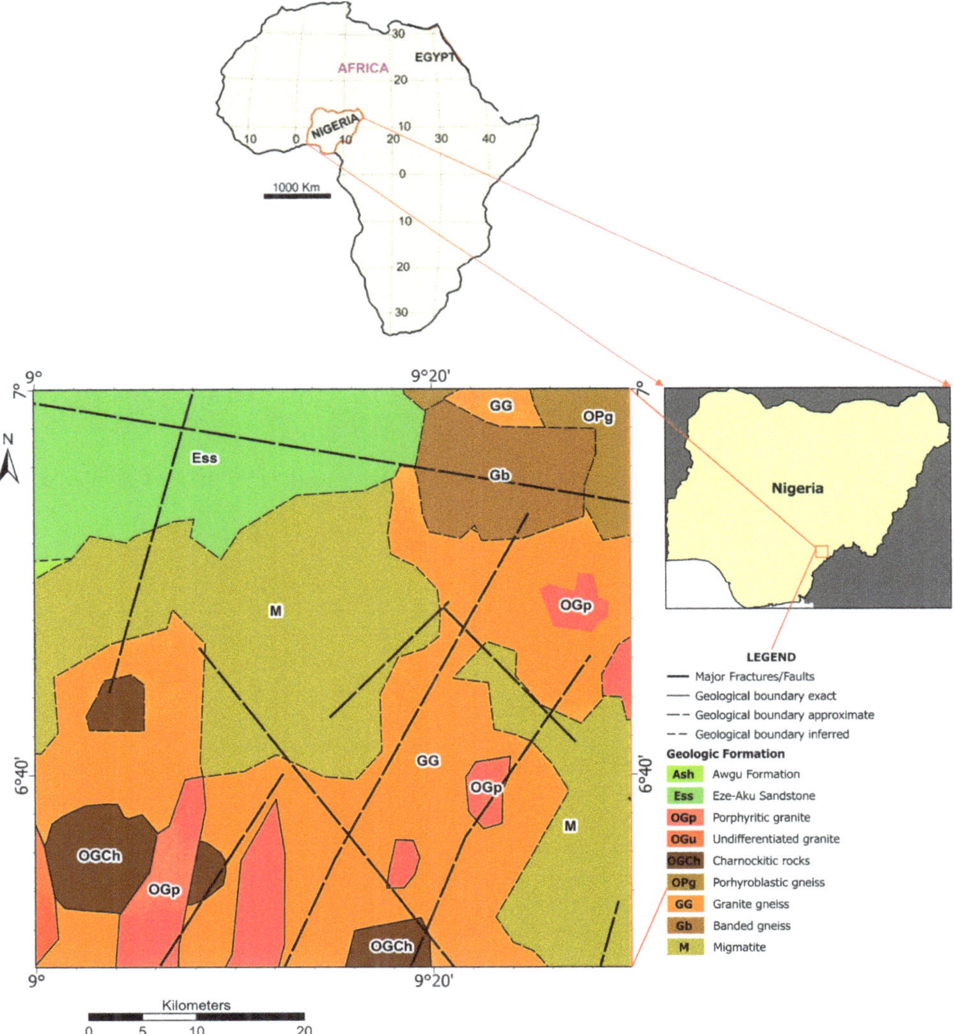

Figure 1. Location and geologic maps of the study area.

There were no magmatic or depositional events that occurred during the middle-to-late Paleozoic era. Younger Jurassic granites are among the alkaline, anorogenic, shallow sub-volcanic intrusive materials that characterize the Mesozoic era. They are found in a north–south thin belt in the western part of the eastern region and extend northward into the Republic of Niger. One of Nigeria's Precambrian basement outcrops is located in the Obudu Plateau, a section of the Bamenda Massif [46]. High-grade metamorphic rocks, including migmatized schists and gneisses that have been affected by granites, quartzo-feldspathic veins, and unmetamorphosed dolerites, are to blame for the region's unusual lithology [46]. Amphibolites, metaquartzites, and metagabbro are also present in these rocks in traces. According to [47], the rocks in this area date to the Archaean, Eburnean, and Pan-African eras. Researchers arrived at the conclusion that the Southeast Nigerian basement evolutionary history is connected to the Pan-African mobile belt in Central Africa because of the relatively close agreement between the lithology of the rocks and ages of the Southeast Nigeria, Northern Cameroon, and Central African Fold Belt basement complexes [47]. The Adamawa-Yade and western Cameroon domain served as the active margin during the continent-to-continent collision that created to the Pan-Central African belt in Central Africa, whereas the northern Congo Craton border served as the passive margin [48]. According to [46], migmatitic gneiss, also known as garnet–sillimanite gneiss, garnet–hornblende gneiss, or simply migmatite gneiss, compose the majority of the OP.

3. Methodology

3.1. Two-Dimensional Magnetic Forward Problem

After carefully examining the magnetic anomaly expressions of horizontal cylinders [49], thin sheets [50], and spheres, [51,52] summarized the general new formula of the two-dimensional magnetic anomaly (T) profile for simple geometric bodies, which is defined as:

$$T_{(x_i,z)} = K \frac{Az^2 + B(x_i - x_0) + C(x_i - x_0)^2}{\left[(x_i - x_0)^2 + z^2\right]^{Sf}}, \ i = 0, 1, 2, 3, \ldots\ldots\ldots, N \tag{1}$$

where K is the amplitude coefficient, z is the depth of the buried body, and A, B, and C are described as follows:

$$A \begin{cases} 3sin^2\theta - 1 \\ 2sin\theta \\ -cos\theta \\ cos\theta \\ \frac{cos\theta}{z} \end{cases} , \ B \begin{cases} -3zsin\theta \\ -3zcos\theta \\ -3zsin\theta, \\ 2zsin\theta \\ -sin\theta \end{cases} \ C \begin{cases} 3cos^2\theta - 1 \\ -sin\theta \\ 2cos\theta \\ 0 \end{cases}$$

- For a sphere (total field);
- For a sphere (vertical field);
- For a sphere (horizontal field);
- For a horizontal cylinder, FHD of thin sheet, and SHD of geological contact (all fields);
- For a thin sheet and FHD of geological contacts (all fields).

θ is the angle of effective magnetization [51,53] in the spheres. Additionally, ref [54] defines the cases of thin sheets and horizontal cylinders as follows: x_0 is the coordinate of the source body's center, FHD and SHD are the first and second horizontal derivatives of the magnetic anomaly, respectively; N is the number of data points; and Sf is the shape factor, with values of 2.5 for spheres, 2 for horizontal cylinders, and 1 for thin sheets [55].

3.2. Magnetic Inverse Problem

In geophysics, the inversion technique is an optimization process aimed at identifying the model parameters of concealed geological features that are most suitable for explaining the measured data. To solve the inverse problem, it is necessary to perform this with an initial model [56]. This preliminary model can be established using drilling, other geophysical methods, and prior geological knowledge [57]. Every iteration step involves progressively refining the initial model until the theoretical and observed data are the best match.

According to the two-dimensional magnetic formula (Equation (1)), the model's unidentified parameters are the origin's location (x_0), depth (z), index angle (θ), shape factor (Sf), and amplitude coefficient (K). Due to its ease of use and speedy operation by merging fewer operators, the PSO technique was utilized to resolving the inverse problem. Furthermore, it is stable in terms of both numbers and mathematics. The model parameter values that would minimize the discrepancies between the collected field data and the theoretical model are determined using the following simple objective function:

$$Q = \frac{2\sum_{i=1}^{N}\left|T_i^m - T_i^c\right|}{\sum_{i=1}^{N}\left|T_i^m - T_i^c\right| + \sum_{i=1}^{N}\left|T_i^m - T_i^c\right|} \tag{2}$$

where N is the number of data points, T_i^m is the observed magnetic anomaly at the point x_i, and T_i^c is the approximated magnetic anomaly at the point x_i.

3.3. Particle Swarm Optimization

Ref [34] suggested the PSO technique as a global optimization algorithm. It depends on creatures (represented as particles) in the environment imitating natural processes, such as fish schooling and birds flocking. A point in M-dimensional space has an equivalent competitive solution for each particle in a PSO algorithm. In the inquiry space of the objective function, the computation is introduced freely and the initial solutions are established arbitrarily [58]. The PSO algorithm effectively guides researchers to achieving a global optimum value. The PSO algorithm's main model structure is that the most likely answers are developed before the best ones. Figure 2 shows a basic flowchart. The PSO algorithm starts by assigning each particle in the swarm a random position and speed in the problem search space. Every bird, which represents a particle or model, has a velocity vector and a position vector that reflect the parameter value. PSO explains a swarm of particles (models) in an M-dimensional space. Each particle retains the position and speed of its previous optimal state. The best location of the swarm and the previous best position, often referred to as the Tbest model inhabited by the particle, are utilized to jointly estimate the speed modification of the particle at each iteration phase. The modified velocity is then utilized to calculate a new position for the particle using the Jbest model [59]. The following equations, according to [60], describe the update:

$$V_i^{k+1} = c_3 V_i^k + c_1 rand\,()(T_{best} - P_i^{k+1}) + c_2 rand\,()[(J_{best} - P_i^{k+1})P_i^{k+1}] = P_i^k + V_i^{k+1} \tag{3}$$

$$x_i^{k+1} = x_i^k + v_i^{k+1} \tag{4}$$

where P_i^k is the current i model at the kth iteration, v_i^k is the speed of the ith particle at the kth iteration, and rand() is an identical random number in the rang (0, 1). c_1 and c_2 are the positive consistent numbers that control individual and social behaviors [61]. c_3 is the inertial coefficient that controls the particle velocity. x_i^k is the position of the particle i at the kth iteration.

The magnetic anomaly from Equation (1) is computed every iterative phase for each x_i using the PSO algorithm. To estimate the quality of the data fit at each iteration phase of the inversion process, the RMS is given as:

$$RMS = \sqrt{\frac{\sum_{i=1}^{N}\left[T_i^m(x_i) - T_i^c(x_i)\right]^2}{N}} \tag{5}$$

This is taken as the misfit between the observed and theoretical anomalies.

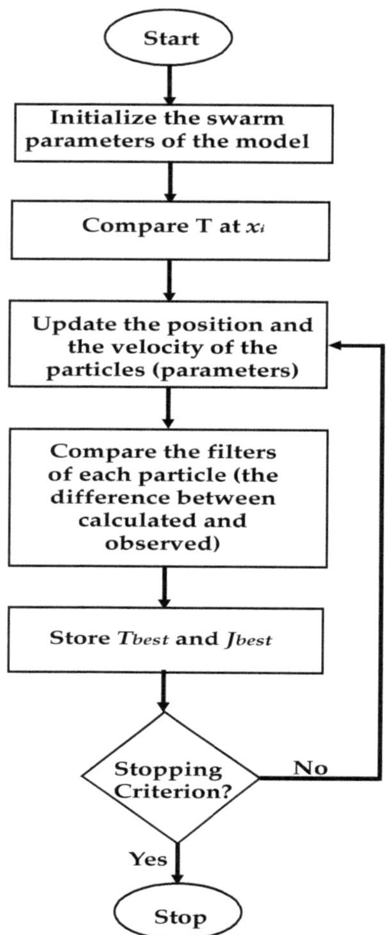

Figure 2. Flowchart of the PSO algorithm applied to magnetic anomalies' interpretations.

4. Data Acquisition and Selection of Profiles

The aero-magnetic data that were used in this work (Figure 3) were collected and filtered between 2005 and 2010 by Fugro Airborne Services, Canada. These data were collected by means of a Flux-Adjusting Surface Data Assimilation System with a flight-line space of 0.1 km, tie line space of 0.5 km, and terrain clearance ranging from 0.08–0.1 km along 826,000 lines. The mean total field, inclination value, and declination value were 32,851.9 nT, $-14.6°$, and $-2.4°$, respectively. The magnetic data have the potential to map small geologic anomalies, and when compared to the aero-magnetic data from 1970, they were found to be of good quality. In order to obtain associated PSO parameters, including the origin position (x_0), depth (z), amplitude coefficient (K), and index angle (θ), the profiles used in the PSO operation were carefully chosen across notable anomalies of the magnetic data (Figure 4).

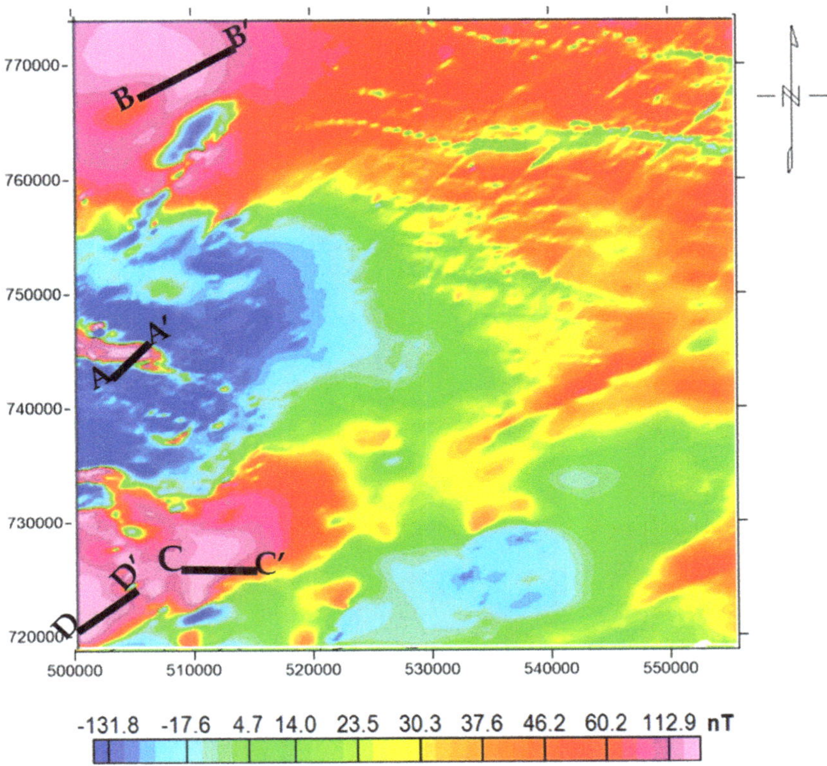

Figure 3. Total magnetic intensity data showing locations of profiles.

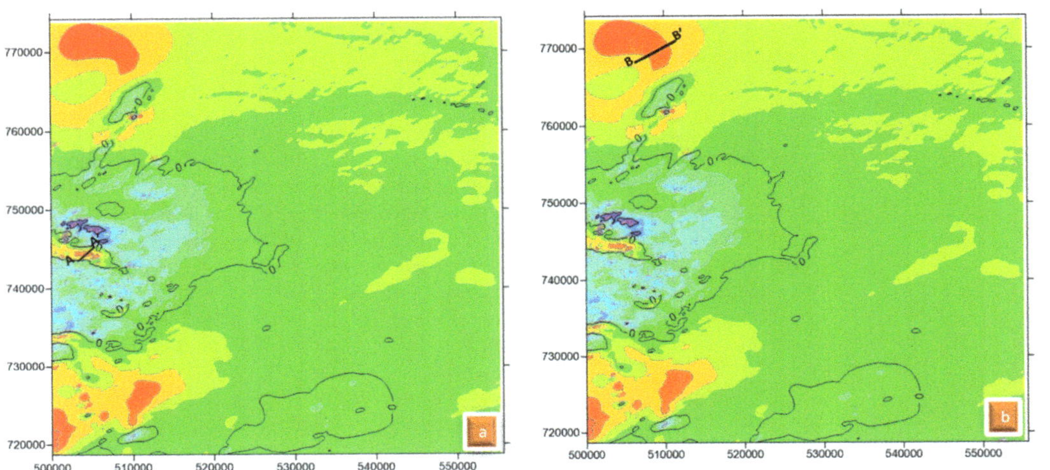

Figure 4. *Cont.*

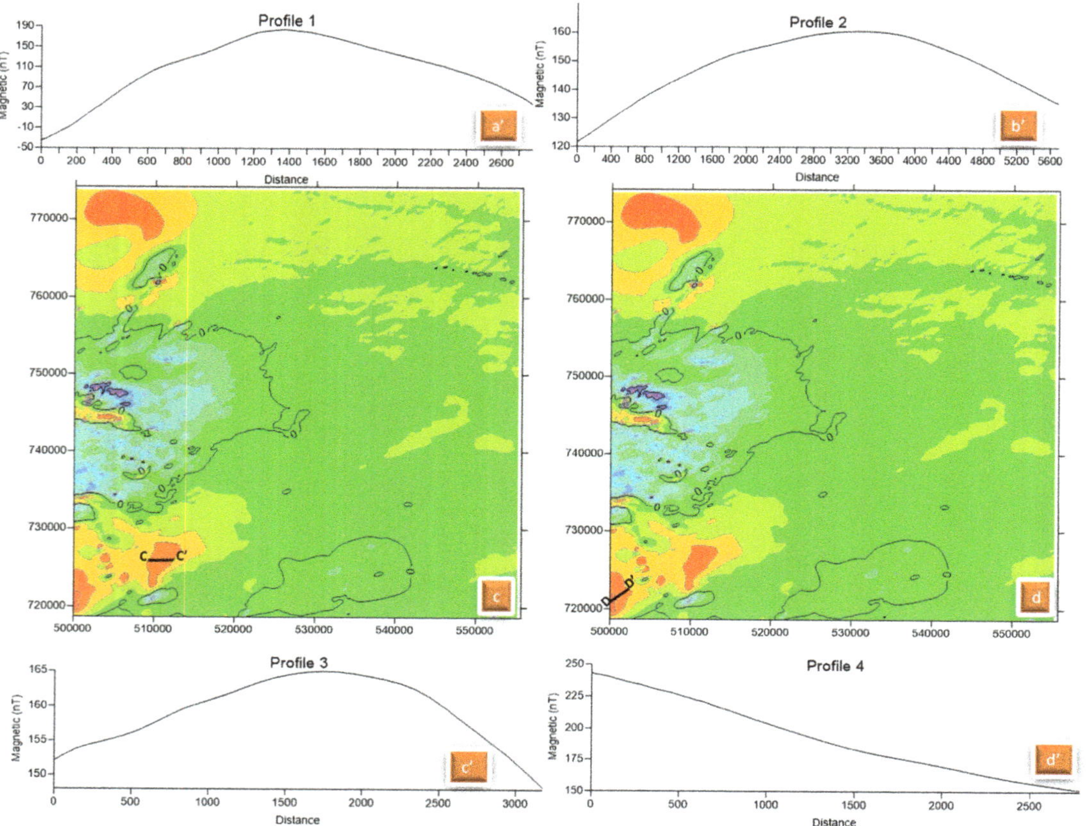

Figure 4. Profiles (**a–a'**), (**b–b'**), (**c–c'**), and (**d–d'**), and related signatures (**a'**), (**b'**), (**c'**), and (**d'**), respectively, obtained from the magnetic data.

5. Interpretation and Discussion of Results

The PSO algorithm was used for magnetic data interpretations over the Obudu Precambrian basement complex in Nigeria. The area that is situated in the eastern flank of Nigeria's border with the Republic of Cameroon is predominantly composed of metamorphic rocks that have been invaded by unaltered granites, dolerites, and quartzo-feldspathic veins [3,46,62–65]. The profile of A–A' (with a distance of 2600 m) and its related signature (Figure 4a') were obtained from the magnetic data. Table 1 displays the results of the inverted model parameters (depth, amplitude coefficient, index angle, shape factor, and location of origin) over A–A'. The optimum results of the model parameters from the magnetic anomaly are observed as K = 315.67 nT, z = 425.34 m, θ = 43°, Sf = 1.15, and x_0 = 1554.86 m.

Profile (B–B') in Figure 4 shows the component of the magnetic anomaly with a profile length of 5600 m. The model parameters are displayed in Table 2, which were obtained from the observed geologic anomaly. Table 2 reveals the evaluated model parameters as K = 257.71 nT, z = 543.75 m, θ = 54°, Sf = 0.96, and x_0 = 3645.42 m. Furthermore, the parameters of the third anomaly (Table 3) obtained from the profile length of 3000 m (Figure 4; profile C–C') were K = 189.53 nT, z = 560.87 m, θ = 48, Sf = 1.2, and x_0 = 1950 m. The fourth magnetic anomaly profile with a length of 2500 m and associated signature (a') (Figure 4) revealed the inverted model parameters (Table 4) of 247.23 nT, 394.16 m,

39°, 1.26, and 165.41 m, respectively, for amplitude coefficient, depth, index angle, shape factor, and location of the origin over D–D′. It can be noticed that profile D–D′ (Figure 4d′) originates from the peak of the noticed magnetic anomaly and, as a result, the profile signature assumes the shape of a quarter cycle. From the obtained shape factor results involving the four models (Sf = 1.15, 0.96, 1.2, and 1.26), it can be established that the four magnetic anomalies are generated by thin sheets. Similar results were reported by [21] (Sf = 0.89 and 0.93) in their comparative study involving the application of PSO to real and synthetic data. On the contrary, Sf values of 2.5 and 2.0 were reported for the sphere and horizontal cylinder, respectively [21]. These geologic structures are created by the invasion of older granite suites by younger granite suites, resulting in a series of metamorphisms [66], folds, faults, and shear zones [67,68] linked to the Pan-African orogeny and succeeding post-orogenic events [67]. According to several published studies, geological structures within tectonically active regions act as depositional zones for igneous-related minerals and migratory pathways for hydrothermal fluids [69–71]. Magmatism and mineralization are linked, according to several investigations [3,45,69,71]. Therefore, it is believed that vast quantities of metallogenic minerals in the study area are under the influence of magmatic intrusions.

Table 1. Numerical results of magnetic anomaly for profile A–A′.

S/N	Parameter	Range	PSO Result	GA Result
1	K (nT)	0–3000	315.67	297.54
2	z (m)	200–1200	425.34	417.69
3	θ (°)	−90–90	43	47
4	Sf (dimensionless)	0.5–2.5	1.15	1.05
5	x_0 (m)	1100–1900	1554.86	1527.97
6	RMS (nT)		6.43	8.59

Table 2. Numerical results of magnetic anomaly for profile B–B′.

S/N	Parameter	Range	PSO Result	GA Result
1	K (nT)	0–3000	257.71	278.28
2	z (m)	200–1200	543.75	549.14
3	θ (°)	−90–90	54	50
4	Sf (dimensionless)	0.5–2.5	0.96	0.89
5	x_0 (m)	2500–3800	3645.42	3654.87
6	RMS (nT)		4.81	6.28

Table 3. Numerical results of magnetic anomaly for profile C–C′.

S/N	Parameter	Range	PSO Result	GA Result
1	K (nT)	0–3000	189.53	208.35
2	z (m)	200–1200	560.87	569.26
3	θ (°)	−90–90	48	52
4	Sf (dimensionless)	0.5–2.5	1.2	1.13
5	x_0 (m)	500–3000	1950	1962
6	RMS (nT)		5.95	9.62

Table 4. Numerical results of magnetic anomaly for profile D–D′.

S/N	Parameter	Range	PSO Result	GA Result
1	K (nT)	0–3000	247.23	219.59
2	z (m)	200–1200	394.16	384.15
3	θ (°)	−90–90	39	45
4	Sf (dimensionless)	0.5–2.5	1.26	0.97
5	x_0 (m)	50–1250	165.41	170.57
6	RMS (nT)		8.92	11.53

The validity of the PSO algorithm was tested on real data obtained from the Pima copper mine, United States of America, and the Hamrawein zone, Egypt [21]. The Pima copper mine field is characterized by minerals related to Laramide Paleozoic igneous activity [72]. The magnetic anomaly profile (696 m length) over a thin sheet-like feature is well-defined by these model parameters (K = 600 nT, z = 71.08 m, h = −47.83, Sf = 0.92, and x_0 = −0.49) [50]. The correlation between the measured and theoretical data, including drilling information [73], indicates a relatively strong match [50]. Similarly, the PSO algorithm was used for magnetic data acquired from Hamrawein, Egypt. This region is located at the western flank of the Red Sea and is dominated by sedimentary and metavolcanic rocks [74]. The observed anomalies were characterized by inverted model parameters of 507.64 nT, 623.05 m, 57.04°, 0.89, and 4255.98 m (amplitude coefficient, depth, index angle, shape factor, and location of origin, respectively) and 427.38 nT, 494.14 m, 37.27°, 0.93, and 14,823.96 m (amplitude coefficient, depth, index angle, shape factor, and location of origin, respectively) for the first and second anomalies [21]. Previous studies have shown the strong correlations of theoretical and observed anomalies [21]. Likewise, the PSO procedure, when compared with the very fast simulated annealing method (VFSAM), generated good results in a shorter time period [75].

The inversion outcomes for the four profiles (A–A′, B–B′, C–C′, and D–D′), presented in Tables 1–4, utilizing the particle swarm optimization method were juxtaposed with those obtained through the genetic algorithm. This comparison reveals that the results achieved via the PSO approach exhibit greater stability and efficiency, primarily attributable to their lower root mean square (RMS) values.

6. Conclusions

In this research, the PSO algorithm was employed in approximating the distinctive model parameters (K, z, θ, x_0, and Sf) of the models. These models were generated from four profiles drawn for high-quality airborne magnetic data obtained from the Precambrian Obudu basement complex in Nigeria. Magnetic anomaly profile A–A′ with a distance of 2600 m generated inverted model parameters of K = 315.67 nT, z = 425.34 m, θ = 43°, Sf = 1.15, and x_0 = 1554.86 m. The second profile (B–B′) of a length of 5600 m had associated model parameters of K = 257.71 nT, z = 543.75 m, θ = 54°, Sf = 0.96, and x0 = 3645.42 m. Likewise, the parameters of the third anomaly obtained from a profile length of 3000 m (profile C–C′) were K = 189.53 nT, z = 560.87 m, θ = 48, Sf = 1.2, and x_0 = 1950 m. The fourth profile (with a length of 2500 m) that originated from the peak of the magnetic anomaly produced inverted model parameters of 247.23 nT, 394.16 m, 39°, 1.26, and 165.41 m, respectively, for the amplitude coefficient, depth, index angle, shape factor, and location of the origin. On the whole, the obtained shape factor values of the four models (Sf = 1.15, 0.96, 1.2, and 1.26) suggest the magnetic anomalies are initiated by thin sheets. The model results show that the PSO procedure is rapid, stable, and proficient for analyzing magnetic data for quantitative interpretations. The observed geological structures from the PSO results reveal depositional zones for igneous-related minerals and migratory pathways for hydrothermal fluids. As a result, it is believed that vast quantities of metallogenic minerals in the study area are associated with magmatic intrusions.

Minerals **2023**, 13, 1209

Author Contributions: Conceptualization, S.E.E., A.M.E. and K.S.E.; methodology S.E.E., A.M.E. and K.S.E.; software, S.E.E., A.M.E. and K.S.E.; validation A.M.E. and K.S.E.; formal analysis, A.M.G.; investigation, A.E.A., K.A., P.A. and M.S.F.; resources, E.I.A.; data curation, S.E.E.; writing—original draft preparation, S.E.E., A.E.A., A.M.G. and E.I.A.; writing—review and editing, A.M.E. and K.S.E. visualization, P.A.; supervision, A.M.E.; project administration, A.M.E.; funding acquisition, K.A. and M.S.F. All authors have read and agreed to the published version of the manuscript.

Funding: Researchers Supporting Project (RSP2023R249), King Saud University, Riyadh, Saudi Arabia.

Data Availability Statement: The will be available under the request from the authors.

Acknowledgments: Sincere thanks and gratitude to the Researchers Supporting Project (RSP2023R249), King Saud University, Riyadh, Saudi Arabia, for funding this research article.

Conflicts of Interest: The authors declare no conflict of interest.

References

1. Ekwok, S.E.; Akpan, A.E.; Achadu, O.I.M.; Eze, O.E. Structural and lithological interpretation of aero-geophysical data in parts of the Lower Benue Trough and Obudu Plateau, Southeast Nigeria. *Adv. Space Res.* **2021**, *68*, 2841–2854. [CrossRef]
2. Ekwok, S.E.; Akpan, A.E.; Ebong, D.E. Enhancement and modelling of aeromagnetic data of some inland basins, southeastern Nigeria. *J. Afr. Earth Sci.* **2019**, *155*, 43–53. [CrossRef]
3. Ekwok, S.E.; Akpan, A.E.; Kudamnya, E.A. Exploratory mapping of structures controlling mineralization in Southeast Nigeria using high resolution airborne magnetic data. *J. Afr. Earth Sci.* **2020**, *162*, 103700. [CrossRef]
4. Ivakhnenkoa, O.P.; Abirova, R.; Logvinenkoc, A. New method for characterisation ofpetroleum reservoir fluid mineral deposits using magnetic analysis. *Energy Procedia* **2015**, *76*, 454–462. [CrossRef]
5. Eldosouky, A.M.; Ekwok, S.E.; Akpan, A.E.; Achadu, O.-I.M.; Pham, L.T.; Abdelrahman, K.; Gómez-Ortiz, D.; Alarifi, S.S. Delineation of structural lineaments of Southeast Nigeria using high resolution aeromagnetic data. *Open Geosci.* **2022**, *14*, 331–340. [CrossRef]
6. Eldosouky, A.M.; Ekwok, S.E.; Ben, U.C.; Ulem, C.A.; Abdelrahman, K.; Gomez-Ortiz, D.; Akpan, A.E.; George, A.M.; Pham, L.T. Appraisal of geothermal potentials of some parts of the Abakaliki Anticlinorium and adjoining areas (Southeast Nigeria) using magnetic data. *Front. Earth Sci.* **2023**, *11*, 1216198. [CrossRef]
7. Alfaifi, H.J.; Ekwok, S.E.; Ulem, C.A.; Eldosouky, A.M.; Qaysi, S.; Abdelrahman, K.; András, P.; Akpan, A.E. Exploratory assessment of geothermal resources in some parts of the Middle Benue Trough of Nigeria using airborne potential field data. *J. King Saud. Univ. Sci.* **2023**, *35*, 102521. [CrossRef]
8. Abdelrahman, K.; Ekwok, S.E.; Ulem, C.A.; Eldosouky, A.M.; Al-Otaibi, N.; Hazaea, B.Y.; Hazaea, S.A.; Akpan, A.E. Exploratory Mapping of Geothermal Anomalies in the Neoproterozoic Arabian Shield, Saudi Arabia Using Magnetic Data. *Minerals* **2023**, *13*, 694. [CrossRef]
9. Gündoğdu, N.Y.; Candansayar, M.E.; Genç, E. Rescue archaeology application: Investigation of Kuriki mound archaeological area (Batman, SE Turkey) by using direct current resistivity and magnetic methods. *J. Environ. Eng. Geophys.* **2017**, *22*, 177–189. [CrossRef]
10. Abo-Ezz, E.R.; Essa, K.S. A least-squares minimization approach for model parameters estimate by using a new magnetic anomaly formula. *Pure Appl. Geophys.* **2016**, *173*, 1265–1278. [CrossRef]
11. Ben, U.C.; Ekwok, S.E.; Akpan, A.E.; Mbonu, C.C.; Eldosouky, A.M.; Abdelrahman, K.; Gómez-Ortiz, D. Interpretation of magnetic anomalies by simple geometrical structures using the manta-ray foraging optimization. *Front. Earth Sci.* **2022**, *10*, 849079. [CrossRef]
12. Ben, U.C.; Mbonu, C.C.; Thompson, C.E.; Ekwok, S.E.; Akpan, A.E.; Akpabio, I.; Eldosouky, A.M.; Abdelrahman, K.; Alzahrani, H.; Gómez-Ortiz, D.; et al. Investigating the applicability of the social spider optimization for the inversion of magnetic anomalies caused by dykes. *J. King Saud. Univ. Sci.* **2023**, *35*, 102569. [CrossRef]
13. Essa, K.S.; Elhussein, M. Interpretation of magnetic data through particle swarm optimization: Mineral exploration cases studies. *Nat. Resour. Res.* **2020**, *29*, 521–537. [CrossRef]
14. Abdelrahman, E.M.; Abo-Ezz, E.R.; Essa, K.S.; El-Araby, T.M.; Soliman, K.S. A new least-squares minimization approach to depth and shape determination from magnetic data. *Geophys. Prospect.* **2007**, *55*, 433–446. [CrossRef]
15. Abdelrahman, E.M.; Abo-Ezz, E.R.; Essa, K.S. Parametric inversion of residual magnetic anomalies due to simple geometric bodies. *Explor. Geophys.* **2012**, *43*, 178–189. [CrossRef]
16. Ku, C.C.; Sharp, J.A. Werner deconvolution for automated magnetic interpretation and its refinement using Marquardt's inverse modeling. *Geophysics* **1983**, *48*, 754–774. [CrossRef]
17. Thompson, D.T. EULDPH—A new technique for making computer-assisted depth estimates from magnetic data. *Geophysics* **1982**, *47*, 31–37. [CrossRef]
18. Pilkington, M. Joint inversion of gravity and magnetic data for two-layer models. *Geophysics* **2006**, *71*, L35–L42. [CrossRef]
19. Abdelrahman, E.M.; El-Araby, H.M.; El-Araby, T.M.; Essa, K.S. A least-squares minimization approach to depth determination from magnetic data. *Pure Appl. Geophys.* **2003**, *160*, 1259–1271. [CrossRef]

20. Tlas, M.; Asfahani, J. Fair function minimization for interpretation of magnetic anomalies due to thin dikes, spheres and faults. *J. Appl. Geophys.* **2011**, *75*, 237–243. [CrossRef]
21. Essa, K.S.; Elhussein, M. PSO (Particle Swarm Optimization) for interpretation of magnetic anomalies caused by simple geometrical structures. *Pure Appl. Geophys.* **2018**, *175*, 3539–3553. [CrossRef]
22. Van den Bergh, F.; Engelbrecht, A.P. A Cooperative approach to particle swarm optimization. *IEEE Trans. Evol. Comput.* **2004**, *8*, 225–239. [CrossRef]
23. Boschetti, F.; Denith, M.C.; List, R.D. Inversion of potential field data by genetic algorithms. *Geophys. Prospect.* **1997**, *45*, 461–478. [CrossRef]
24. Biswas, A. Interpretation of residual gravity anomaly caused by a simple shaped body using very fast simulated annealing global optimization. *Geosci. Front.* **2015**, *6*, 875–893. [CrossRef]
25. Ekinci, Y.L.; Balkaya, Ç.; Göktürkler, G.; Turan, S. Model parameter estimations from residual gravity anomalies due to simple-shaped sources using differential evolution algorithm. *J. Appl. Geophys.* **2016**, *129*, 133–147. [CrossRef]
26. Balkaya, C.; Ekinci, Y.L.; Göktürkler, G.; Turan, S. 3D non-linear inversion of magnetic anomalies caused by prismatic bodies using differential evolution algorithm. *J. Appl. Geophys.* **2017**, *136*, 372–386. [CrossRef]
27. Biswas, A. Inversion of source parameters from magnetic anomalies for mineral/ore deposits exploration using global optimization technique and analysis of uncertainty. *Nat. Resour. Res.* **2018**, *27*, 77–107. [CrossRef]
28. Essa, K.S.; Elhussein, M. Gravity Data Interpretation Using New Algorithms: A Comparative Study. In *Gravity-Geoscience Applications, Industrial Technology and Quantum Aspect*; Zouaghi, Z., Ed.; InTech: Rijeka, Croatia, 2018.
29. Srivastava, S.; Datta, D.; Agarwal, B.N.P.; Mehta, S. Applications of ant colony optimization in determination of source parameters from total gradient of potential fields. *Near Surf. Geophys.* **2014**, *12*, 373–389. [CrossRef]
30. Di Maio, R.; Rani, P.; Piegari, E.; Milano, L. Selfpotential data inversion through a genetic-price algorithm. *Comput. Geosci.* **2016**, *94*, 86–95. [CrossRef]
31. Holland, J.H. *Adaptation in Natural and Artificial Systems*; University of Michigan Press: Ann Arbor, MI, USA, 1975.
32. Kirkpatrick, S.; Gelatt, C.D.; Vecchi, M.P. Optimization by Simulated Annealing. *Science* **1983**, *220*, 671–680. [CrossRef]
33. Storn, R.; Price, K. Differential Evolution—A Simple and Efficient Heuristic for Global Optimization over Continuous Spaces. *J. Glob. Optim.* **1997**, *11*, 341–359. [CrossRef]
34. Kennedy, J.; Eberhart, R. Particle Swarm Optimization. In Proceedings of the IEEE International Conference on Neural Networks, Perth, WA, Australia, 27 November–1 December 1995; IEEE Service Center: PiscatawaY, NJ, USA, 1995; IV, pp. 1942–1948.
35. Mishra, S.; Pant, M. Particle Swamp Optimization: A Hybrid Swarm Intelligence Algorithm. *Procedia Comput. Sci.* **2017**, *115*, 453–460.
36. Eberhart, R.C.; Shi, Y. Particle Swarm Optimization: Developments, Applications and Resources. In Proceedings of the Congress on Evolutionary Computation, Seoul, Republic of Korea, 27–30 May 2001; pp. 81–86.
37. Toushmalani, R. Gravity inversion of a fault by particle swarm optimization (PSO). *Springer Plus* **2013**, *2*, 315. [CrossRef]
38. Juang, C.F. A hybrid genetic algorithm and particle swarm optimization for recurrent network design. *IEEE Trans. Syst. Man Cybern. Part B* **2004**, *34*, 997–1006. [CrossRef]
39. Robinson, J.; Rahamat-Samii, Y. Particle swarm optimization in electromagnetic. *IEEE Trans. Antennas Propag.* **2004**, *52*, 397–407. [CrossRef]
40. Cedeno, W.; Agrafiotis, D.K. Using particle swarms for the development of QSAR models based on K-nearest neighbor and kernel regression. *J. Comput. Aided Mol. Des.* **2003**, *17*, 255–263. [CrossRef]
41. Donelli, M.; Franceschini, G.; Martini, A.; Mass, A. An integrated multiscaling strategy based on a particle swarm algorithm for inverse scattering problems. *IEEE Trans. Geosci. Remote Sens.* **2006**, *44*, 298–312. [CrossRef]
42. Wachowiak, M.P.; Smolıkova, R.; Zheng, Y.; Zurada, J.M.; Elmaghraby, A.S. An approach to multimodal biomedical image registration utilizing particle swarm optimization. *IEEE Trans. Evol. Comput.* **2004**, *8*, 289–301. [CrossRef]
43. Chau, W.K. Application of a Particle Swarm Optimization Algorithm to Hydrological Problems. In *Water Resources Research Progress*; Robinson, L.N., Ed.; Nova Science Publishers Inc.: New York, NY, USA, 2008; pp. 3–12.
44. Rahaman, M.A.; Lancelot, J.R. Continental Crust Evolution in SW Nigeria: Constraints from U/Pb dating of Pre-Pan-African Gneisses. In *Rapport D'activite 1980–1984 Documents et Travaux du Centre Geologique et Geophysique de Montpellier*; Universitédes Sciences et Techniques du Languedoc: Montpellier, France, 1984; pp. 41–50.
45. Haruna, I.V. Review of the Basement Geology and Mineral Belts of Nigeria. *J. Appl. Geol. Geophys.* **2017**, *5*, 37–45.
46. Agbi, I.; Ekwueme, B.N. Preliminary review of the geology of the hornblende biotite gneisses of Obudu Plateau Southeastern Nigeria. *Glob. J. Geol. Sci.* **2018**, *17*, 75–83. [CrossRef]
47. Ukwang, E.E.; Ekwueme, B.N.; Kröner, A. Single zircon evaporation ages: Evidence for the Mesoproterozoic crust in the southeastern Nigerian basement complex. *Chin. J. Geochem.* **2012**, *31*, 48–54. [CrossRef]
48. Toteu, S.F.; Penaye, J.; Djomani, Y.P. Geodynamic evolution of the Pan-African belt in Central Africa with special reference to Cameroon. *Can. J. Earth Sci.* **2004**, *41*, 73–85. [CrossRef]
49. Rao, T.K.S.P.; Subrahmanyam, M.; Srikrishna Murthy, A. Nomograms for the direct interpretation of magnetic anomalies due to long horizontal cylinders. *Geophysics* **1986**, *51*, 2156–2159. [CrossRef]
50. Gay, P. Standard curves for interpretation of magnetic anomalies over long tabular bodies. *Geophysics* **1963**, *28*, 161–200. [CrossRef]

51. Rao, T.; Subrahmanyam, M. Characteristic curves for the inversion of magnetic-anomalies of spherical ore bodies. *Pure Appl. Geophys.* **1988**, *126*, 69–83.
52. Abdelrahman, E.M.; Essa, K.S. A new method for depth and shape determinations from magnetic data. *Pure Appl. Geophys.* **2015**, *172*, 439–460. [CrossRef]
53. Rao, B.S.R.; Murthy, I.V.R.; Rao, C.V. A computer program for interpreting vertical magnetic anomalies of spheres and horizontal cylinders. *Pure Appl. Geophys.* **1973**, *110*, 2056–2065. [CrossRef]
54. Gay, P. Standard curves for the interpretation of magnetic anomalies over long horizontal cylinders. *Geophysics* **1965**, *30*, 818–828. [CrossRef]
55. Abdelrahman, E.M.; Essa, K.S. Magnetic interpretation using a least-squares, depth-shape curves method. *Geophysics* **2005**, *70*, L23–L30. [CrossRef]
56. Lines, L.R.; Treitel, S. A review of least-squares inversion and its application to geophysical problems. *Geophys. Prospect.* **1984**, *32*, 159–186. [CrossRef]
57. Zhdanov, M.S. *Geophysical Inversion Theory and Regularization Problems*; Elsevier: Amsterdam, The Netherlands, 2002; p. 633.
58. He, J.; Guo, H. A modified particle swarm optimization algorithm. *Telkomnika* **2013**, *11*, 6209–6215. [CrossRef]
59. Sen, M.K.; Stoffa, P.L. *Global Optimization Methods in Geophysical Inversion*; Cambridge University Press: Cambridge, UK, 2013; p. 279.
60. Sweilam, N.H.; El-Metwally, K.; Abdelazeem, M. Self potential signal inversion to simple polarized bodies using the particle swarm optimization method: A visibility study. *J. Appl. Geophys.* **2007**, *6*, 195–208.
61. Parsopoulos, K.E.; Vrahatis, M.N. Recent approaches to global optimization problems through particle swarm optimization. *Nat. Comput.* **2002**, *1*, 235–306. [CrossRef]
62. Ekwok, S.E.; Akpan, A.E.; Achadu, O.-I.M.; Thompson, C.E.; Eldosouky, A.M.; Abdelrahman, K.; András, P. Towards understanding the source of brine mineralization in Southeast Nigeria: Evidence from high-resolution airborne magnetic and gravity data. *Minerals* **2022**, *12*, 146. [CrossRef]
63. Ekwok, S.E.; Akpan, A.E.; Achadu, O.I.M.; Ulem, C.A. Implications of tectonic anomalies from potential feld data in some parts of Southeast Nigeria. *Environ. Earth Sci.* **2022**, *81*, 6. [CrossRef]
64. Ekwok, S.E.; Akpan, A.E.; Ebong, E.D.; Eze, O.E. Assessment of depth to magnetic sources using high resolution aeromagnetic data of some parts of the Lower Benue Trough and adjoining areas, Southeast Nigeria. *Adv. Space Res.* **2021**, *67*, 2104–2119. [CrossRef]
65. Ekwok, S.E.; Akpan, A.E.; Kudamnya, E.A.; Ebong, D.E. Assessment of groundwater potential using geophysical data: A case study in parts of Cross River State, south-eastern Nigeria. *Appl. Water Sci.* **2020**, *10*, 144. [CrossRef]
66. Nwankwo, L.I. Structural styles in the Precambrian Basement Complex of southeastern Nigeria. *J. Afr. Earth Sci.* **2009**, *53*, 73–86.
67. Ajibade, A.C.; Oyawoye, M.O.; Rahaman, M.A.; Adekeye, O.A. Petrology of migmatites of the Obudu Plateau, Nigeria. *J. Afr. Earth Sci.* **2010**, *56*, 23–30.
68. Asouzu, E.C.; Onyeagocha, A.C. Geology and mineralization of the Obudu area of southeastern Nigeria. *J. Afr. Earth Sci.* **2013**, *86*, 20–34.
69. Airo, M.L. Aeromagnetic and aeroradiometric response to hydrothermal alteration. *Surv. Geophys.* **2002**, *23*, 273–302. [CrossRef]
70. United States Geological Survey (USGS). Setting and Origin of Iron Oxide-Copper-Cobalt-Gold-Rare Earth Element Deposits of Southeast Missouri. 2013. Available online: http://minerals.usgs.gov/east/semissouri/index.html (accessed on 5 June 2023).
71. Geological Survey of Canada (GSC). *Airborne Geophysical Survey, Mount Milligan area, British Columbia (NTS 93 O/4W, N/1, N/2E)*; GSC Open File 2535; GSC: Montreal, Canada, 1992.
72. Shafiqullah, M.; Langlois, J.D. The Pima Mining District Arizona—A Geochronologic Update. In *New Mexico Geological Society Guidebook 29th Annual Fall Field Conference Guidebook*; New Mexico Geological Society: Boulder, CO, USA, 1978; pp. 321–327.
73. Abdelrahman, E.M.; Sharafeldin, S.M. An iterative least-squares approach to depth determination from residual magnetic anomalies due to thin dikes. *J. Appl. Geophys.* **1996**, *34*, 213–220. [CrossRef]
74. Salem, A.; Aboud, E.; Elsirafy, A.; Ushijima, K. Structural mapping of Quseir area, northern Red Sea, Egypt, using high-resolution aeromagnetic data. *Earth Planets Space* **2005**, *57*, 761–765. [CrossRef]
75. Biswas, A.; Parija, M.P.; Kumar, S. Global nonlinear optimization for the interpretation of source parameters from otal gradient of gravity and magnetic anomalies caused by thin dyke. *Ann. Geophys.* **2017**, *60*, G0218. [CrossRef]

Article

Preferred Orientations of Magnetic Minerals Inferred from Magnetic Fabrics of Hantangang Quaternary Basalts

Jong Kyu Park [1], Ji Young Shin [2], Seungwon Shin [3,*] and Yong-Hee Park [3,*]

[1] Department of Geophysics, Kangwon National University, Chuncheon 24341, Republic of Korea; fckhd100@kangwon.ac.kr
[2] Laboratory of Orogenic Belts and Crustal Evolution, School of Earth and Space Sciences, Peking University, Beijing 100871, China; jyshin@pku.edu.cn
[3] Division of Geology and Geophysics, Kangwon National University, Chuncheon 24341, Republic of Korea
* Correspondence: ssw7304@kangwon.ac.kr (S.S.); aegis@kangwon.ac.kr (Y.-H.P.); Tel.: 82-33-250-8588 (Y.-H.P.)

Abstract: This paper presents a comprehensive analysis of the anisotropy of magnetic susceptibility (AMS) and paleomagnetic data from Quaternary basalt outcrops along the Hantangang River, Korea. A total of 554 samples were collected from 20 sites, representing three distinct units, Unit I, Unit II, and Unit III. Paleomagnetic data reveal a difference in the timing of eruptions between Units I and II, suggesting distinct periods by volcanic episodes. The mineral magnetic analysis identified titanomagnetite as the dominant magnetic carrier in the samples. AMS results indicated weak anisotropy and scattered AMS directions, indicating a low degree of preferred orientation of grains within the basalt rocks. The inverse AMS fabrics observed at specific sites are attributed to single-domain (SD) grains. Comparing the AMS data with the anisotropy of anhysteretic remanent magnetization (AARM) data, three distinct types of magnetic fabrics (normal, intermediate, and inverse) were discerned. The magnetic fabric was utilized to ascertain the flow direction based on the findings obtained from the AMS results. The findings suggest that the Quaternary basalts in this study's area were primarily confined to the Hantangang River channel and its immediate vicinity during lava flow. However, distinct flow patterns are observed in the southwestern region, implying the presence of unknown volcanic sources.

Keywords: magnetic minerals; anisotropy of magnetic susceptibility; Quaternary basalt; lava flows

check for updates

Citation: Park, J.K.; Shin, J.Y.; Shin, S.; Park, Y.-H. Preferred Orientations of Magnetic Minerals Inferred from Magnetic Fabrics of Hantangang Quaternary Basalts. *Minerals* **2023**, *13*, 1011. https://doi.org/10.3390/min13081011

Academic Editors: Luan Thanh Pham, Saulo Pomponet Oliveira and Le Van Anh Cuong

Received: 9 July 2023
Revised: 27 July 2023
Accepted: 27 July 2023
Published: 29 July 2023

1. Introduction

The Hantangang River, located in the central part of the Korean Peninsula, is unique in South Korea and known for its distinct geologic and geomorphologic features resulting from basalt erosion. Despite extensive research, the exact eruptive origin of the Quaternary Hantangang River Volcanic Field (HRVF) [1] remains uncertain, with two proposed possibilities [2–8]. Ori Mountain (38°23′25″N, 127°16′01″E) and Upland (680 m) in North Korea have been identified as potential sources, suggesting that low-viscosity basaltic lava may have erupted multiple times and flowed through ancient river channels, resulting in the formation of the HRVF, which extends for 110 km and varies in thickness from 3 to 40 m [2–4,6,7]. The erupted lava undergoes cooling processes, resulting in the formation of columnar jointed structures and pillow lavas. Subsequently, these formations undergo erosion by reformed river systems, giving rise to waterfalls and other distinct features along the channel walls. These geologic valuations led to the establishment of the Hantangang River National Geopark in 2015, later certified as a UNESCO Global Geopark in 2020.

The HRVF has been the subject of ongoing research to determine the origin of the lava [2,6,7,9]. However, accurate research on the origin has been challenging due to the presumed location in North Korea. In such cases, one of the methods employed to infer the origin of lava flows is the analysis of anisotropy of magnetic susceptibility (AMS). AMS is a technique that examines the preferred orientation of minute magnetic minerals in volcanic or sedimentary rocks, allowing for the estimation of the direction of lava flow or paleocurrents. This technique has been proven valuable in geologic studies [10,11] and has been extensively utilized in many research effects [12–26].

Although several previous studies have examined the paleomagnetic characteristics of the Hantangang River [3,27–29], none have utilized AMS analysis. In this study, we conducted a rigorous AMS analysis of Quaternary basalt outcrops along the Hantangang River to determine the source of the lava and gain insights into the paleo-river channels in this study's area.

2. Geological Setting

This study's area, where the Hantangang River is situated within the Chugaryeong rift valley, also known as the Chugaryeong Fault Zone, is a geological feature in the central part of the Korean Peninsula. It consists of several parallel fault lines trending from northeast to southwest (Figure 1). The Chugaryeong rift valley formed approximately 25–30 million years ago during the late Cenozoic era as a result of the movement of the Eurasian and Philippine Sea plates [30–32]. Stretching for about 250 km in length and varying in width from 20 to 60 km, the Chugaryeong rift valley holds significant geologic and tectonic importance for the Korean Peninsula. It has profoundly influenced the region's landscape, geology, and natural resources. The valley is also an active seismic zone, occasionally experiencing earthquakes along the fault lines.

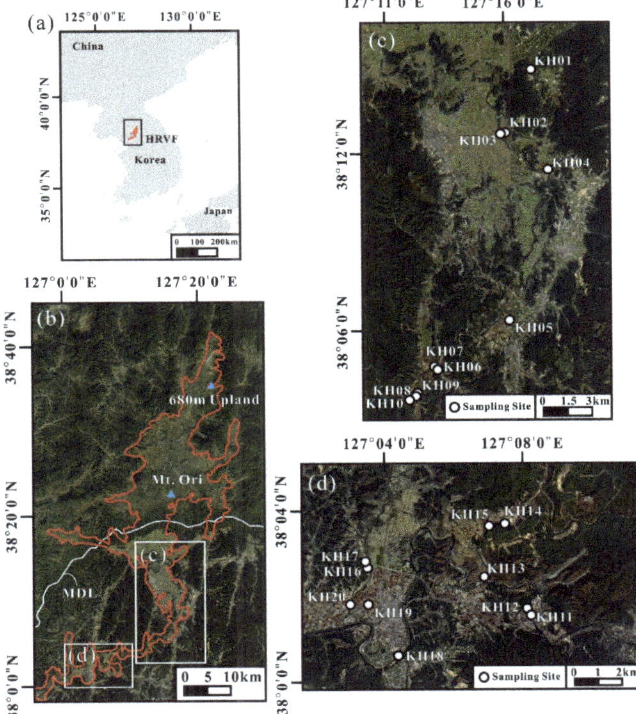

Figure 1. (**a**) A map of Korean Peninsula with its vicinity showing the location of the Hantangang River Volcanic Field (HRVF) [1]. (**b**) A satellite map showing the distribution of the HRVF (inside of the red closed curve) with Mt. Ori and 680 uplands. MDL: Military Demarcation Line between South and North Korea. (**c**,**d**) Satellite maps of this study's area with sampling site locations.

The basement rock in this study's area consists of a Precambrian metamorphic complex intruded by Jurassic and Cretaceous granite, where the fault zones have intersected [4]. The Quaternary alkali olivine basalts in this study's area erupted during the latest volcanic activity in the Chugaryeong rift valley, following the volcanic eruptions of acidic volcanisms during the Cretaceous period [3]. The Cretaceous silicic volcanic rocks occupy an oval-shaped region bound by the Dongducheon and Dongsong faults. At the same time, the tholeiitic basalt outcrops are sparsely distributed in the lower part of the acidic volcanic rock outcrops. The Quaternary basalts exhibit a narrow and elongated distribution along the Hantangang River, covering the process of the Dongducheon and Dongsong faults.

In this study, the term "Hantangang basalts" will be used to refer to all the basalts distributed around the Hantangang River, even though these Quaternary basalts are commonly known as Jeongok basalts in the Jeongok region. The Hantangang basalts are characterized by short eruption intervals, as inferred from their chemical composition. However, their precise eruption history remains unknown [2,3].

Based on the volcanic stratigraphy analyzed in this study's area, the Hantangang basalt has been classified into three units. Unit I represents the lowermost part of the basalt outcrop along the Hantangang River, with an observed thickness of about 3 to 4 m. This unit displays either weakly developed or poorly observed columnar fractures. Pillow lava is also a typical feature of this unit. Evidence suggests that the lava unit experienced a relatively slow cooling process, indicated by the development of vertical and horizontal segregation textures within the lava flow. The lava flow appears massive and void-free exposures, with a K-Ar dating estimate of 0.51 ± 0.07 Ma [33] (Danhara et al., 2002 [33]).

Unit II is characterized by prominently exposed lava flows along the Hantangang River, exhibiting well-developed vertical columnar fractures. This unit can be further classified into lower, middle, and upper parts. The upper part of Unit II is distinguished by the presence of multiple layers of lava, with thicknesses ranging from 20 to 30 cm and a total thickness from 1 to 1.5 m.

During a volcanic eruption, the lower part of the lava flow is in contact with the Earth's surface, while the upper part is exposed to the atmosphere and gradually cools. As a result, it is likely that the relatively dense lower part of Unit II primarily cooled through heat conduction in contact with the Earth's surface. In contrast, the upper part of Unit II represents the surface portion of the lava flow that cooled upon contact with the atmosphere. This cooling process, combined with fluctuations in the thickness of the lava flow, led to the development of multiple layers of lava. The absence of distinct features, such as paleosols, sedimentary layers, or erosion surfaces at the upper part of Unit II, indicates a temporal gap, making it challenging to differentiate between the upper part of Unit II and Unit III in the field.

Several age estimates have been reported for this basalt unit, including 1.08 ± 0.158 Ma [34], 0.40 Ma [35], and 0.51 ± 0.01 Ma [36]. The lower part of Unit II was estimated to have a burial age of 0.48 Ma [37] based on cosmic-origin isotopes of the Baekui-ri layer, while the upper part was estimated to be $0.09 \pm 0.03 \sim 0.18 \pm 0.03$ Ma [36] based on K-Ar dating.

Unit III is a lava flow unit that typically rests atop a scarp. Although thinner than Unit II, its thickness varies considerably between outcrops and locations, measuring up to 5 to 6 m in some areas. The lava flows in Unit III generally lack columnar fractures but display well-defined pores in the upper and lower parts, along with clear segregation structures. Lee et al. [38] estimated that this unit formed approximately 40,000 years ago, likely representing the most recent eruption of Hantangang basalt.

3. Methods

A total of 554 samples were collected from 20 sites in this study's area, five sites from Unit I, thirteen sites from Unit II, and three sites from Unit III (Figure 1). All samples were cored with a gasoline-powered portable rock drill and oriented with a Brunton compass in the field. In the laboratory, the samples were subsequently sliced into 25 mm diameter and

22 mm long cylinders and stored in mu-metal shield boxes to prevent the acquisition of viscous remanence caused by the external magnetic field, such as the Earth's field.

AMS data for each sample were measured in fifteen positions [39] using a Bartington magnetic susceptibility meter (Model MS-2) connected with an MS-2B susceptibility bridge. Eigen parameters for each AMS data were calculated using the Hext statistic [40]. The data were visualized by an ellipsoid with the principal susceptibility axes labeled $K_{max} > K_{int} > K_{min}$, which were calculated using the PmagPy program [41]. The mean magnetic susceptibility (K_m) for a single sample is calculated by $K_m = (K_{max} + K_{int} + K_{min})/3$. The magnitude and shape of the susceptibility ellipsoid are expressed by the corrected degree of anisotropy (P') and the symmetry of shape on the vertical axis (T), respectively, proposed by Jelinek [42].

The anisotropy of anhysteretic remanent magnetization (AARM) of pilot samples from each site was measured in fifteen different orientations on a Molspin spinner magnetometer. This was carried out to determine the orientation distribution of remanence-bearing or ferromagnetic (sensu lato) minerals and to test the possibility of an inverse fabric influenced by single-domain magnetite (SD effect; [43]). The ARM was imparted using a Molspin AF demagnetizer and an ARM attachment with a 0.05 m TDC bias field and a peak alternating field of 90 mT. The tensor of the AARM was then calculated using the Hext statistic [40].

Magnetic remanence measurements for paleomagnetic analysis were performed using a Molspin spinner magnetometer. Natural remanent magnetization (NRM) was measured and demagnetized stepwise up to a peak field of 90 mT in 5~10 mT intervals. The palaeomagnetic data from all specimens were projected onto the orthogonal vector diagram [44]. The characteristic remanent magnetization (ChRM) direction of each specimen was determined using principal component analysis with the anchored line fit method [45] from at least three or more points.

Low-field thermomagnetic measurements (K-T curves) at low and high temperatures were conducted on bulk rock samples using a Bartington MS2 susceptibility meter equipped with an MS2WF furnace. Twenty cylindrical subsamples of 15 mm in height were made from residual fragments of core samples. Low-temperature measurements were carried out by heating the samples cooled to about $-170\ ^\circ$C in liquid nitrogen to room temperature. Then, the samples underwent a gradual heating process, reaching temperatures of up to $700\ ^\circ$C at a rate of $20\ ^\circ$C/min, followed by cooling to room temperature at the same rate. Magnetic susceptibility measurements were taken at intervals of $1\ ^\circ$C during the process.

To determine the grain size of magnetic minerals, hysteresis parameters were measured on 80 cylindrical subsamples (8 mm in diameter and 10 mm in height) using a Molspin vibrating sample magnetometer (Model VSM Nuvo) calibrated by a paramagnetic standard sample containing 0.6 g of $FeSO_4$.

4. Results

4.1. Paleomagnetic Results

The NRMs of the basalt samples accepted by the selection criteria [46] vary from 140.8 to 5386.2 mA/m in intensity, with predominantly northerly positive directions. Most samples display a simple decay of a northerly and moderately steep down component toward the origin above 15 mT or 20 mT, indicating the effective isolation of the ChRM component through the AF demagnetization method (Figure 2). This observation suggests that a low-coercivity ferrimagnetic mineral, such as titanomagnetite, is the predominant magnetic mineral. Field observations reveal that the lava flows in this study's area exhibit very gentle dips from 1° to 6° toward the flow direction, indicating a topographic origin rather than a tectonic origin. Previous petrographic investigations also support these findings, reporting no evidence of tectonic movement following lava eruptions [3]. Therefore, in this study, we did not correct for tilting in the paleomagnetic directions.

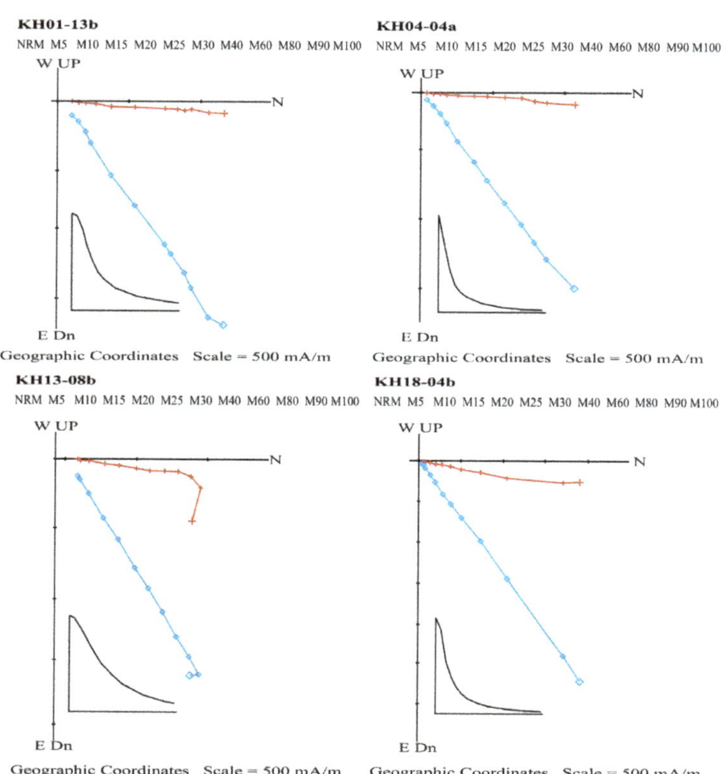

Figure 2. Demagnetization behaviors of representative samples. Orthogonal vector diagrams with normalized intensity are shown in geographic coordinates. Steps of AF demagnetization are labeled with M below the sample number. The horizontal and vertical projections are plotted with cross (red) and solid (blue) symbols, respectively.

The site-mean paleomagnetic directions were calculated from 427 samples, excluding 35 samples that showed anomalous directions likely due to the acquisition of isothermal remanence or chemical alteration (Table 1). The ChRM directions within each site cluster demonstrate high precision parameters ($k \geq 118.5$) and narrow 95% confidence limits around the mean direction ($\alpha_{95} \leq 3.2°$). The unit-mean directions are as follows: Unit I, D/I: 6.8°/60.4° ($\alpha_{95} = 1.8°$); Unit II, D/I: 2.3°/58.7° ($\alpha_{95} = 1.5°$); Unit III, D/I: 0.2°/60.8° ($\alpha_{95} = 5.9°$) (Figure 3). F-test results indicate that the directions of Unit I and Unit II are statistically distinguished from each other at a 5% confidence level, suggesting different timing of remanence acquisition for these two units. The direction of Unit III is not statistically distinguishable from those of Units I and II at a 5% significance level due to the large 95% confidence circle of Unit III. However, it is interpreted that the acquisition of remanent magnetization in Units I and III could not have occurred simultaneously.

Table 1. Paleomagnetic results in this study's area.

Unit	Site	n/N	Dec (°)	Inc (°)	k	α_{95} (°)	Sampling Site Location		
							Lat (°N)	Long (°E)	Elev (m)
I	KH01	17/18	6.0	59.6	162.2	2.6	38.2466	127.2864	188.1
	KH06	33/33	10.8	60.8	640.6	1.0	38.0803	127.2236	106.0
	KH07	22/24	8.5	62.4	227.7	2.0	38.0789	127.2178	113.0
	KH09	20/21	1.2	60.5	118.5	2.9	38.0628	127.2069	143.0
	KH10	26/26	7.4	58.5	761.1	1.0	38.0617	127.2003	143.0
	Unit Mean		6.8	60.4	1262.2	1.8			
II	KH02	18/20	0.1	59.2	277.5	2.0	38.2105	127.2658	150.2
	KH03	11/12	352.8	57.4	584.3	1.8	38.2105	127.2658	150.2
	KH04	17/18	6.6	58.2	246.2	2.2	38.1907	127.2887	125.3
	KH08	14/16	356.9	61.2	350.4	2.0	38.0655	127.2047	96.7
	KH11	15/16	3.8	61.1	131.1	3.2	38.0241	127.1375	61.6
	KH12	8/10	6.9	60.0	249.0	3.1	38.0256	127.1372	61.8
	KH13	5/17	4.6	58.1	741.0	2.3	38.0405	127.1134	50.7
	KH15	28/31	8.5	59.1	497.2	1.2	38.0614	127.1166	49.3
	KH16	25/25	357.3	55.7	373.1	1.5	38.0449	127.0577	29.1
	KH18	19/23	1.8	56.7	476.1	1.5	38.0112	127.0727	39.1
	KH19	24/24	8.1	58.3	713.7	1.1	38.0300	127.0602	43.4
	KH20	32/32	1.1	58.5	131.1	2.2	38.0283	127.0481	54.3
	Unit Mean		2.3	58.7	700.1	1.5			
III	HK05	35/35	9.5	62.4	671.2	0.9	38.1072	127.2688	124.6
	KH14	30/30	347.4	59.5	694.8	1.0	38.0625	127.1201	80.5
	KH17	28/31	4.4	59.5	317.1	1.5	38.0455	127.0575	63.1
	Unit Mean		0.2	60.8	187.0	5.9			

n/N: number of accepted/measured samples; Dec: declination; Inc: inclination; α95: 95% confidence level; Lat: latitude; Long: longitude; Elev: elevation.

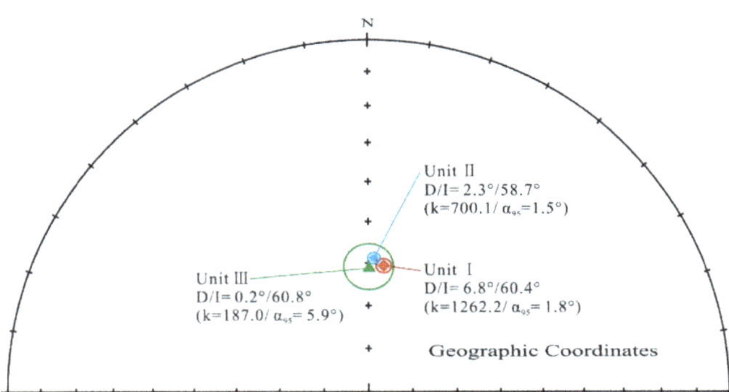

Figure 3. Unit-mean paleomagnetic directions with 95% confidence circle on an equal area projection; D/I: declination/inclination; k: precision parameter; α_{95}: 95% confidence level. Red, blue and green symbols are Unit I, II and III, respectively.

4.2. Mineral Magnetic Results

4.2.1. Thermomagnetic (K-T) Curves

Figure 4 displays representative thermomagnetic curves (K-T curves). The low-temperature K-T curves exhibit a gradual increase in susceptibility as the temperature rises from −200 °C, with a sharp increase occurring at the inflection point ranging from −42 °C to −74 °C. This behavior agrees with the findings of [47] using synthetic titanomagnetite ($Fe_{3-x}Ti_xO_4$) samples with a Ti-rich composition ($x > 0.5$). The high-temperature K-T curves of all samples reveal irreversible heating and cooling curves, indicating a magnetic phase alteration during the thermal treatment. In the heating curve, the susceptibility peaks around 0 °C, followed by a rapid decrease within the range of 64 °C to 88 °C, suggesting a relatively low Curie temperature (Figure 4). These observations strongly support that Ti-rich titanomagnetites ($x \approx 0.7$) are major magnetic carriers in the samples [48]. Furthermore, the heating curves exhibit a gradual rise above ~300 °C, followed by two subsequent decreases at 467~505 °C and 553~565 °C, indicating the possible presence of Ti-poor titanomagnetite or metastable titanomaghemite resulting from the low-temperature oxidation of titanomagnetite during the heating process [19,49]. During the cooling cycle, a significant increase in susceptibility is observed at approximately 550 °C (Figure 4), suggesting the formation of new magnetic minerals such as Ti-poor titanomagnetite through the transformation of Ti-rich magnetite during the heating process [48,50].

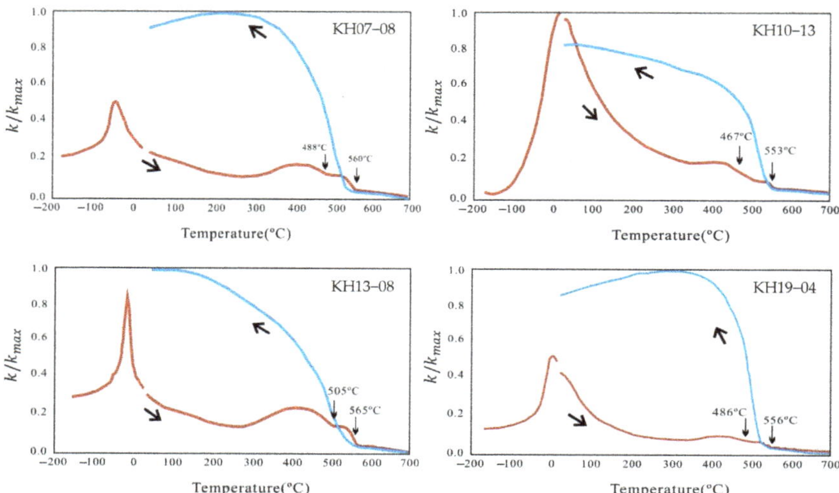

Figure 4. Thermomagnetic curves at low and high temperatures for representative samples. Heating curve and cooling curve are in red and blue, respectively.

4.2.2. Hysteresis Parameters

Magnetic hysteresis loops were measured for 80 samples, and paramagnetic corrections were applied using the best-fitting high field slope (K_{HF}). Hysteresis parameters, such as coercivity of remanence (B_{cr}), coercive force (B_c), saturation remanence (M_{rs}), and saturation magnetization (M_s), were obtained (Table S1). For the magnetic grain size analysis, the hysteresis ratio of M_{rs}/M_s versus B_{cr}/B_c was plotted on a Day diagram, as shown in Figure 5 [51,52].

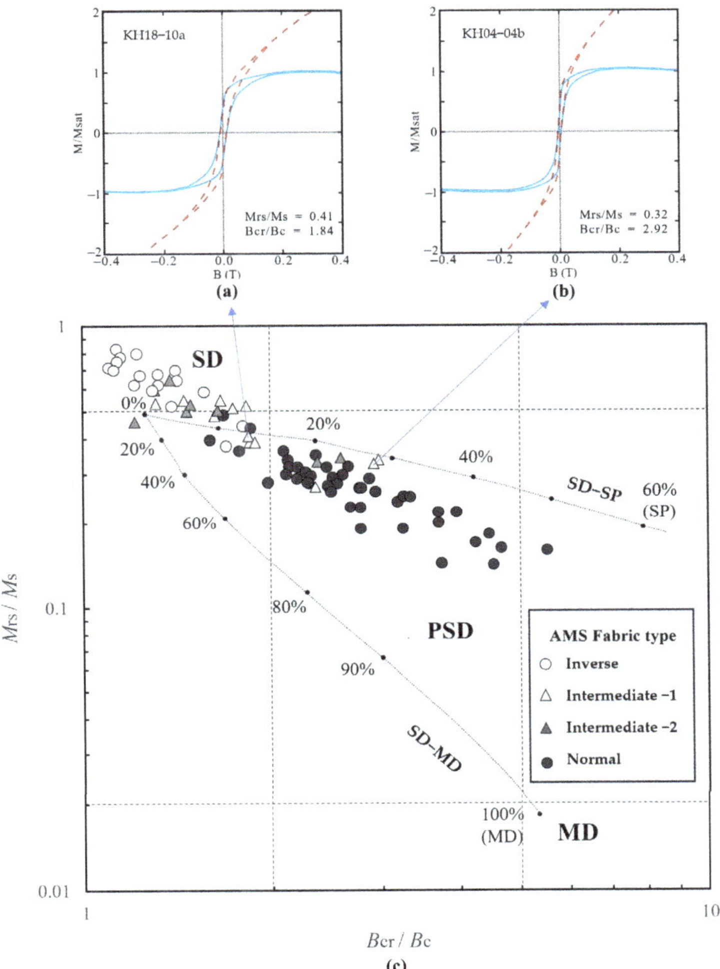

Figure 5. (**a,b**) Examples of slight wasp-wasted hysteresis loop. A blue solid line is paramagnetic corrected curve and a red dotted line is uncorrected curve. Samples KH04-04b and KH18-10a correspond to the Intermediate Group I of AMS fabric type. (**c**) Hysteresis ratios (M_{rs}/M_s versus B_{cr}/B_c) for 20 sites (four samples for each site) classified by AMS fabric type are plotted on the modified Day diagram [51,52]. The theoretical curves indicated by the solid lines represent the mixing lines of single-domain (SD), multi-domain (MD), and SD- superparamagnetic (SP) mixtures [52].

Day diagrams can be ambiguous due to the presence of various variables [53]; nonetheless, we have chosen to adopt them for quantitative rock magnetism interpretation. The distribution of these hysteresis ratios ranges from single domain (SD) to pseudo single domain (PSD) particle size. More than half of the data points fall in the PSD region, indicating a predominance of PSD particles in samples.

On the other hand, about a quarter of the data points fall in the SD region. The rest of the points are close to the SD/PSD boundary and the theoretical curve for a mixture of SD and 10 nm superparamagnetic (SP) particles proposed by [52] (Figure 5c). These results suggest that the magnetic particles in most samples are predominantly composed of SD or PSD grains.

In contrast, a subset of samples contains a dominant population of SD grains with some presence of SP grains. Hysteresis loops in these samples are slightly wasp-waisted (Figure 5a,b), supporting the presence of SP grains [54]. Because the data distribution is far from the multi-domain region or the theoretical curve for the mixture of SD-MD grains [52], the content of MD grains in samples is negligible.

4.3. Anisotropy of Magnetic Susceptibility (AMS) Results

The site-mean magnetic susceptibility (K_m) ranges from 0.97×10^{-3} SI to 6.53×10^{-3} SI, with an overall average of 2.36×10^{-3} SI (Table 2). These values are relatively lower than the expected range for basaltic rocks ($> 5 \times 10^{-3}$ SI). The lower K_m values can be attributed to two factors, the small grain size falling within the SD-PSD range and the mineralogy of the magnetic grains, primarily composed of Ti-rich titanomagnetite [10,55]. Sites dominated by SD grains (KH01, KH02, KH03, KH04, KH8, KH11, KH12, KH18, KH20) generally exhibit lower K_m values compared to other sites (Table 2), providing further evidence of the influence of grain size on the observed lower K_m values.

Table 2. AMS results in this study's area.

Site (Unit)	N	K_m (10^{-3} SI)	Mean AMS parameters		K_{max}		K_{int}		K_{min}	
			P'	T	Dec (°)	Inc (°)	Dec (°)	Inc (°)	Dec (°)	Inc (°)
KH01(I)	30	1.282	1.016	−0.4366	207.2	38.5	14.6	50.8	112.3	6.2
KH02(II)	24	1.437	1.015	−0.2760	133.4	39.3	248.9	27.7	3.3	38.2
KH03(II)	22	1.296	1.010	−0.0893	182.0	52.7	57.7	23.2	314.8	27.4
KH04(II)	18	1.493	1.009	0.0885	198.1	35.1	108.1	0.1	17.9	54.9
KH05(III)	35	3.300	1.010	0.2804	337.1	22.7	69.0	4.4	169.4	66.9
KH06(I)	33	3.147	1.009	−0.1903	312.3	34.0	52.2	14.2	161.4	52.3
KH07(I)	24	1.903	1.011	−0.3391	151.2	12.5	56.8	19.0	272.6	67.0
KH08(II)	27	0.974	1.012	−0.6346	234.6	43.3	355.2	28.5	106.2	33.4
KH09(I)	23	3.783	1.010	−0.3054	65.2	23.8	155.6	1.0	247.9	66.2
KH10(I)	33	6.532	1.006	−0.1616	15.6	33.6	281.2	6.6	181.6	55.6
KH11(II)	30	1.293	1.012	−0.2887	158.3	47.8	336.4	42.2	67.2	1.0
KH12(II)	22	1.112	1.009	−0.1227	173.4	47.4	353.3	42.6	83.4	0.1
KH13(II)	22	1.881	1.008	−0.0523	284.5	18.1	17.8	10.1	135.7	69.1
KH14(III)	30	2.315	1.007	−0.0719	318.8	42.0	225.2	4.1	130.7	47.7
KH15(II)	31	2.060	1.006	0.1137	295.4	13.4	203.3	8.8	81.1	73.9
KH16(II)	25	4.856	1.017	−0.2017	342.4	17.9	249.3	9.7	132.2	69.5
KH17(III)	31	2.315	1.007	−0.0924	186.2	12.1	94.4	8.4	330.4	75.2
KH18(II)	23	2.075	1.009	0.0528	11.0	19.5	117.3	38.4	260.1	45.1
KH19(II)	24	2.130	1.005	0.0696	170.3	9.7	263.8	19.7	55.4	67.8
KH20(II)	32	2.140	1.011	−0.3096	84.2	2.2	174.2	0.6	279.7	87.7

N: number of samples measured; K_m: mean magnetic susceptibility; P'/T: degree of magnetic anisotropy/shape parameter; K_{max}/K_{int}/K_{min}: mean AMS eigenvectors corresponding to the maximum/intermediate/minimum susceptibility; Dec/Inc: declination/Inclination.

Lava flows typically exhibit a low degree of anisotropy and a scattered distribution of AMS directions due to the weak preferred orientation of grains [56,57]. Table 2 presents the mean values of the degree of anisotropy (P'), which range from 1.005 to 1.017, indicating very low anisotropy. When projected on a stereonet, the AMS principal directions display significant scattering (Figure 6). Despite the weak anisotropy and scattered directions, the axial clusters of AMS ellipsoids can provide valuable information on the flow direction [58]. The mean directions for the principal eigenvectors (K_{max}, K_{int}, and K_{min}) calculated using bootstrap statistics are listed in Table 2. In certain sites (KH15, KH17, and KH20), the K_{max} axes are nearly horizontal, while the K_{min} axes are predominantly vertical (Figure 6). Although the mean Kmax and K_{min} directions exhibit slight deviations (less than ~20 degrees) from the horizontal and vertical orientations, respectively (Table 2), these deviations do not appear statistically significant, considering the relatively large scatter in the principal

AMS directions. For these sites, the flow directions of lava flows can be inferred from the magnetic lineation.

On the other hand, the remaining sites show mean K_{min} axes plunging at angles of 50 to 70 degrees from the horizontal axis (Table 2), indicating magnetic imbrications. The flow directions for these sites can be deduced from the imbricated sense of elongated grains. However, some of these sites (KH01, KH02, KH03, KH04, KH08, KH11, and KH12) exhibit anomalously horizontal to sub-horizontal K_{min} axes (Figure 6 and Table 2). We attributed this phenomenon to the interchanged principal axes resulting from inverse components in the AMS fabrics, as discussed further in Section 5.2.

The site-mean shape parameters (T) are plotted on the Jelinek diagram [42] in Figure 7. Some sites (KH01, KH02, KH06, KH07, KH08, KH09, and KH16) are predominantly prolate (T < 0), indicating an elongated ellipsoidal shape. Only KH05 exhibits a predominantly oblate shape (T > 0), indicating a flattened ellipsoidal shape. The remaining sites exhibit a minimal site-mean T value due to a wide range of T values. This result reflects the significant scatter in the AMS ellipsoid axes and the measurement uncertainties resulting from the low anisotropy.

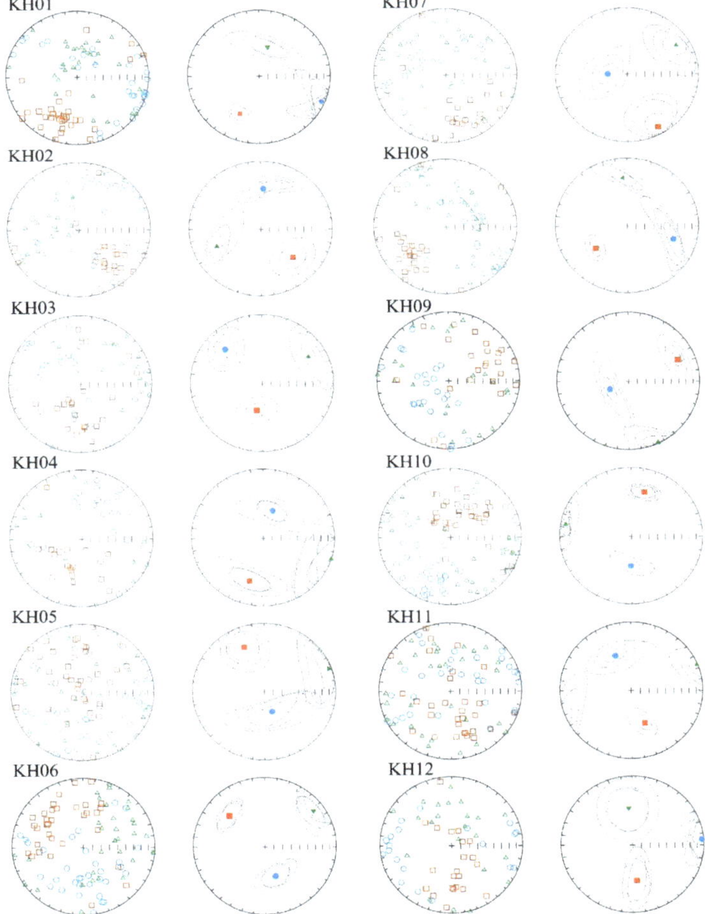

Figure 6. *Cont.*

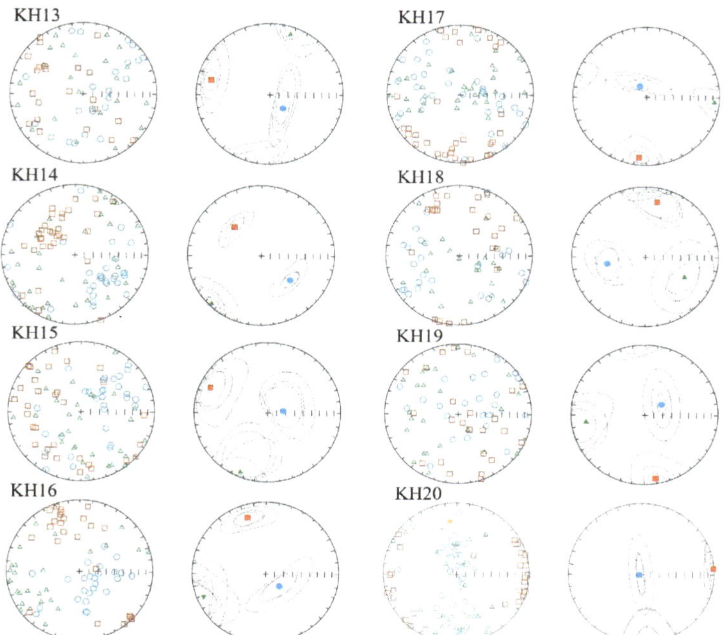

Figure 6. Equal area, lower-hemisphere projection of principal directions of AMS fabric in geographic coordinates. Squares (red), triangles (green), and circles (blue) represent maximum, intermediate, and minimum eigenvectors, respectively. The site-mean directions of three eigenvectors and their corresponding error ovals, calculated using the bootstrap method, are shown on the right side.

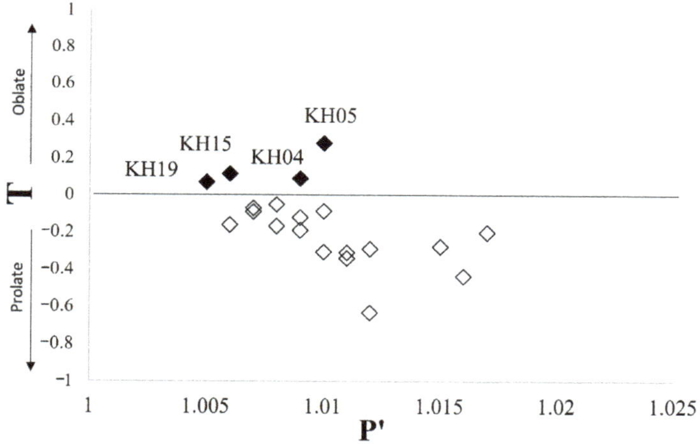

Figure 7. Jelinek diagram showing the site-mean values of degree of anisotropy (P') and shape parameter (T). Closed and open symbols are oblate (T > 0) and prolate (T < 0) shapes, respectively.

4.4. Anisotropy of Anhysteretic Remanent Magnetization (AARM) Results

Four representative samples from each of the 20 sites were subjected to AARM measurements, and the results are shown in Table 3. The site-mean degree of anisotropy (P') of the AARM ellipsoids ranges from 1.072 to 1.230 (Table 3), which is higher than that of the AMS ellipsoids (from 1.005 to 1.017). This result supports the general observation that remanence is generally more anisotropic than susceptibility [43,59,60]

Table 3. AARM results in this study's area.

Site (Unit)	N	Mean AARM Parameters		$AARM_{max}$		$AARM_{int}$		$AARM_{min}$		AMS Fabric Type
		P'	T	Dec (°)	Inc (°)	Dec (°)	Inc (°)	Dec (°)	Inc (°)	
KH01(I)	4	1.117	0.0419	351.6	29.4	95.2	22.8	216.9	51.4	Int (Group 2)
KH02(II)	4	1.078	0.2234	0.7	26.9	262.4	15.7	145.4	58.2	Inverse
KH03(II)	4	1.144	0.0351	296.8	32.0	45.5	27.2	167.2	45.6	Inverse
KH04(II)	4	1.117	0.1781	108.3	3.1	199.5	21.3	10.4	68.4	Int (Group 1)
KH05(III)	4	1.119	−0.1809	322.3	26.1	227.4	9.9	118.3	61.8	Normal
KH06(I)	4	1.078	0.0130	336.4	14.5	240.5	21.8	97.5	63.4	Normal
KH07(I)	4	1.134	0.1216	161.7	4.5	69.8	22.6	262.3	66.9	Normal
KH08(II)	4	1.204	0.4252	108.6	18.0	10.3	24.1	231.7	59.2	Inverse
KH09(I)	4	1.157	0.2560	50.8	5.7	141.5	6.8	281.4	81.1	Normal
KH10(I)	4	1.154	0.1694	342.9	21.6	251.1	4.5	149.9	67.9	Normal
KH11(II)	4	1.199	0.0820	0.8	38.7	261.2	11.7	157.5	48.9	Inverse
KH12(II)	4	1.134	0.0102	2.9	30.1	272.0	1.6	179.3	59.9	Int (Group 2)
KH13(II)	4	1.095	−0.2983	121.7	4.5	29.8	22.6	222.3	66.9	Normal
KH14(III)	4	1.142	−0.4803	321.0	17.7	53.5	7.9	166.7	70.5	Normal
KH15(II)	4	1.112	−0.3199	279.4	4.6	188.1	15.6	25.6	73.7	Normal
KH16(II)	4	1.124	0.1711	191.8	5.1	283.6	19.8	88.1	69.5	Normal
KH17(III)	4	1.079	−0.3168	165.4	14.6	73.3	8.1	315.1	73.2	Normal
KH18(II)	4	1.230	−0.1969	113.5	17.0	11.0	35.3	224.6	49.6	Int (Group 1)
KH19(II)	4	1.072	−0.0241	138.2	9.6	45.9	13.5	262.6	73.3	Int (Group 1)
KH20(II)	4	1.154	−0.0587	74.0	22.3	164.8	2.0	259.7	67.6	Normal

N: number of samples measured; P'/T: degree of magnetic anisotropy/shape parameter; $AARM_{max}/AARM_{int}/AARM_{min}$: mean AARM eigenvectors corresponding to the maximum/intermediate/minimum intensity; Dec/Inc: declination/Inclination; Int: Intermediate.

As with the AMS data, the principal eigenvectors ($AARM_{max}$, $AARM_{int}$, and $AARM_{min}$) for all sites were plotted on a stereonet (Figure 8), and the mean directions were calculated (Table 3). The fabric types of AMS were determined by comparing the directional data between AMS and AARM. Most sites (KH05, KH06, KH07, KH08, KH09, KH10, KH13, KH14, KH15, KH16, KH17, KH20) were classified as having normal AMS fabrics, as the mean principal axes of the AMS ellipsoid were nearly coaxial with those of the AARM ellipsoid. Four sites (KH02, KH03, KH8, and KH11) were identified as having inverse AMS fabrics, as the mean K_{max} and K_{min} orientations were sub-parallel to those of $AARM_{min}$ and $AARM_{max}$, respectively. These inverse components of AMS fabrics can mix with normal components, resulting in an "intermediate" AMS fabric [61,62]. In our samples, intermediate AMS fabrics are observed in five sites (KH01, KH04, KH12, KH18, KH19). Two sites (KH01, KH12) exhibit an interchange of all three principal axes of the AMS ellipsoid compared to the AARM fabrics, while three sites (KH04, KH18, KH19) show an interchange between two axes (K_{int} and K_{max}). When considering both AMS and AARM results, anomalous AMS fabrics (KH01, KH02, KH03, KH8, KH11, KH12) with highly tilted K_{min} axes from the vertical orientation were found to be associated with the inverse components of the AMS ellipsoid.

In most sites, the site-mean T values indicate a tendency toward oblate AARM ellipsoid shapes (Table 3). Some sites exhibit prolate AARM ellipsoid shapes (KH05, KH13, KH14, KH15, KH17, KH18, KH19, KH20). However, when comparing the T values between AMS and AARM, no consistent relationship was observed with the AMS fabric types. It is likely due to the weak preferred orientation of the magnetic grains and the interference from the inverse components of the AMS ellipsoids.

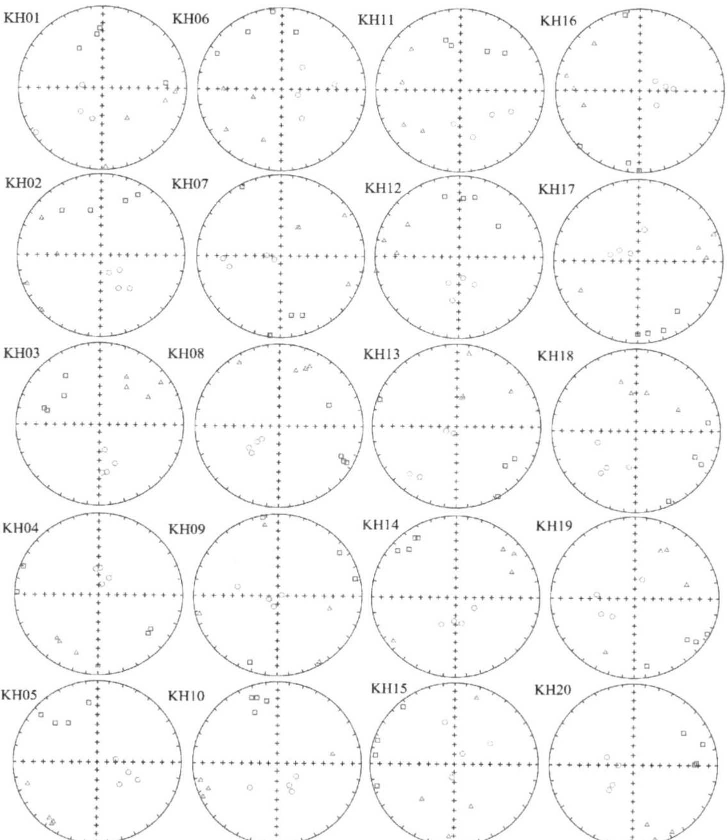

Figure 8. Equal area, lower-hemisphere projection of principal directions of AARM data for four representative samples from each site. Squares (red), triangles (green), and circles (blue) represent maximum, intermediate, and minimum eigenvectors, respectively.

5. Discussion

5.1. Paleomagnetic Implications for the Age of Eruption

The paleomagnetic data from each site met the data selection criteria proposed by McElhinny and McFadden [63]. These criteria included obtaining mean ChRM components from more than five samples per site, maintaining a sufficiently small radius of the 95% confidence circle ($\alpha_{95} \leq 3.2°$), and ensuring that the virtual geomagnetic pole (VGP) latitude for each site exceeded 50°.

As discussed in Section 4.1, the mean directions of Unit I and Unit II were statistically distinguished. At the same time, Unit III did not pass the significance test with the other units due to the limited number of sites and a larger α_{95} value. However, the F-test results comparing Unit I and Unit II indicate different lava eruption times for these two units.

The VGP positions for each unit are as follows: Unit I, 83.8°N/183.0°E ($A_{95} = 2.5°$); Unit II, 87.7°N/179.0°E ($A_{95} = 1.7°$). These VGPs are located close to the geographic North Pole, suggesting that any significant tectonic movement has not influenced the basalts in this study's area since the time of the lava eruption. Furthermore, the primary flow fabrics (which will be discussed in Section 5.2) support that tectonic correction is not necessary for determining the ChRM directions.

Previous studies on eruption ages for Quaternary basalts in this study's area reported the following: Unit I, 0.51 Ma [33]; Unit II, 1.08 Ma to 0.09 Ma [34–37]; Unit III, 0.04 Ma [38]. Notably, these basalts exhibit only normal polarity, suggesting eruption during the Brunhes Epoch, namely, after 0.78 Ma. Additionally, the significantly large k-value (700.1) and small α_{95} value (1.5°) of the mean direction of Unit II indicate that the sampled basalts erupted over a relatively short geologic time. The paleomagnetic directional data derived from this study are expected to contribute to a global geomagnetic field database as more detailed and precise dating data on basalts in this study's area are accumulated in the future.

5.2. Single Domain Effect on Magnetic Fabrics

The occurrence of inverse AMS fabrics, characterized by the interchange of K_{max} and K_{min} axes, can be attributed to mainly two causes: (1) secondary processes such as hydrothermal alteration [64] and tectonic overprinting [65]; (2) the presence of SD grains [66,67]. Our samples exhibit predominantly sub-horizontal $AARM_{max}$ axes and sub-vertical $AARM_{min}$ axes (Figure 8 and Table 3), indicating that the original fabrics are associated with the primary lava flow and have no significant subsequent alteration. Although some sites show inclined $AARM_{min}$ axes of up to 50° from the horizontal axis (Table 3), this can be attributed to the imbrication of the magnetic foliation plane.

Our mineral magnetic results suggest that SD grains are responsible for the observed inverse AMS fabrics. Figure 5c presents the grain sizes of the dominant AMS carriers for each site, classified according to the AMS fabric type. Sites exhibiting inverse AMS fabrics display hysteresis parameters (white circles) within the SD range (Figure 5c), providing evidence of SD grains' influence on the inverse fabric. Conversely, sites with normal AMS fabrics predominantly exhibit hysteresis ratios (grey circles) within the PSD range, indicating that SD grains did not significantly contribute to the normal fabric.

The hysteresis ratios vary across the SD to PSD range for sites displaying intermediate AMS fabrics (white and grey triangles). They are distributed near the boundary between the SD and PSD regions (Figure 5c). These data points align closely with the SD-SP mixing line proposed by Dunlop [52] (Figure 5c), indicating the presence of SD grains mixed with some SP grains, as indicated by slightly wasp-waisted hysteresis loops (Figure 5a,b). Thus, the intermediate fabrics observed in our samples might arise from a combination of inverse components resulting from the SD effect mixed with the normal components.

5.3. Inferred Lava Flow Direction

Flow directions were determined for sites with normal AMS fabric. For these sites, the flow direction is inferred from the magnetic lineation. One of them, site KH20, exhibits a mean K_{min} axis nearly vertical with a slight tilt of 2.3° (Table 2), indicating minimal imbrication. In contrast, the remaining sites with normal AMS fabrics show an oblique orientation of the mean K_{min} axes, ranging from 14.8° to 42.3° from the vertical (Table 2). Most sites have mean K_{min} axes plunging approximately at 20° (14.8°~24.6°) from the vertical axis, while a few sites (KH06, KH10, KH14) exhibit highly inclined mean K_{min} axes, deviating by around 40° (34.4°~42.3°).

In this study, mean K_{min} axes plunging at 70° are considered nearly vertical based on the significant scattering of the principal AMS directions and the poor grouping of K_{min} axes. However, mean K_{min} axes with high obliquity (20°~45°) from the vertical indicate magnetic imbrication of elongated grains. This imbrication is likely caused by a parabolic velocity profile within the lava flow between the ground surface and a solidified roof [13,14]. As the roof solidifies, the flow velocity in the upper and lower parts of the lava becomes lower compared to the central part, resulting in the opposite imbrication of elongated grains [13,14,68,69]. In opposite imbrications, the plunge direction of the magnetic foliation pole (K_{min} azimuth) indicates the flow direction [14,70]. Given that our samples were collected from the lower half of the lava flow, the azimuth of the K_{min} axes could be parallel to the flow vector [13,17,70].

The determined lava flow directions were shown on the geologic map (Figure 9). Three sites (KH15, KH17, KH20) used bidirectional lineation as the flow direction indicator. For the remaining sites, flow directions were determined using magnetic imbrication (Figure 9). The inferred flow directions, pointing toward the south–southwest, are consistent with previous work that utilized vesicle structure in the field [3]. This agreement is observed at sites KH09 and KH20.

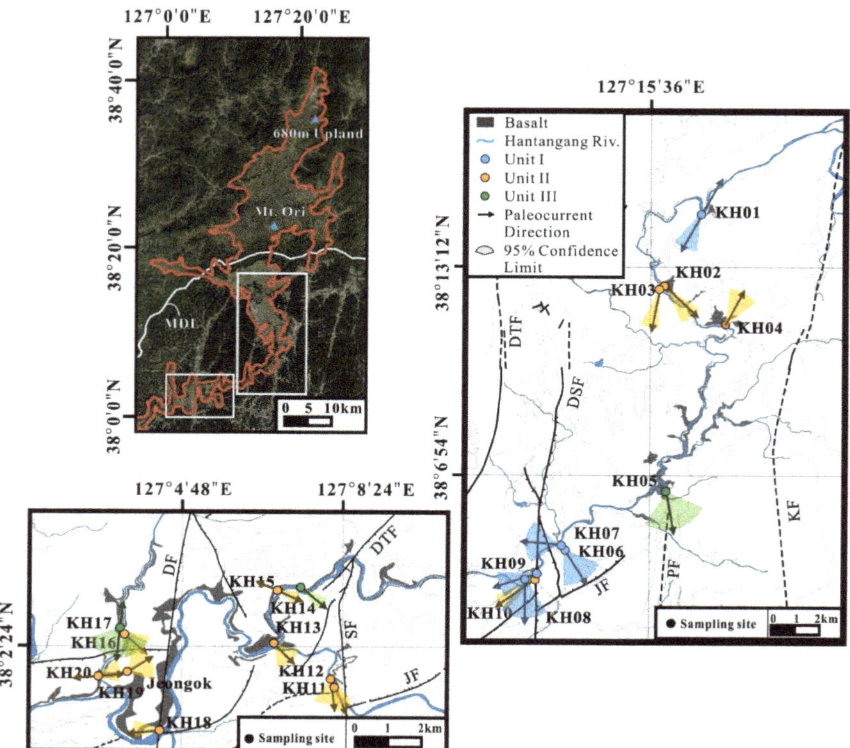

Figure 9. (upper left) A satellite map showing this study's area. MDL: Military Demarcation Line between South and North Korea. (lower left/right) Inferred lava flow directions with 95% confidence level on the map in the downstream/upstream regions of Hantangang River. DF: Dongduchen Fault, DSF: Dongsong Fault, JF: Jeongok Fault, KF: Kimhwa Fault, DTF: Damteo Fault, SF: Singo Fault, and PF: Pochen Fault.

The lava flow directions were determined for four sites (KH02, KH03, KH08, KH11) exhibiting an inverse fabric (K_{max}-K_{min} interchange). The direction of K_{max} axes, which deviates from the vertical axis, was taken as the direction of imbrication. For three sites (KH04, KH18, KH19) showing intermediate fabric (group 1; K_{max}-K_{int} interchange), the direction of the K_{min} axes deviated from the vertical axis by from 22.2° to 44.9°. Therefore, this deviation was considered magnetic imbrication. In the case of two sites (KH01, KH12) with intermediate fabric (group 2), where all three axes interchanged, the axes' distribution was considered. In the case of site KH01, the direction of K_{max} axes was estimated as the magnetic lineation (Figure 9).

In this study, lava flow directions were determined from a total of 20 sites. However, due to the limited number of sampling sites, it has not been possible to identify specific lava sources definitively. Nonetheless, the lava flow directions obtained for each Unit provide valuable information about the past topography during the lava eruption.

Unit I, the oldest basalt in this study's area, is exclusively found in the upstream region of the Hantangang River (the northeastern part). The lava flow directions observed at five sites (KH01, KH06, KH07, KH09, and KH10) consistently point toward the southwest or southeast, indicating a downstream flow direction. These observations suggest that the Unit I lavas originated from a previously proposed source within North Korea, such as Mt. Ori and the 680 m Upland [2]. Similarly, both Unit II (KH02 and KH03) and Unit III (KH05) demonstrate a strong alignment with the current channel of the Hantangang River (Figure 9). These findings imply that the paleo-channel and surrounding topography of the Hantangang River during the time of lava flows were largely consistent with the present-day configuration, providing further support for the work by Hakim et al. [9].

However, the Jeongok region in the southwestern part of this study's area shows a different pattern compared to the northeastern part. Although most lava flows from Units II and III follow the Hantangang River channel, the flow directions suggest that they may have originated from other directions to the north or northwest rather than upstream in the northeastern region (e.g., KH14, KH15, KH16, KH19). Therefore, the possibility of potentially unknown volcanic sources in the surrounding area cannot be ruled out.

In summary, the distribution of Quaternary basalts in this study's area is predominantly confined to the channel of the Hantangang River and its immediate vicinity during lava flow. The sampling sites are limited to the outcrops located within the current channel of the Hantangang River, as much of the larger area has been eroded or buried. Furthermore, the complete stratigraphic relationship of the lava flows has yet to be fully elucidated, emphasizing the need for a more robust eruption history related to the source regions. Given several faults with a north–south orientation in this study's area (as shown in Figure 9), further comprehensive investigations are required to identify potential volcanic sources throughout the area.

Supplementary Materials: The following supporting information can be downloaded at: https://www.mdpi.com/article/10.3390/min13081011/s1, Table S1: Magnetic hysteresis parameter of a representative sample of the site.

Author Contributions: Conceptualization, J.K.P., J.Y.S., S.S. and Y.-H.P.; methodology, J.Y.S. and Y.-H.P.; software, J.K.P. and J.Y.S.; validation, J.K.P. and Y.-H.P.; formal analysis, J.K.P. and J.Y.S.; investigation, J.K.P., J.Y.S., S.S. and Y.-H.P.; resources, S.S. and Y.-H.P.; data curation, J.Y.S., J.K.P. and Y.-H.P.; writing—original draft preparation J.K.P., J.Y.S. and Y.-H.P.; writing—review and editing, J.K.P., J.Y.S., S.S. and Y.-H.P.; visualization, J.K.P., J.Y.S. and Y.-H.P.; supervision, Y.-H.P.; project administration, S.S. and Y.-H.P.; funding acquisition, J.K.P., J.Y.S., S.S. and Y.-H.P. All authors have read and agreed to the published version of the manuscript.

Funding: This research was funded by the National Research Foundation of Korea, grant number 2019R1A6A1A03033167 and 2022R1F1A1073012(Y.-H.P.), and RS-2023-00211265 (S.S.) and by Kangwon National University in 2019 (Y.-H.P.).

Acknowledgments: The authors would like to thank the referees for their constructive reviews. The authors also sincerely thank Seong-Jae Doh at Korea University, Daekyo Cheong at Kangwon National University and Yongjae Yu at Chungnam National University for their valuable advice on this study This research was also supported by the 2021 Research Project for the UNESO Hantangang River Global Geopark and the Gyeonggi Provincial Office.

References

1. Kil, Y.; Ahn, K.S.; Woo, K.S.; Lee, K.C.; Jwa, Y.J.; Jung, W.; Sohn, Y.K. Geoheritage values of the Quaternary Hantangang river volcanic field in the central Korean Peninsula. *Geoheritage* **2019**, *11*, 765–782. [CrossRef]
2. Won, J.K. A study on the Quaternary volcanism in the Korean peninsula. *J. Geol. Soc. Korea* **1983**, *19*, 159–168, (In Korean with English abstract).
3. Kim, K.H.; Kim, O.J.; Min, K.D.; Lee, Y.S. Structural, paleomagnetic and petrological studies of the Chugaryeong Rift Valley. *Econ. Environ. Geol.* **1984**, *17*, 215–230, (In Korean with English abstract).
4. Won, C.K.; Kim, Y.K.; Lee, M.W. The study on the geochemistry of Choogaryong Alkali Basalt. *J. Geol. Soc. Korea* **1990**, *26*, 70–81, (In Korean with English abstract).
5. Song, M.Y.; Shin, K.S. Satellite image analysis for the geologic structure and land surface environments in the Chugaryung rift valley near Cholwon. *J. Korean Earth Sci. Soc.* **1998**, *19*, 675–683.
6. Lee, M.B.; Lee, G.R.; Kim, N.S. Drainage derangement and revision by the formation of Cheolwon-Pyeonggang lava plateau in Chugaryeong rift valley, central Korea. *J. Korean Geogr.* **2004**, *39*, 833–844, (In Korean with English abstract).
7. Won, C.K.; Lee, M.W.; Jin, M.S.; Choi, M.J.; Jeong, B.H. *Geological Survey in the Hantangang River*; Jisungsa: Seoul, Republic of Korea, 2010; p. 232. (In Korean)
8. Lee, M.B.; Lee, G.R. A study of regional geomorphology in the Chugaryeong tectonic valley, central Korea. *J. Korean Geogr.* **2016**, *51*, 473–490, (In Korean with English abstract).
9. Hakim, W.L.; Ramayanti, S.; Park, S.; Ko, B.; Cheong, D.K.; Lee, C.W. Estimating the Pre-Historical volcanic eruption in the Hantangang River volcanic field: Experimental and simulation study. *Remote Sens.* **2022**, *14*, 894.
10. Tarling, D.; Hrouda, F. (Eds.) *Magnetic Anisotropy of Rocks*; Springer Science & Business Media: London, UK, 1993; p. 218.
11. Henry, B.; Jordanova, D.; Jordanova, N.; Souque, C.; Robion, P. Anisotropy of magnetic susceptibility of heated rocks. *Tectonophysics* **2003**, *366*, 241–258.
12. Cañón-Tapia, E.; Walker, G.P.; Herrero-Bervera, E. Magnetic fabric and flow direction in basaltic pahoehoe lava of Xitle volcano, Mexico. *J. Volcanol.* **1995**, *65*, 249–263.
13. Cañón-Tapia, E.; Walker, G.P.; Herrero-Bervera, E. The internal structure of lava flows—Insights from AMS measurements I: Near-vent a'a. *J. Volcanol.* **1996**, *70*, 21–36.
14. Cañón-Tapia, E.; Walker, G.P.; Herrero-Bervera, E. The internal structure of lava flows—Insights from AMS measurements II: Hawaiian pahoehoe, toothpaste lava and 'a'ā. *J. Volcanol.* **1997**, *76*, 19–46.
15. Cañón-Tapia, E.; Pinkerton, H. The anisotropy of magnetic susceptibility of lava flows: An experimental approach. *J. Volcanol.* **2000**, *98*, 219–233.
16. Cañón-Tapia, E. Anisotropy of magnetic susceptibility of lava flows and dykes: A historical account. *Geol. Soc. Spec. Publi.* **2004**, *238*, 205–225.
17. Herrero-Bervera, E.; Cañon-Tapia, E.; Walker, G.P.; Tanaka, H. Magnetic fabrics study and inferred flow directions of lavas of the Old Pali Road, O' ahu, Hawaii. *J. Volcanol.* **2002**, *118*, 161–171.
18. Gurioli, L.; Zanella, E.; Pareschi, M.T.; Lanza, R. Influences of urban fabric on pyroclastic density currents at Pompeii (Italy): 1. Flow direction and deposition. *J. Geophys. Res. Solid Earth* **2007**, *112*, B05214. [CrossRef]
19. Loock, S.; Diot, H.; de Vries, B.V.W.; Launeau, P.; Merle, O.; Vadeboin, F.; Petronis, M.S. Lava flow internal structure found from AMS and textural data: An example in methodology from the Chaîne des Puys, France. *J. Volcanol.* **2008**, *177*, 1092–1104.
20. Maffione, M.; Pucci, S.; Sagnotti, L.; Speranza, F. Magnetic fabric of Pleistocene continental clays from the hanging-wall of an active low-angle normal fault (Altotiberina Fault, Italy). *Int. J. Earth Sci.* **2012**, *101*, 849–861.
21. Závada, P.; Kratinová, Z.; Kusbach, V.; Schulmann, K. Internal fabric development in complex lava domes. *Tectonophysics* **2009**, *466*, 101–113.
22. Boiron, T.; Bascou, J.; Camps, P.; Ferré, E.C.; Maurice, C.; Guy, B.; Launeau, P. Internal structure of basalt flows: Insights from magnetic and crystallographic fabrics of the La Palisse volcanics, French Massif Central. *Geophys. J. Int.* **2013**, *193*, 585–602.
23. Hrouda, F.; Verner, K.; Kubinova, S.; Burianek, D.; Faryad, S.W.; Chlupacova, M.; Holub, F. Magnetic fabric and emplacement of dykes of lamprophyres and related rocks of the Central Bohemian Dyke Swarm (Central European Variscides). *J. Geosci.* **2016**, *61*, 335–354. [CrossRef]
24. Bradák, B.; Seto, Y.; Chadima, M.; Kovács, J.; Tanos, P.; Újvári, G.; Hyodo, M. Magnetic fabric of loess and its significance in Pleistocene environment reconstructions. *Earth Sci. Rev.* **2020**, *210*, 103385.
25. Moncinhatto, T.R.; Haag, M.B.; Hartmann, G.A.; Savian, J.F.; Poletti, W.; Sommer, C.A.; Trindade, R.I. Mineralogical control on the magnetic anisotropy of lavas and ignimbrites: A case study in the Caviahue-Copahue field (Argentina). *Geophy. J. Int.* **2020**, *220*, 821–838.
26. Hrouda, F.; Chadima, M.; Ježek, J. Anisotropy of out-of-phase magnetic susceptibility and its potential for rock fabric studies: A review. *Geosciences* **2022**, *12*, 234.
27. Suk, D.W. A Paleomagnetic Study on Basalt Distributed in the Jeongok Area and Hantangan River Basins. Master's Thesis, Yonsei University, Seoul, Republic of Korea, 2 February 1982. (In Korean with English abstract).
28. Lee, Y.S.; Min, K.D.; Hwang, J.H. The Geodynamic evolution of the Chugaryeong fault valley in a view point of paleomagnetism. *Econ. Environ. Geol.* **2001**, *34*, 555–571, (In Korean with English abstract).

29. Kim, B.C.; Hwang, J.H.; Lee, Y.S.; Nahm, W.H. Paleomagnetic and Soil Chemical Studies on the Quaternary Paleosol Around the Hantan River. *Econ. Environ. Geol.* **2004**, *37*, 325–334, (In Korean with English abstract).

30. Lee, Y.S.; Torii, M.; Nishimura, S.; Min, K.D. Apparent polar wander path for the Southern part of Korean Peninsula. In Proceedings of the 29th International Geological Congress, Kyoto, Japan, 24 August–3 September 1992.

31. Lee, Y.S.; Nishimura, S.; Min, K.D. Paleomagnetotectonics of east Asia in the Proto-Tethys Ocean. *Tectonophysics* **1997**, *270*, 157–166.

32. Min, K.D.; Lee, Y.S. Phanerozoic Geodynamics of the Korean Peninsula. *Econ. Environ. Geol.* **2006**, *39*, 353–368, (In Korean with English abstract).

33. Danhara, T.; Bae, K.; Okada, T.; Matsufuji, K.; Hwang, S. What is the real age of the Chongokni Paleolithic site? A new approach by fission track dating, K-Ar dating and tephra analysis. In *Paleolithic Archaeology in Northeast Asia*; Bae, K., Lee, J.C., Eds.; Yeoncheon County & The Institute of Cutural Properties: Seoul, Republic of Korea, 2002; pp. 77–116.

34. Takayanagi, M. Dating of the basalt rock. In *Report on the excavation of Chongokri site*; National Research Institute of Cultural Heritage: Seoul, Republic of Korea, 1983; pp. 586–588, (In Korean and Japanese).

35. Yi, S.; Lee, Y.; Lim, H. *Basic study on geological development of Paleolithic sites along Imjin Basin (I)*; Seoul National University Museum: Seoul, Republic of Korea, 2005; p. 104. (In Korean)

36. Ryu, S.; Oka, M.; Yagi, K.; Sakuyama, T.; Itaya, T. K-Ar ages of the Quaternary basalts in the Jeongok area, the central part of Korean Peninsula. *Geosci. J.* **2011**, *15*, 1–8.

37. Seong, Y.B. Burial age dating of geomorphic event using multiple cosmogenicradioactive nuclides: A case study on the first flow of Jeongok basalt over Baekeuri formation. *J. Korean Geomorphol. Assoc.* **2007**, *14*, 1–7, (In Korean with English abstract).

38. Lee, M.B.; Seong, Y.B.; Lee, G.R. Formative age and process on basalt of lava plateau in the Cheolwon and Yeoncheon areas, central Korea. *J. Korean Geomorphol. Assoc.* **2020**, *27*, 41–51.

39. Jelínek, V. *The Statistical Theory of Measuring Anisotropy of Magnetic Susceptibility of Rocks and Its Application*; Geofyzika: Brno, Czech Republic, 1977; Volume 29, pp. 1–87.

40. Hext, G.R. The estimation of second-order tensors, with related tests and designs. *Biometrika* **1963**, *50*, 353–373.

41. Tauxe, L. PmagPy, Software Package. 2011. Available online: https://pmagpy.github.io/PmagPy-docs/intro.html (accessed on 6 July 2023).

42. Jelinek, V. Characterization of the magnetic fabric of rocks. *Tectonophysics* **1981**, *79*, 63–67. [CrossRef]

43. Jackson, M. Anisotropy of magnetic remanence: A brief review of mineralogical sources, physical origins, and geological applications, and comparison with susceptibility anisotropy. *Pure Appl. Geophys.* **1991**, *136*, 1–28.

44. Zijderveld, J.D.A. AC demagnetization of rocks: Analysis of results. In *Methods in Paleomagnetism*; Collinson, D.W., Creer, K.M., Runcorn, S.K., Eds.; Elsevier: New York, NY, USA, 1967; pp. 254–286.

45. Kirschvink, J.L. The least-squares line and plane and the analysis of palaeomagnetic date. *Geophys. J. R. Astron. Soc.* **1980**, *62*, 699–718.

46. Tauxe, L.; Constable, C.; Johnson, C.L.; Koppers, A.A.P.; Miller, W.R.; Staudigel, H. Paleomagnetism of the southwestern U.S.A. recorded by 0-5 Ma igneous rocks. *Geochem. Geophys. Geosys.* **2003**, *4*, 8802. [CrossRef]

47. Moskowitz, B.M.; Jackson, M.; Kissel, C. Low-temperature magnetic behavior of titanomagnetites. *Earth Plant. Sci. Lett.* **1998**, *157*, 141–149. [CrossRef]

48. Dunlop, D.J.; Özdemir, Ö. *Rock Magnetism: Fundamentals and Frontiers*; Cambridge University Press: Cambridge, UK, 1997; p. 573.

49. Bascou, J.; Camps, P.; Dautria, J.M. Magnetic versus crystallographic fabrics in a basaltic lava flow. *J. Volcanol.* **2005**, *145*, 119–135.

50. Vlag, P.; Alva-Valdivia, L.; De Boer, C.B.; Gonzalez, S.; Urrutia-Fucugauchi, J. A rock-and paleomagnetic study of a Holocene lava flow in central Mexico. *Phys. Earth Planet. Inter.* **2000**, *118*, 259–272.

51. Day, R.; Fuller, M.; Schmidt, V.A. Hysteresis properties of titanomagnetites: Grain-size and compositional dependence. *Phys. Earth Planet. Inter.* **1977**, *13*, 260–267.

52. Dunlop, D.J. Theory and application of the Day plot (Mrs/Ms versus Hcr/Hc) 2. Application to data for rocks, sediments, and soils. *J. Geophys. Res. Solid Earth* **2002**, *107*, 2057. [CrossRef]

53. Roberts, A.P.; Tauxe, L.; Heslop, D.; Zhao, X.; Jiang, Z. A critical appraisal of the "Day" diagram. *J. Geophys. Res. Solid Earth* **2018**, *123*, 2618–2644.

54. Tauxe, L.; Mullender, T.A.T.; Pick, T. Potbellies, wasp-waists, and superparamagnetism in magnetic hysteresis. *J. Geophys. Res. Solid Earth* **1996**, *101*, 571–583. [CrossRef]

55. Jackson, M.; Moskowitz, B.; Rosenbaum, J.; Kissel, C. Field-dependence of AC susceptibility in titanomagnetites. *Earth Planet. Sci. Lett.* **1998**, *157*, 129–139. [CrossRef]

56. Hrouda, F. Magnetic anisotropy of rocks and its application in geology and geophysics. *Geophys. Surv.* **1982**, *5*, 37–82.

57. Tauxe, L. *Paleomagnetic Principles and Practice*; Kluwer Academic Publishers: Dordrecht, The Netherlands, 1998; pp. 230–240.

58. Ellwood, B.B. Flow and emplacement direction determined for selected basaltic bodies using magnetic susceptibility anisotropy measurements. *Earth Planet. Sci. Lett.* **1978**, *41*, 254–264.

59. McCabe, C.; Jackson, M.; Ellwood, B.B. Magnetic anisotropy in the trenton limestone: Results of a new technique, anisotropy of anhysteretic susceptibility. *Geophys. Res. Lett.* **1985**, *12*, 333–336.

60. Stephenson, A.; Sadikun, S.T.; Potter, D.K. A theoretical and experimental comparison of the anisotropies of magnetic susceptibility and remanence in rocks and minerals. *Geophys. J. Int.* **1986**, *84*, 185–200.

61. Rochette, P.; Jackson, M.; Aubourg, C. Rock magnetism and the interpretation of anisotropy of magnetic susceptibility. *Rev. Geophys.* **1992**, *30*, 209–226. [CrossRef]
62. Rochette, P.; Aubourg, C.; Perrin, M. Is this magnetic fabric normal? A review and case studies in volcanic formations. *Tectonophysics* **1999**, *307*, 219–234.
63. McElhinny, M.W.; McFadden, P.L. Palaeosecular variation over the past 5 Myr based on a new generalized database. *J. Geophys. Res. Solid Earth* **1997**, *131*, 240–252.
64. Krása, D.; Herrero-Bervera, E. Alteration induced changes of magnetic fabric as exemplified by dykes of the Koolau volcanic range. *Earth. Planet. Sci. Lett.* **2005**, *240*, 445–453.
65. Park, J.K.; Tanczyk, E.I.; Desbarats, A. Magnetic fabric and its significance in the 1400 Ma Mealy diabase dykes of Labrador, Canada. *J. Geophys. Res. Solid Earth* **1988**, *93*, 13689–13704.
66. Potter, D.K.; Stephenson, A. Single-domain particles in rocks and magnetic fabric analysis. *Geophys. Res. Lett.* **1988**, *15*, 1097–1100. [CrossRef]
67. Rochette, P. Inverse magnetic fabric in carbonate-bearing rocks. *Earth Planet. Sci. Lett.* **1988**, *90*, 229–237. [CrossRef]
68. Knight, M.D.; Walker, G.P. Magma flow directions in dikes of the Koolau Complex, Oahu, determined from magnetic fabric studies. *J. Geophys. Res. Solid Earth* **1988**, *93*, 4301–4319.
69. Tauxe, L.; Gee, J.S.; Staudigel, H. Flow directions in dikes from anisotropy of magnetic susceptibility data: The bootstrap way. *J. Geophys. Res. Solid Earth* **1998**, *103*, 17775–17790.
70. Hillhouse, J.W.; Wells, R.E. Magnetic fabric, flow directions, and source area of the lower Miocene Peach Springs Tuff in Arizona, California, and Nevada. *J. Geophys. Res. Solid Earth* **1991**, *96*, 12443–12460.

minerals

Article

Low-Dimensional Multi-Trace Impedance Inversion in Sparse Space with Elastic Half Norm Constraint

Nanying Lan [1], Fanchang Zhang [1,*], Kaipan Xiao [1], Heng Zhang [2] and Yuhan Lin [3]

[1] National Key Laboratory of Deep Oil and Gas, China University of Petroleum (East China), Qingdao 266580, China; nanyinglan@126.com (N.L.); xkpzcy@163.com (K.X.)
[2] China National Offshore Oil Corporation (CNOOC) Limited Tianjin Branch, Tianjin 300450, China; zhangheng18@cnooc.com.cn
[3] Changqing Geophysical Department, BGP Inc., China National Petroleum Corporation, Xi'an 710021, China; lyh201966@163.com
* Correspondence: zhangfch@upc.edu.cn

Abstract: Recently, multi-trace impedance inversion has attracted great interest in seismic exploration because it improves the horizontal continuity and fidelity of the inversion results by exploiting the lateral structure information of the strata. However, computational inefficiency affects its practical application. Furthermore, in terms of vertical constraints on the model parameters, it only considers smooth features while ignoring sharp discontinuity features. This leads to yielding an over-smooth solution that does not accurately reflect the distribution of underground rock. To deal with the above situations, we first develop a low-dimensional multi-trace impedance inversion (LMII) framework. Inspired by compressed sensing, this framework utilizes low-dimensional measurements in sparse space containing the maximum information of the signal to construct the objective function for multi-trace inversion, which can significantly reduce the size of the inversion problem and improve the inverse efficiency. Then, we introduce the elastic half (EH) norm as a vertical constraint on the model parameters in the LMII framework and formulate a novel constrained LMII model for impedance inversion. Because the introduced EH norm takes into account both the smoothness and blockiness of rock impedance, the constrained LMII model can effectively raise the inversion accuracy of complex strata. Finally, an efficient alternating multiplier iteration algorithm is derived based on the variable splitting technique to optimize the constrained LMII model. The performance of the developed approaches is tested using synthetic and practical data, and the results prove their feasibility and superiority.

Keywords: multi-trace impedance inversion; low-dimensional space; elastic half norm; compressed sensing; variable splitting

Citation: Lan, N.; Zhang, F.; Xiao, K.; Zhang, H.; Lin, Y. Low-Dimensional Multi-Trace Impedance Inversion in Sparse Space with Elastic Half Norm Constraint. *Minerals* **2023**, *13*, 972. https://doi.org/10.3390/min13070972

Academic Editors: Luan Thanh Pham, Saulo Pomponet Oliveira and Le Van Anh Cuong

Received: 25 May 2023
Revised: 18 July 2023
Accepted: 20 July 2023
Published: 22 July 2023

1. Introduction

Impedance is a critical parameter to describe the properties of underground rock, and its accurate estimation is essential for mineral zone prediction and deposit characterization [1–3]. In mineral exploration, seismic inversion is a major tool to obtain the impedance of subsurface rocks. Affected by the band-limited nature of seismic data, incomplete data coverage, and noise interference, seismic inversion is usually ill-conditioned. Generally, it is difficult to obtain a stable and unique impedance profile [4,5]. To overcome the ill-conditions of impedance inversion and obtain reliable inversion results, regularized inversion techniques that introduce additional prior information have been extensively studied by exploration geophysicists.

Currently, the regularized inversion methods available for impedance prediction can be divided into two main categories: Type one is Tikhonov regularization inversion. This kind of method employs the L2 norm [6,7] or Gaussian prior distribution [8] to constrain the solution and force it to satisfy the smoothness condition. Unfortunately, these methods can blur sharp impedance edges and produce over-smooth solutions [9,10], which is detrimental to fine reservoir characterization. Another type is sparsity regularized inversion. This type of approach constrains the solution by using sparse norms or long-tailed distributions, imposing it to exhibit blocky features. Typical sparse regularization for impedance inversion includes total-variation regularization [11–14], L0 norm [15], L1 norm [16,17], L1-2 norm [18,19], Lp ($0 < p < 1$) norm [20], and Cauchy distribution [21,22]. Although sparse regularization preserves important edge information, it is inadequate for describing the prior information of rock impedance. The reason is that subsurface strata usually contain edge discontinuities that separate two smooth regions [9,10], while sparse regularization can only identify a single sharp property and cannot effectively characterize the smooth structures of stratigraphy. Similar to sparse regularization, Tikhonov regularization is also unable to model both smooth and sharp features of subsurface strata. Therefore, the regularized inversion methods mentioned above are ineffective in obtaining unbiased impedance estimation [14,23]. Additionally, these methods perform a trace-by-trace inversion, which ignores the lateral coherence between seismic traces [24–28].

Recently, several multi-trace impedance inversion (MII) methods were developed to eliminate the lateral instability problem caused by trace-by-trace inversion. Hamid and Pidlisecky present a laterally constrained MII algorithm [24], which constructs a constraint term in the horizontal direction using a second-order derivative operator to improve the lateral continuity of the inverted impedance. Karimi discusses the limitations of the MII method in the presence of tilted layers [29]. Hamid and Pidlisecky integrate seismic dip information into the impedance inversion and develop an MII method with structural constraints [26]. This method rotates the derivative operator by employing dip information to force the inversion results to honor the local structure, which enhances the adaptability of the MII method to high-steep layers. Yin et al. propose a cross-correlation-driven MII method that uses cross-correlation to describe the structural characteristics of stratigraphic reflections and raises the stability of impedance inversion [30]. Zhang et al. extract the local structural direction using the structure tensor and design a structure-oriented regularization method for impedance inversion [31]. Although the above MII techniques have made important contributions to improving the spatial continuity and stability of inversion results, there are still two critical issues to be addressed. One is the efficiency issue. The above MII approaches need to expand the forward operator via the Kronecker product to construct the objective function. Since the size of the forward operator is exponentially related to the number of traces and sampling points, the inverse problem has a large scale when both are enormous. Solving the large-scale inverse problem is so time-consuming that the MII method is challenged in practical application. Another one is the accuracy issue. The above MII approaches all construct the constraint terms with the L2 norm, which implies these methods only consider the smoothness and ignore the sharp discontinuity features in terms of vertical constraints on the model parameters. Thus, these methods cannot produce an optimal solution that accurately reflects the impedance distribution.

To address the above issues as much as possible, we propose a low-dimensional multi-trace impedance inversion (LMII) method with elastic half (EH) norm constraints in this paper. Specifically, we first develop a novel LMII inversion framework. Enlightened by compressed sensing [32,33], this framework uses low-dimensional measurements in a sparse domain containing the maximum information of the signal to construct the objective function for multi-trace inversion, which can effectively reduce the size of the inversion problem and improve the inversion efficiency. Subsequently, we introduce the sum of L1/2 norm and L2 norm, called EH norm [34], as a vertical constraint on the model parameters in the LMII framework to obtain an unbiased impedance estimation. As a combination of sparse regularization and Tikhonov regularization, the EH norm is able to characterize both blocky and smooth features of underground rock. Therefore, the LMII model constrained by the EH norm can effectively raise the inversion accuracy compared with the traditional MII method that only considers smoothness. Then, based on the variable splitting technique [35,36], we derive an efficient alternating multiplier iteration algorithm to optimize the constrained LMII problem. Finally, the effectiveness and superiority of the developed methods are verified by synthetic and field data.

2. Methodology

2.1. MII Method

According to the Robinson convolution model [37], seismic records can be represented as a convolution of source wavelet and reflection coefficient series:

$$\mathbf{s} = \mathbf{w} * \mathbf{r} + \mathbf{n}, \tag{1}$$

where $\mathbf{s} \in \mathbb{R}^M$ denotes the seismic record, $\mathbf{w} \in \mathbb{R}^M$ denotes the source wavelet, $\mathbf{r} \in \mathbb{R}^M$ denotes the reflection coefficient series, and $\mathbf{n} \in \mathbb{R}^M$ denotes the noise. Reformulating Equation (1) as matrix-vector product yields:

$$\mathbf{s} = \mathbf{W}\mathbf{r} + \mathbf{n}, \tag{2}$$

where $\mathbf{W} \in \mathbb{R}^{M \times M}$ is the circular matrix formed by source wavelet time-shifting. Under the small reflection coefficient hypothesis [37], the following linearized relationship exists between the subsurface impedance and reflection coefficient:

$$\mathbf{r}(t) = \frac{1}{2} \ln\left(\frac{\overline{\mathbf{z}}(t+1)}{\overline{\mathbf{z}}(t)} \right) = \frac{1}{2}(\ln(\overline{\mathbf{z}}(t+1)) - \ln(\overline{\mathbf{z}}(t))), \tag{3}$$

where $\overline{\mathbf{z}}$ denotes impedance and t represents time. By arranging the reflection coefficients along the time axis, one obtains:

$$\mathbf{r} = \frac{1}{2}\mathbf{D}\mathbf{z}, \tag{4}$$

where $\mathbf{D}$ is the first-order derivative operator in the time direction and $\mathbf{z}$ represents the natural logarithm of the stratigraphic impedance. Combining Equations (2) and (4) produces the following forward problem

$$\mathbf{s} = \frac{1}{2}\mathbf{W}\mathbf{D}\mathbf{z} + \mathbf{n} = \mathbf{G}\mathbf{z} + \mathbf{n}. \tag{5}$$

In the multi-trace case, Equation (5) can be expanded to the following form

$$\underbrace{\begin{bmatrix} \mathbf{s}_1 \\ \mathbf{s}_2 \\ \vdots \\ \mathbf{s}_N \end{bmatrix}}_{S} = \underbrace{\begin{bmatrix} \mathbf{G}_1 & 0 & 0 & 0 \\ 0 & \mathbf{G}_2 & 0 & 0 \\ 0 & 0 & \ddots & 0 \\ 0 & 0 & 0 & \mathbf{G}_N \end{bmatrix}}_{G} \underbrace{\begin{bmatrix} \mathbf{z}_1 \\ \mathbf{z}_2 \\ \vdots \\ \mathbf{z}_N \end{bmatrix}}_{Z} + \underbrace{\begin{bmatrix} \mathbf{n}_1 \\ \mathbf{n}_2 \\ \vdots \\ \mathbf{n}_N \end{bmatrix}}_{N}. \tag{6}$$

To obtain the subsurface impedance, a common method is to formulate a least-square optimization problem:

$$\mathbf{Z}^* = \arg\min_{Z} \frac{1}{2} \|\mathbf{S} - \mathbf{G}\mathbf{Z}\|_2^2, \tag{7}$$

where $\mathbf{Z}^*$ denotes the optimal estimation of natural logarithmic impedance $\mathbf{Z}$. Due to the ill-conditioning feature of $\mathbf{G}$ and the presence of noise, Equation (7) is often ill-posed. Thus, Equation (7) needs to introduce prior information through regularization terms to obtain a stable and reliable solution. In the conventional MII method [24], lateral structure as well as reference model constraints are inserted to improve the inversion accuracy. Specifically, the objective function of the conventional MII method can be expressed as

$$\begin{aligned}
\mathbf{Z}^* &= \arg\min_{\mathbf{Z}} \tfrac{1}{2}\|\mathbf{S} - \mathbf{G}\mathbf{Z}\|_2^2 + \lambda\|\mathbf{Z}_{ref} - \mathbf{Z}\|_2^2 + \mu\|\mathbf{C}\mathbf{Z}\|_2^2, \\
&= \arg\min_{\mathbf{Z}} \tfrac{1}{2}\|\overline{\mathbf{S}} - \overline{\mathbf{G}}\mathbf{Z}\|_2^2
\end{aligned} \tag{8}$$

where $\overline{\mathbf{S}} = \begin{bmatrix} \mathbf{S} \\ \sqrt{2\lambda}\mathbf{Z}_{ref} \\ 0 \end{bmatrix} \in \mathbb{R}^{3MN}$, $\overline{\mathbf{G}} = \begin{bmatrix} \mathbf{G} \\ \sqrt{2\lambda}\mathbf{I} \\ -\sqrt{2\mu}\mathbf{C} \end{bmatrix} \in \mathbb{R}^{3MN\times MN}$, $\mathbf{Z}_{ref}$ is the low-frequency reference model constructed from geological knowledge or well log, $\mathbf{I}$ is the identity matrix, $\mathbf{C}$ is the second-order difference operator in the spatial direction, λ and μ are the weight factors. Equation (8) has an analytical solution, that is:

$$\mathbf{Z}^* = \left(\overline{\mathbf{G}}^T \overline{\mathbf{G}} \right)^{-1} \overline{\mathbf{G}}^T \overline{\mathbf{S}}, \tag{9}$$

where $\overline{\mathbf{G}}^T$ represents the transpose of $\overline{\mathbf{G}}$. Equation (9) involves the inverse operation of large-scale matrices and using it directly for impedance inversion is very time-consuming. Alternatively, the conjugate gradient method [38], which does not involve an inverse matrix, is usually adopted to solve $\mathbf{Z}^*$. However, it still suffers from computational inefficiency due to the limitation of matrix scale. In order to address this issue, a novel LMII framework is developed in this paper.

2.2. LMII Framework

Given an orthogonal transform matrix $\boldsymbol{\Phi}$, whose columns are orthogonal atoms $\{\varphi_i\}_{i=1}^{M}$, any M-dimensional discrete signal can be represented as a linear combination of these atoms:

$$\mathbf{x} = \sum_{i=1}^{M} \varphi_i \alpha_i = \boldsymbol{\Phi}\boldsymbol{\alpha}, \tag{10}$$

where $\alpha_i = \langle \varphi_i, \mathbf{x} \rangle$ is the projection coefficient of signal $\mathbf{x}$ on the i-th atom and $\langle \bullet, \bullet \rangle$ denotes the Euclidean inner product operator. Compressive sensing [32,33] proved that, if the projection $\boldsymbol{\alpha}$ is sparse, then K ($K \ll M$) non-adaptive measurements of $\boldsymbol{\alpha}$ observed by a measurement matrix satisfying certain conditions are sufficient to accurately recover $\mathbf{x}$. Mathematically, the above K-dimensional measurements can be expressed as:

$$\mathbf{f} = \boldsymbol{\Psi}\boldsymbol{\alpha}, \tag{11}$$

where $\mathbf{f} \in \mathbb{R}^K$ is the low-dimensional measurement and $\boldsymbol{\Psi} \in \mathbb{R}^{K\times M}$ is the measurement matrix. Theoretically, the condition to be satisfied by the measurement matrix can be described by the restricted isometry (RIP) property [39] as follows:

$$(1 - \delta_K)\|\boldsymbol{\alpha}\|_2^2 \leq \|\boldsymbol{\Psi}\boldsymbol{\alpha}\|_2^2 \leq (1 + \delta_K)\|\boldsymbol{\alpha}\|_2^2, \tag{12}$$

where $\delta_k \in [0, 1]$ is the isometry constant of $\boldsymbol{\Psi}$. Essentially, δ_k quantitatively describes the preservation for the information contained in the original signal by low-dimensional measurements. $\delta_K = 0$ indicates complete preservation, while $\delta_K = 1$ indicates no preservation. To recover the signal exactly, δ_K needs to be as small as possible, which means that the subsets of K columns of $\boldsymbol{\Psi}$ are close to a set of standard orthogonal basis [40].

Inspired by the above idea, we develop a novel LMII framework in sparse space. Concretely, by employing orthogonal transform to convert the multi-trace inversion objective function from the spatio-temporal domain to sparse space, one can obtain:

$$\mathbf{Z}^* = \arg\min_{\mathbf{Z}} \frac{1}{2}\|\mathbf{S}' - \mathbf{F}_1\mathbf{G}\mathbf{Z}\|_2^2 + \lambda\|\mathbf{Z}'_{ref} - \mathbf{F}_2\mathbf{Z}\|_2^2 + \mu\|\mathbf{C}\mathbf{Z}\|_2^2, \tag{13}$$

where $\mathbf{F}_1 = kron(\mathbf{I}_{N \times N}, \boldsymbol{\Phi}_1)$, $\mathbf{F}_2 = kron(\mathbf{I}_{N \times N}, \boldsymbol{\Phi}_2)$, $kron(\bullet, \bullet)$ denotes the Kronecker product operator, $\boldsymbol{\Phi}_1$ and $\boldsymbol{\Phi}_2$ denote the employed sparse transform, and $\mathbf{S}' = \mathbf{F}_1 \mathbf{S}$ and $\mathbf{Z}'_{ref} = \mathbf{F}_2 \mathbf{Z}_{ref}$ denote projections of the seismic profile and the low-frequency reference model in sparse space, respectively. Afterwards, the measurement matrix is designed according to the RIP property and the low-dimensional measurement is performed in the sparse space to obtain the following LMII framework:

$$
\begin{aligned}
\mathbf{Z}^* &= \arg\min_{\mathbf{Z}} \tfrac{1}{2} \|\hat{\mathbf{S}} - \mathbf{M}_1 \mathbf{F}_1 \mathbf{G} \mathbf{Z}\|_2^2 + \lambda \|\hat{\mathbf{Z}}_{ref} - \mathbf{M}_2 \mathbf{F}_2 \mathbf{Z}\|_2^2 + \mu \|\mathbf{C}\mathbf{Z}\|_2^2 \\
&= \arg\min_{\mathbf{Z}} \tfrac{1}{2} \|\widetilde{\mathbf{S}} - \widetilde{\mathbf{G}} \mathbf{Z}\|_2^2
\end{aligned}
\tag{14}
$$

where $\mathbf{M}_1 = kron(\mathbf{I}_{N \times N}, \boldsymbol{\Psi}_1)$, $\mathbf{M}_2 = kron(\mathbf{I}_{N \times N}, \boldsymbol{\Psi}_2)$, $\boldsymbol{\Psi}_1$, and $\boldsymbol{\Psi}_2$ denote the designed measurement matrix and $\hat{\mathbf{S}} = \mathbf{M}_1 \mathbf{S}'$ and $\hat{\mathbf{Z}}_{ref} = \mathbf{M}_2 \mathbf{Z}'_{ref}$ represent the low-dimensional measurements of seismic profile and low-frequency reference model in the sparse space,

$$
\widetilde{\mathbf{S}} = \begin{bmatrix} \hat{\mathbf{S}} \\ \sqrt{2\lambda}\hat{\mathbf{Z}}_{ref} \\ 0 \end{bmatrix} \in \mathbb{R}^{(\kappa_1 + \kappa_2 + 1)MN}, \quad \widetilde{\mathbf{G}} = \begin{bmatrix} \mathbf{M}_1 \mathbf{F}_1 \mathbf{G} \\ \sqrt{2\lambda}\mathbf{M}_2 \mathbf{F}_2 \\ -\sqrt{2\mu}\mathbf{C} \end{bmatrix} \in \mathbb{R}^{(\kappa_1 + \kappa_2 + 1)MN \times MN}, \quad \kappa_1, \kappa_2 \in (0,1]
$$

denote the measurement ratios of seismic profile and reference model, respectively. Since both κ_1 and κ_2 are less than 1, the scale of the LMII problem is significantly smaller than that of the traditional MII problem. Obviously, this provides a reliable guarantee for the efficient inversion of rock impedance.

2.3. Constrained LMII Model via EH Norm

Although the developed LMII framework effectively improves solution efficiency, it still has limitations in accuracy (the conventional MII method also has this problem). Concretely, in terms of vertical constraints on the model parameters, this method only considers smooth features and ignores sharp discontinuity features, which leads to its solution being too smooth to accurately reflect the distribution of underground rock. To address this issue, we introduce the EH norm [34] in the LMII framework as a vertical constraint on the model parameters and formulate a constrained LMII model for impedance inversion. Since the introduced EH norm takes into account both the smoothness and blockiness of the rock impedance, the constrained LMII model can raise the inversion accuracy of complex stratigraphy. Formulaically, the constrained LMII model can be written as:

$$
\mathbf{Z}^* = \arg\min_{\mathbf{Z}} \frac{1}{2} \|\widetilde{\mathbf{S}} - \widetilde{\mathbf{G}} \mathbf{Z}\|_2^2 + \mu \|\mathbf{C}\mathbf{Z}\|_2^2 + \gamma \|\mathbf{B}\mathbf{Z}\|_{1/2}^{1/2} + \beta \|\mathbf{B}\mathbf{Z}\|_2^2,
\tag{15}
$$

where $\widehat{\mathbf{S}} = \begin{bmatrix} \hat{\mathbf{S}} \\ \sqrt{2\lambda}\hat{\mathbf{Z}}_{ref} \end{bmatrix}$, $\widehat{\mathbf{G}} = \begin{bmatrix} \mathbf{M}_1 \mathbf{F}_1 \mathbf{G} \\ \sqrt{2\lambda}\mathbf{M}_2 \mathbf{F}_2 \end{bmatrix}$, $\mathbf{B} = kron(\mathbf{I}_{N \times N}, \mathbf{D})$, $\|\bullet\|_{1/2}^{1/2}$ denotes the $L_{1/2}$ norm and γ and β are weighting factors. Equation (15) can be equivalently converted into the following optimization problem:

$$
\mathbf{Z}^* = \arg\min_{\mathbf{Z}} \frac{1}{2} \|\widehat{\mathbf{S}} - \widehat{\mathbf{G}} \mathbf{Z}\|_2^2 + \mu \|\mathbf{A}\mathbf{Z}\|_2^2 + \gamma \|\mathbf{B}\mathbf{Z}\|_{1/2}^{1/2},
\tag{16}
$$

where $\mathbf{A} = \mathbf{C} + \sqrt{\beta/\mu}\mathbf{B}$. To optimize Equation (16), we derive an efficient alternating multiplier iteration algorithm based on variable splitting technique. Specifically, with the aid of the variable splitting technique [35,36], the following constrained minimization problem is generated by introducing an auxiliary variable $\mathbf{T} = \mathbf{B}\mathbf{Z}$ into Equation (16)

$$
(\mathbf{Z}^*, \mathbf{T}^*) = \arg\min_{\mathbf{Z}, \mathbf{T}} \frac{1}{2} \|\widehat{\mathbf{S}} - \widehat{\mathbf{G}} \mathbf{Z}\|_2^2 + \mu \|\mathbf{A}\mathbf{Z}\|_2^2 + \gamma \|\mathbf{T}\|_{1/2}^{1/2} \text{ s.t. } \mathbf{T} = \mathbf{B}\mathbf{Z}.
\tag{17}
$$

Equation (17) can be minimized in an unconstrained manner by formulating it as an augmented Lagrangian function. To be specific, its augmented Lagrangian function is:

$$\mathcal{L}(\mathbf{Z}, \mathbf{T}, \mathbf{c}) = \min_{\mathbf{Z}, \mathbf{T}, \mathbf{c}} \frac{1}{2} \|\widehat{\mathbf{S}} - \widehat{\mathbf{G}}\mathbf{Z}\|_2^2 + \mu\|\mathbf{A}\mathbf{Z}\|_2^2 + \gamma\|\mathbf{T}\|_{1/2}^{1/2} + \langle \mathbf{BZ} - \mathbf{T}, \mathbf{c}\rangle + \frac{\varsigma}{2}\|\mathbf{BZ} - \mathbf{T}\|_2^2, \quad (18)$$

where $\mathbf{c}$ is a Lagrangian multiplier and ς is a non-negative parameter that controls the convergence rate of the algorithm. The minimizer of Equation (18) is the saddle point of function $\mathcal{L}(\mathbf{Z}, \mathbf{T}, \mathbf{c})$, which can be computed by the following iterative steps:

$$\mathbf{Z}^{(k+1)} = \underset{\mathbf{Z}}{\arg\min}\,\mathcal{L}\left(\mathbf{Z}, \mathbf{T}^{(k)}, \mathbf{u}^{(k)}\right) = \underset{\mathbf{Z}}{\arg\min}\,\frac{1}{2}\|\widehat{\mathbf{S}} - \widehat{\mathbf{G}}\mathbf{Z}\|_2^2 + \mu\|\mathbf{A}\mathbf{Z}\|_2^2 + \frac{\varsigma}{2}\|\mathbf{BZ} - \mathbf{T}^{(k)} + \mathbf{u}^{(k)}\|_2^2, \quad (19)$$

$$\mathbf{T}^{(k+1)} = \underset{\mathbf{T}}{\arg\min}\,\mathcal{L}\left(\mathbf{Z}^{(k+1)}, \mathbf{T}, \mathbf{u}^{(k)}\right) = \underset{\mathbf{T}}{\arg\min}\,\gamma\|\mathbf{T}\|_{1/2}^{1/2} + \frac{\varsigma}{2}\|\mathbf{BZ}^{(k+1)} - \mathbf{T} + \mathbf{u}^{(k)}\|_2^2, \quad (20)$$

$$\mathbf{u}^{(k+1)} = \mathbf{u}^{(k)} + \mathbf{BZ}^{(k+1)} - \mathbf{T}^{(k+1)}, \quad (21)$$

where $\mathbf{u} = \mathbf{c}/\varsigma$ is the scaled Lagrangian multiplier and k denotes the iteration index. Equation (19) is a quadratic equation whose solution can be obtained by first order necessary conditions. Specifically, letting its derivative of variable $\mathbf{Z}$ be zero yields

$$-\widehat{\mathbf{G}}^T\left(\widehat{\mathbf{S}} - \widehat{\mathbf{G}}\mathbf{Z}\right) + 2\mu\mathbf{A}^T\mathbf{A}\mathbf{Z} + \varsigma\mathbf{B}^T\left(\mathbf{BZ} - \mathbf{T}^{(k)} + \mathbf{u}^{(k)}\right) = 0. \quad (22)$$

By simplifying Equation (22), the optimal solution of Equation (19) can be obtained:

$$\mathbf{Z}^{(k+1)} = \left(\widehat{\mathbf{G}}^T\widehat{\mathbf{G}} + 2\mu\mathbf{A}^T\mathbf{A} + \varsigma\mathbf{B}^T\mathbf{B}\right)^{-1}\left(\widehat{\mathbf{G}}^T\widehat{\mathbf{S}} + \varsigma\mathbf{B}^T\left(\mathbf{T}^{(k)} - \mathbf{u}^{(k)}\right)\right). \quad (23)$$

Equation (20) is a typical $L1/2$ norm optimization problem [41], whose solution can be computed by the following half-threshold function:

$$\mathbf{T}^{(k+1)} = \mathcal{H}_\lambda\left(\mathbf{BZ}^{(k+1)} + \mathbf{u}^{(k)}\right), \quad (24)$$

where

$$\mathcal{H}_\lambda(\mathbf{x}) = \begin{cases} f_\lambda(\mathbf{x}_i) & , |\mathbf{x}_i| > \frac{3\sqrt[3]{2}}{4}\lambda^{\frac{2}{3}} \\ 0, & \text{otherwise} \end{cases}, \quad (25)$$

$$f_\lambda(\mathbf{x}_i) = \frac{2\mathbf{x}_i}{3}\left(1 + \cos\left(\frac{2\pi}{3} - \frac{2\varphi_\lambda(\mathbf{x}_i)}{3}\right)\right), \quad (26)$$

$$\varphi_\lambda(\mathbf{x}_i) = \arccos\left(\frac{\lambda}{8}\left(\frac{|\mathbf{x}_i|}{3}\right)^{-3/2}\right), \quad (27)$$

where $\lambda = \gamma/\varsigma$. Updating the variables $\mathbf{Z}, \mathbf{T}, \mathbf{u}$ by Equations (21), (23), and (24) until the change of objective function is below the threshold level, we attain the optimal natural logarithmic impedance $\mathbf{Z}^*$. Note that the computational bottleneck of Equation (23) is the matrix inversion. Similar to solving Equation (9), this equation also needs to be solved by the simple and efficient conjugate gradient algorithm. Finally, perform the following exponential operation:

$$\bar{\mathbf{z}} = \exp(\mathbf{Z}^*), \quad (28)$$

and the final predicted rock impedance can be obtained.

3. Examples

In this section, we implement two experiments (one synthetic and one field) to evaluate the feasibility and effectiveness of our presented methods. For our methods, it is a critical task to select appropriate sparse transforms and measurement DHT matrices. In all examples, we employ the discrete Hartley transform (DHT) [42] as a tool for domain conversion because it can concentrate signal energy in the low-frequency part of the spectrum (see Figure 1). Meanwhile, we employ a binary matrix with orthogonal rows as the measurement matrix, which greatly avoids the loss of information contained in the original signal and can effectively maintain the accuracy of impedance inversion. For comparison, the conventional MII method is used as a benchmark. To be fair, the parameters of comparison methods are carefully tuned several times to yield optimal inversion results in each case.

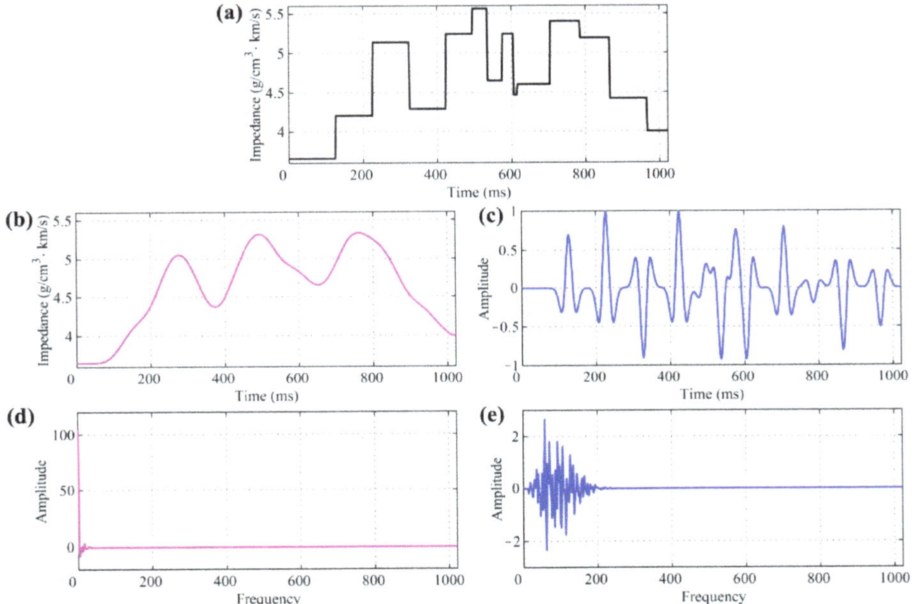

Figure 1. Verification of the validity of DHT domain conversion: (**a**) Impedance model; (**b**) Low-frequency reference model; (**c**) Synthetic seismic data; (**d,e**) DHT coefficients of low-frequency reference model and synthetic seismic data, respectively. The DHT coefficients for both the low-frequency reference model and the synthetic data are distributed at the low-frequency end. This indicates that DHT can concentrate signal energy in the low-frequency part of the spectrum.

3.1. Synthetic Example

First, we check the effectiveness of the developed methods with the SEG/EAGE over-thrust impedance model (Figure 2a). Figure 2b depicts the noise-free seismic profile synthesized by convolving a 30 Hz Ricker wavelet with the reflection coefficients calculated in Figure 2a. This profile contains 801 traces and 373 sampling points with a sampling interval of 2 ms. The conventional MII, LMII, and constrained LMII methods are used for impedance inversion, and their inversion results are presented in Figure 3a–c, respectively. For the above methods, they are set to the same stopping criterion, i.e., the algorithm stops when the number of iterations is greater than 500 or the tolerance error is less than 10^{-6}. It should be noted that the low-frequency reference model (Figure 2c) utilized in the above three methods is obtained by performing 10 Hz low-pass filtering on Figure 2a. In addition, the measurement ratios of seismic data and reference model are set to 0.45 and 0.1 in the LMII and constrained LMII methods, respectively. As shown in Figure 3a–c, all these methods effectively recover small-scale geological information such as thin layers, lenses, pinch-outs, and channels. Figure 3d–f depicts the difference between the results inverted by the above three methods and the true impedance. As observed from Figure 3d–f, the residuals of the conventional MIII method and the LMII method are basically the same, while the residuals of the constrained LMII method are significantly smaller than the other two methods. This indicates that introducing the EH norm effectively improves inversion accuracy. In other words, the EH norm is more accurate than the conventional Tikhonov regularization in characterizing the subsurface structure. To visually compare the advantages and disadvantages of the three methods, Figure 4 displays the zoomed-in views of the true impedance and the inversion results of Figure 3. Figure 4 demonstrates that the inversion results obtained by the MII and LMII methods are oversmooth, blurring the stratigraphic boundaries and producing additional pseudo-layer interferences (as shown by the arrows). In contrast, the inversion results yielded by the constrained LMII method show stratigraphic boundaries more clearly, while greatly avoiding erroneous pseudo-layer interferences.

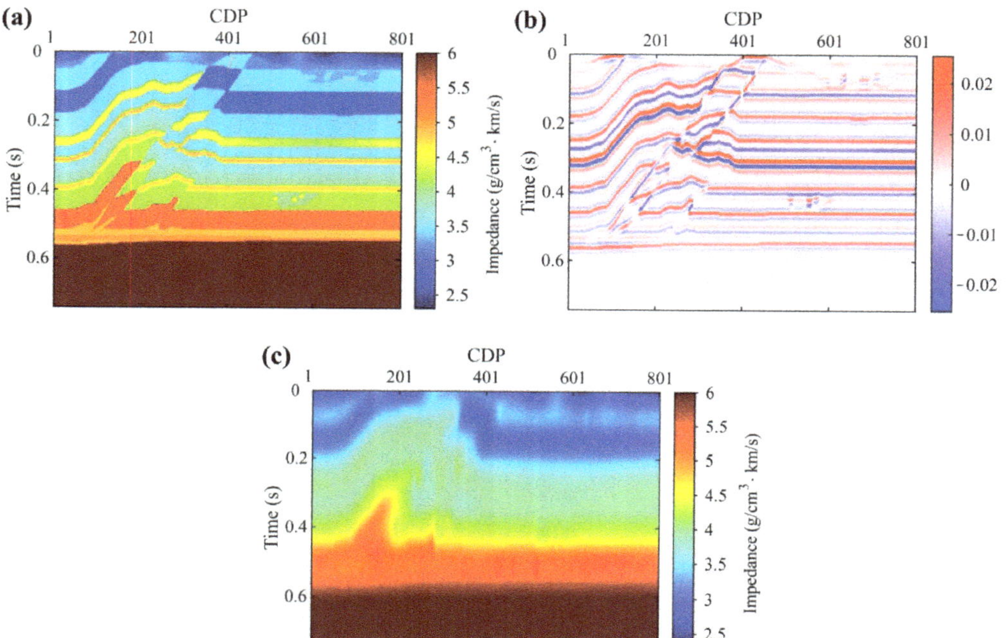

Figure 2. Synthetic example: (**a**) Theoretical overthrust impedance model; (**b**) Synthetic noise-free seismic profile; (**c**) Low-frequency reference model.

In order to quantitatively evaluate the reliability of these inverted results, root mean square error (RMSE) defined by the following equation is used as the evaluation index:

$$RMSE = \sqrt{\frac{1}{M \times N} \sum_{i=1}^{M} \sum_{j=1}^{N} \left(\overline{\mathbf{z}}_{ij} - \mathbf{y}_{ij} \right)^2} \tag{29}$$

where $\mathbf{y}$ denotes the true impedance, $\overline{\mathbf{z}}$ denotes the predicted impedance, M and N represent the number of sampling points and the number of seismic traces, respectively. Note that a smaller RMSE means higher reliability of the predicted impedance. The quantitative comparison of the inverted results of Figure 3 is listed in Table 1. As seen in Table 1, the constrained LMII method achieves a smaller RMSE value than the other two methods. This further demonstrates that the constrained LMII method can recover the impedance of subsurface rock more precisely. Moreover, the RMSE values obtained by the MII and LMII methods are basically consistent, but the calculation time of the conventional MII method is obviously longer than that of the LMII method. This provides reliable proof that the LMII method can effectively improve inversion efficiency while maintaining inversion accuracy. It should be pointed out that although the constrained LMII method uses an efficient alternating multiplier iteration algorithm to solve its objective function, it is slightly longer than the LMII method in terms of computational time due to the non-convex optimization involved. Notwithstanding, the constrained LMII method still has an advantage in computational efficiency compared to the conventional MII method.

Table 1. Quantitative evaluation index and running time of the inversion results of Figure 3.

Method	RMSE ($\times 10^{-3}$)	Time (s)
MII	46.10	261.89
LMII	46.11	135.13
Constrained LMII	29.06	161.17

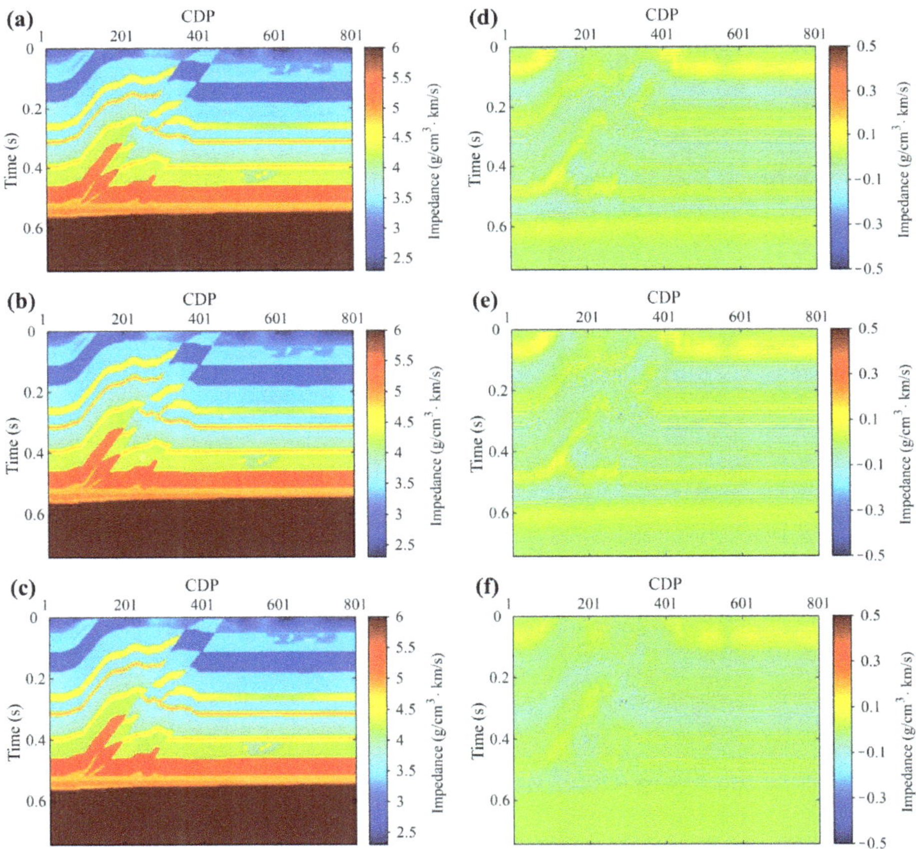

Figure 3. Comparison of the inversion results obtained by different methods in the noise-free case: (**a**) Inversion results by MII; (**b**) Inversion results by LMII; (**c**) Inversion results by constrained LMII; (**d**) Difference between (**a**) and the true impedance; (**e**) Difference between (**b**) and the true impedance; (**f**) Difference between (**c**) and the true impedance.

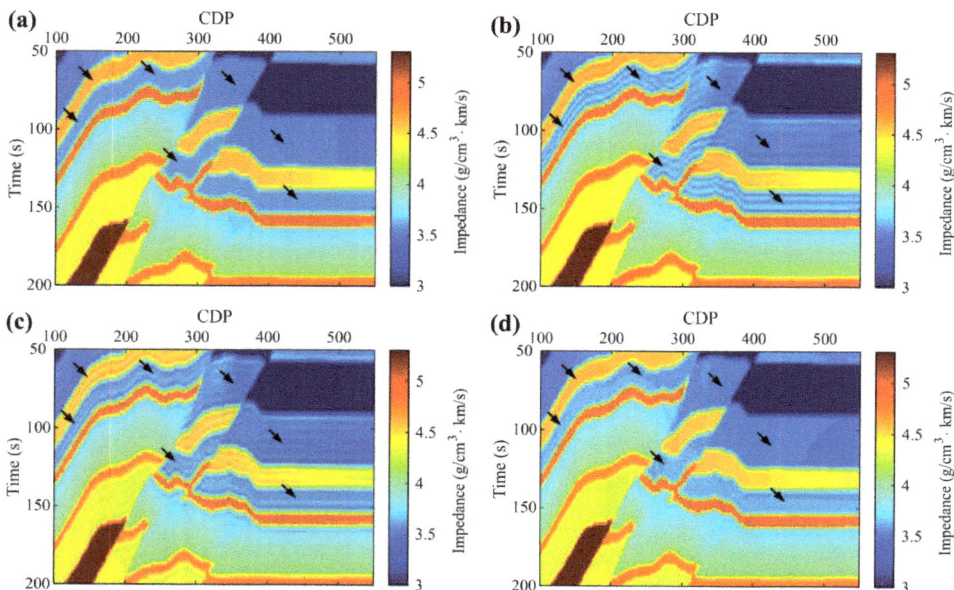

Figure 4. Zoomed-in view of the inversion results shown in Figure 3: (**a**) Zoomed-in view of the true model; (**b–d**) Zoomed-in views of Figure 3a–c, respectively.

To test the stability of the developed LMII and constrained LMII methods, we add Gaussian noise with a signal-to-noise ratio (SNR) of 3 (whose mean is 0 and standard deviation is one-third of the root mean square of the signal) to Figure 2b and obtain a noisy seismic profile for impedance inversion (Figure 5). In this case, the reference model and the setting of measurement ratios are the same as in the noise-free experiment. Figure 6a–c presents the results inverted by MII, LMII, and constrained LMII methods, respectively. Figure 6d–f presents the residuals between these inversion results and the true impedance, respectively. Note that the maximum number of iterations and tolerance error for the above methods are set to 500 and 10^{-3}. As observed from Figure 6, all inversion methods can recover impedance from noise-contaminated seismic data. Nevertheless, the constrained LMII method is significantly better than the other two methods in terms of accuracy since the constrained LMII method achieves a smaller residual. For better comparison, a zoomed window of the true impedance and the inverted results of Figure 6 are shown in Figure 7. At the same time, Figure 8 compares the impedance curves of the 300th trace of the inversion results shown in Figure 6. It can be seen from Figures 7 and 8 that the impedance interfaces of the MII and LMII methods are blurred, while the pseudo-layer interferences are obvious (as shown by arrows). In contrast, the impedance inverted by the constrained LMII method is much better than the other two methods, where the prediction errors are lower, and the layer interfaces are clear. To quantitatively evaluate the inversion quality, we calculate the RMSE values of the three inversion results and show them in Table 2. Also, the elapsed time of the above three inversion methods is listed in Table 2. From Table 2, it can be found that the inversion accuracy of the LMII method is essentially the same as that of the traditional MII methods, while the elapsed time is reduced by nearly half. This further illustrates the advantage of the LMII method in improving inversion efficiency. Besides, it can also be found from Table 2 that the constrained LMII method has a smaller RMSE value than the other two methods, which again effectively proves that the constrained LMII method is effective in improving the inversion quality. At the same time, this result also validates that the constrained LMII method has better robustness under noise circumstances.

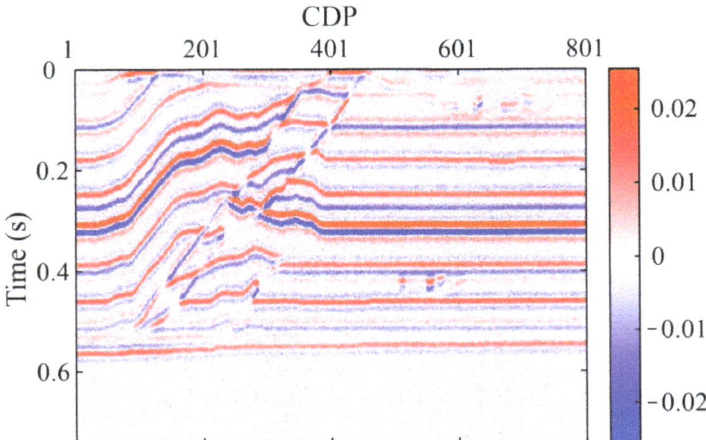

Figure 5. Seismic profile with SNR = 3.

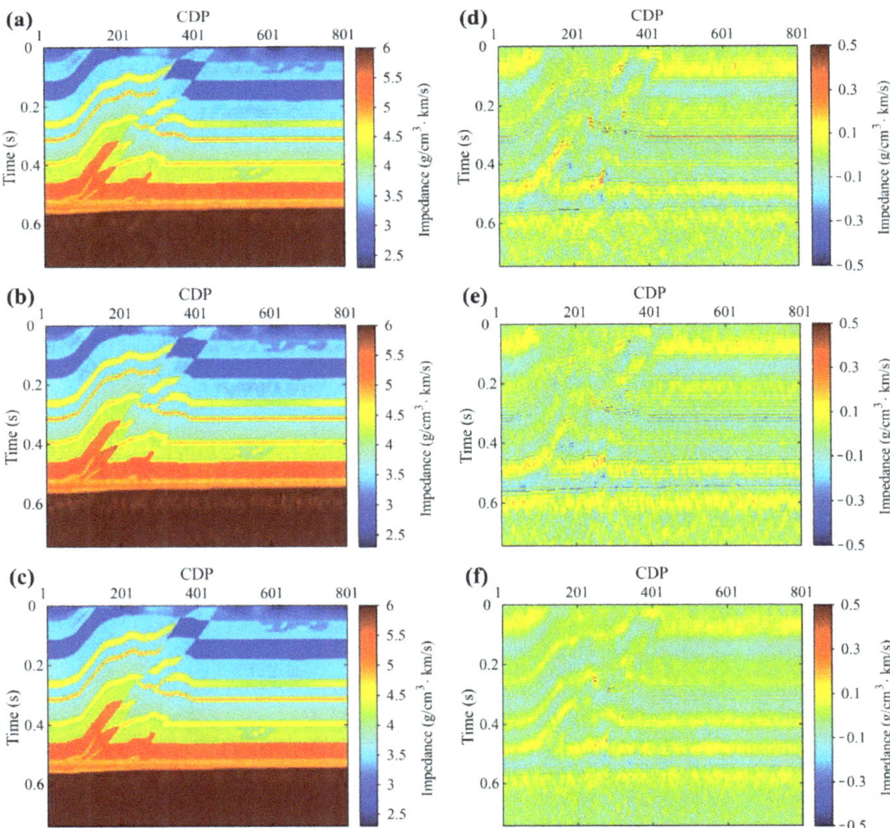

Figure 6. Comparison of inversion results by different methods in the noisy case: (**a–c**) present inversion results by MII, LMII, and constrained LMII, respectively; (**d–f**) present their residuals with the true impedance, respectively.

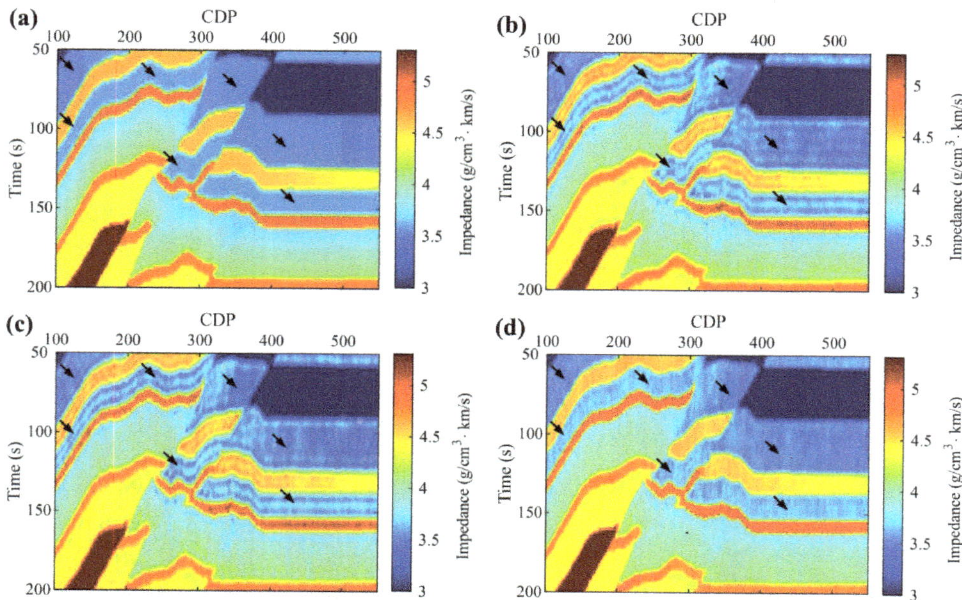

Figure 7. Zoomed-in comparison of the inversion results of Figure 6: (**a**) Zoomed-in view of Figure 2a; (**b–d**) Zoomed-in views of Figure 6a–c, respectively.

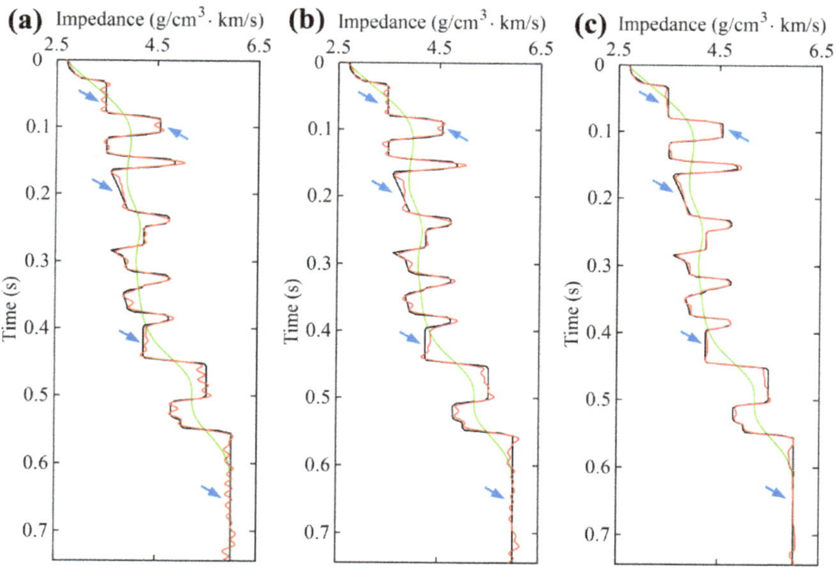

Figure 8. Comparison of the 300th trace from the inversion results of Figure 6: (**a–c**) present inversion results by MII, LMII, and constrained LMII, respectively. The green, black, and red lines represent the low-frequency reference model, true model, and inverted impedance, respectively.

Table 2. Quantitative evaluation index and running time of the inversion results of Figure 6.

Method	RMSE ($\times 10^{-3}$)	Time (s)
MII	79.93	269.31
LMII	79.94	138.28
Constrained LMII	57.92	165.80

3.2. Field Example

In this subsection, we apply the developed algorithms to the field data to test their practicability for real data inversion. Figure 9a shows 2D field post-stack data, which consists of 901 traces, each containing 300 sampling points with an interval of 1 ms. Therein, a blind well at the 288th trace (shown as the gray line) to judge the reliability of the predicted result. Note that this blind well does not participate in the inversion procedure. Figure 9b displays the low-frequency reference model used for impedance inversion, which is derived by interpolation from the 12–14 Hz high-cut filtered log data within the work area. Figure 10a–c presents the results predicted by the MII, LMII, and constrained LMII methods, respectively. In this case, the maximum number of iterations and the tolerance error of the above methods are set to 500 and 10^{-5}, respectively. In addition, the measurement ratios of the seismic record and reference model are set to 0.25 and 0.1 in the LMII and constrained LMII methods, respectively. It is evident from Figure 10 that the inversion results obtained by all three methods well reflect the lateral extension and lithological variation patterns of the geological bodies. Nevertheless, one notices that the constrained LMII method performs better than the other two methods in identifying the stratigraphic boundaries. Specifically, the results inverted by the constrained LMII method (Figure 10c) have obvious blocky characteristics and show clear geological boundaries, while the results inverted by the other two are over-smooth, blurring the lithological boundary features and leading to difficulties in layer identification. Additionally, the conventional MII method and the LMII method also produce additional pseudo-layer interferences (as shown in the elliptical zone). To better illustrate this point, Figure 11a–c presents the measured well logging curve at the 288th trace and the predicted impedance by the above three inversion methods, respectively. Herein, the green line represents the low-frequency reference curve, the black line represents the logging data, and the red line represents the predicted impedance curve. It is clear from Figure 11 that these inversion results show similar trends to the well log, but the inverted curve by the constrained LMII approach is more consistent with the well log (as shown by the arrows). In addition, as seen in the logging curve, a low-impedance shale layer exists near 3.23 s. After inversion by the conventional MII method and the LMII method, pseudo-layer interferences are generated in their results, which led to the appearance of false sand-shale interbeds. In contrast, the constrained LMII method effectively avoids the pseudo-layer interferences and correctly identifies this stratum. It reveals that the constrained LMII method outperforms the other two methods in terms of inversion reliability, which is consistent with the synthetic examples. In this case, the running times of the MII method, the LMII method, and the constrained LMII method are 298.74, 138.19, and 152.21 s, respectively. Clearly, both LMII and constrained LMII methods have smaller computing costs compared with the MII method, which further demonstrates that it is effective to compress the size of the inverse problem by low-dimensional measurements in sparse space.

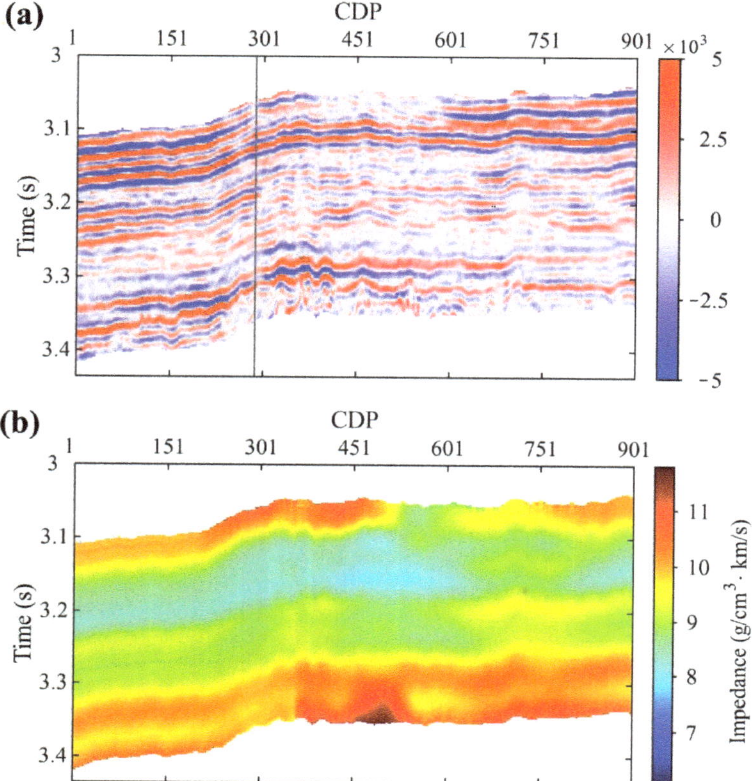

Figure 9. Field example: (**a**) 2D post-stack seismic profile; (**b**) Low-frequency reference model.

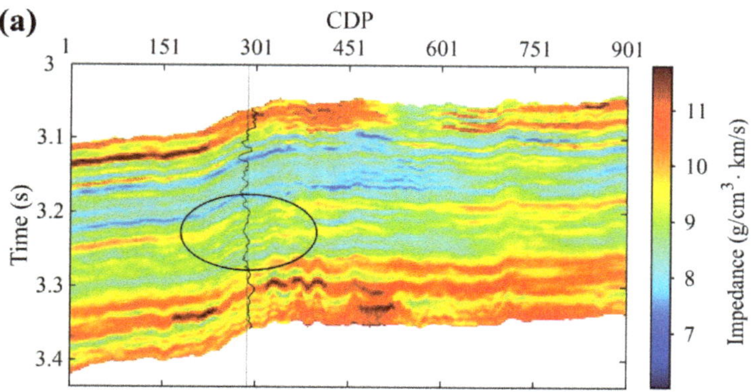

Figure 10. *Cont.*

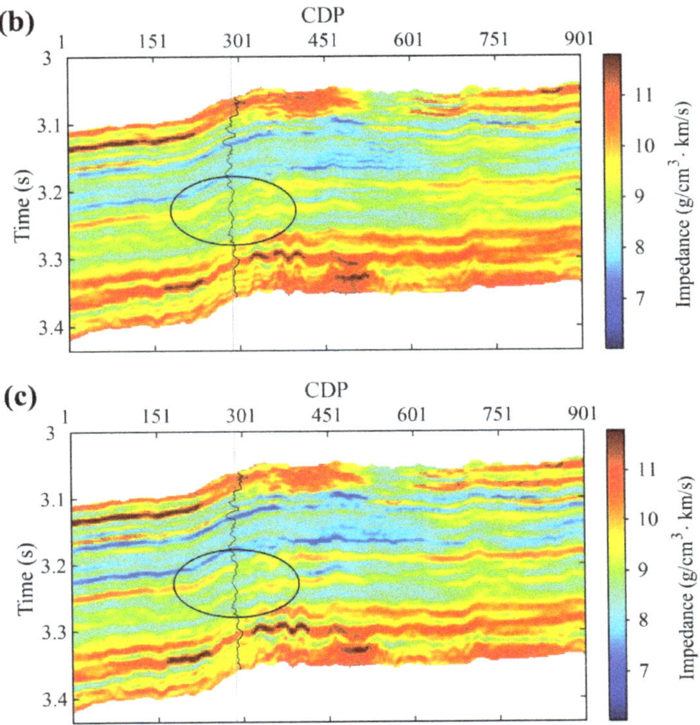

Figure 10. Comparison of inversion results by different methods for the field data: (**a–c**) present inversion results by MII, LMII, and constrained LMII, respectively. The grey and black lines represent the location of the blind well as well as the well log, respectively.

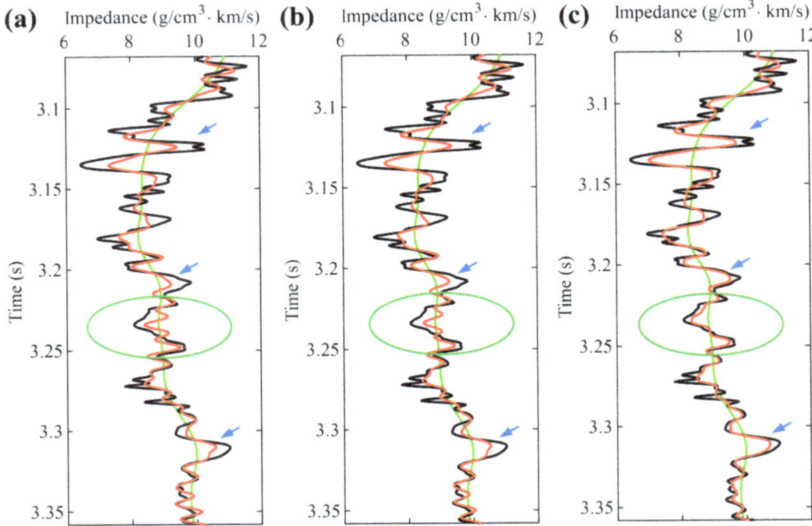

Figure 11. Comparison of the 288th trace from the inversion results of Figure 10: (**a–c**) present inversion results by MII, LMII, and constrained LMII, respectively. The green, black, and red lines represent the

low-frequency reference model, well log, and inverted impedance curve, respectively. The well log shows a low-impedance shale layer at the green circle. The MII and LMII methods produce pseudo-layer interferences, resulting in false sand-shale interbeds at this location. The constrained LMII method effectively avoids the pseudo-layer interferences and correctly identifies this formation.

4. Discussion

Investigating the sensitivity of the impedance inversion algorithm to the estimation accuracy of the source wavelet has important guiding significance for its subsequent practical application. Herein, we take Figure 2 as an example to study the sensitivity of the proposed methods to the source wavelet. Figure 12 shows the relationship curves between the quantitative evaluation index (RMSE) of inversion results by the proposed algorithms and wavelet estimation accuracy. As shown in Figure 12, the smallest RMSE value is obtained when the exact wavelet (i.e., frequency of 30 Hz) is used as the input of the LMII and constrained LMII algorithms. This means that both LMII and constrained LMII algorithms achieve the highest inversion accuracy at this time. With the decrease in the wavelet estimation accuracy, the RMSE value of inversion results yielded by the LMII and constrained LMII methods gradually increases. This result demonstrates that the performance of the LMII and constrained LMII methods has a strong dependence on the estimation accuracy of the source wavelet. Therefore, in order to achieve a satisfactory inversion result, the exact source wavelet must be extracted before applying the LMII and constrained LMII methods for impedance inversion.

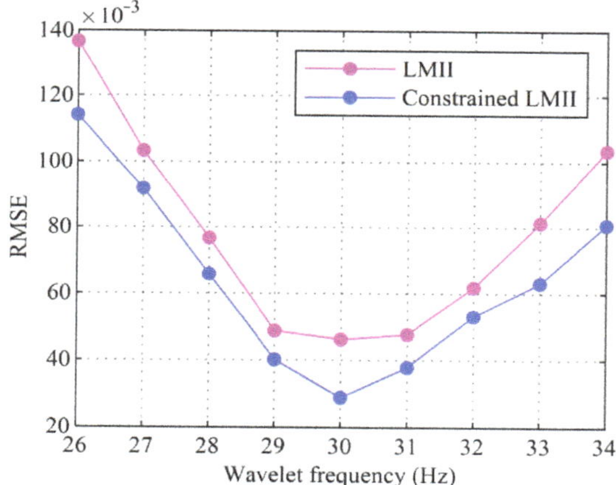

Figure 12. Relationship curves between the quantitative evaluation index (RMSE) of inversion results and wavelet estimation accuracy.

Although the developed method effectively improves the computational accuracy and efficiency of the multi-trace impedance inversion, it is not flawless and is prone to suffering from lateral smearing when dealing with highly dipping structures. The reason is that the lateral constraint term constructed by the second-order difference operator only forces lateral smoothing and ignores the morphology of underground structures [26]. Recently, structure-oriented multi-trace impedance inversion methods [26,30,31] have effectively solved the lateral smearing problem and can be better adapted to inversion work in regions with large dips. However, similar to the traditional multi-trace impedance method, these methods also suffer from insufficient inversion efficiency and accuracy. Fortunately, the inversion framework presented in this paper has excellent expansibility. Therefore, combining the proposed inversion framework with structure-oriented multi-trace inversion methods to develop a series of novel impedance inversion algorithms that can adapt to highly dipping structures with high computational efficiency and accuracy will be a future research direction.

5. Conclusions

To address the issues existing in the conventional MII method with low efficiency and accuracy, a low-dimensional multi-trace inversion method with elastic half norm constraints is proposed. First, we develop an LMII inversion framework that efficiently reduces the scale of the MII inversion problem and thus achieves efficient inversion. Concretely, this framework converts the seismic data and reference impedance into the sparse space by an orthogonal transformation and constructs the objective function with a smaller size for multi-trace inversion using the low-dimensional measurements in the sparse domain. Subsequently, on the basis of the developed LMII framework, we introduce the EH norm as a vertical constraint on the model parameters and propose a novel constrained LMII model for inverting the subsurface impedance. Since this inversion procedure considers both the smoothness and blockiness of the subsurface strata, it can significantly improve the inversion precision compared with the traditional MII method, which considers smoothness only. Finally, based on the variable splitting technique, we derive an alternating multiplier iteration algorithm to efficiently solve the constrained LMII problem. Two experiments are conducted to investigate the performance of the developed algorithms. The results indicate that the LMII framework can effectively improve the multi-trace inversion efficiency while keeping the inversion accuracy constant. Meanwhile, it is also shown that the constrained LMII model can evidently promote the quality of impedance inversion while inheriting the efficiency of the LMII framework. In addition, it should be pointed out that the proposed methods have good scalability and can be extended to gravity inversion, magnetic inversion, and other fields to provide new insights for solving more geophysical problems.

Author Contributions: Conceptualization, N.L. and F.Z.; methodology, N.L., F.Z. and K.X.; software, N.L., F.Z. and K.X.; validation, N.L., F.Z. and H.Z.; formal analysis, N.L., F.Z. and Y.L.; investigation, N.L., F.Z. and Y.L.; resources, N.L., H.Z. and K.X.; data curation, N.L., H.Z. and Y.L.; writing—original draft preparation, N.L.; writing—review and editing, N.L., F.Z., H.Z. and Y.L.; funding acquisition, F.Z. All authors have read and agreed to the published version of the manuscript.

Funding: This research is supported by the Laoshan National Laboratory of Science and Technology Foundation (No. LSKJ202203400) and the National Natural Science Foundation of China (No. 42030103).

Data Availability Statement: Not applicable.

Acknowledgments: We are grateful to the editor and all the reviewers for their constructive comments on this paper. We would also like to thank the Laoshan National Laboratory of Science and Technology Foundation and the National Natural Science Foundation of China for their financial support during this research.

Conflicts of Interest: The authors declare no conflict of interest.

References

1. Harrison, C.B. Feasibility of Rock Characterization for Mineral Exploration Using Seismic Data. Ph.D. Thesis, Curtin University, Perth, Australia, 2009.
2. Kieu, D.T.; Kepic, A. Seismic-impedance inversion with fuzzy clustering constraints: An example from the Carlin Gold district, Nevada, USA. *Geophys. Prospect.* **2020**, *68*, 103–128. [CrossRef]
3. Donoso, G.A.; Bautista, C.; Malehmir, A.; Araujo, V. Characterizing volcanogenic massive sulphides using acoustic impedance inversion, Neves-Corvo, southern Portugal. In *NSG2022 4th Conference on Geophysics for Mineral Exploration and Mining*; European Association of Geoscientists and Engineers: Belgrade, Serbia, 2022; pp. 1–5.
4. Russell, B.; Hampson, D. Comparison of poststack seismic inversion methods. In *SEG Technical Program Expanded Abstracts 1991*; Society of Exploration Geophysicists: Houston, TX, USA, 1991; pp. 876–878.
5. Dai, R.; Yin, C.; Zaman, N.; Zhang, F. Seismic inversion with adaptive edge-preserving smoothing preconditioning on impedance model. *Geophysics* **2019**, *84*, R25–R33. [CrossRef]
6. Tikhonov, A. On the solution of ill-posed problems and the method of regularization. *Dokl. Akad. Nauk. SSSR* **1963**, *151*, 501–504.
7. VanDecar, J.C.; Snieder, R. Obtaining smooth solutions to large, linear, inverse problems. *Geophysics* **1994**, *59*, 818–829. [CrossRef]
8. Sacchi, M.D. Reweighting strategies in seismic deconvolution. *Geophys. J. Int.* **1997**, *129*, 651–656. [CrossRef]

9. Gholami, A.; Siahkoohi, H.R. Regularization of linear and nonlinear geophysical ill-posed problems with joint sparsity constraints. *Geophys. J. Int.* **2010**, *180*, 871–882. [CrossRef]
10. Aghamiry, H.S.; Gholami, A.; Operto, S. Compound regularization of full waveform inversion for imaging piecewise media. *IEEE Trans. Geosci. Romote Sens.* **2020**, *58*, 1192–1204. [CrossRef]
11. Zhang, F.; Dai, R.; Liu, H. Seismic inversion based on L1-norm misfit function and total variation regularization. *J. Appl. Geophys* **2014**, *109*, 111–118. [CrossRef]
12. Gholami, A. Nonlinear multichannel impedance inversion by total-variation regularization. *Geophysics* **2015**, *80*, R217–R224. [CrossRef]
13. Li, C.H.; Zhang, F.C. Amplitude-versus-angle inversion based on the L1-norm-based likelihood function and the total variation regularization constraint. *Geophysics* **2017**, *82*, R173–R182. [CrossRef]
14. Guo, S.; Wang, H. Seismic absolute acoustic impedance inversion with L1 norm reflectivity constraint and combined first- and second-order total variation regularizations. *J. Geophys. Eng.* **2019**, *16*, 773–788. [CrossRef]
15. Dai, R.; Yang, J. An alternative method based on region fusion to solve L0-norm constrained sparse seismic inversion. *Explor. Geophys.* **2021**, *52*, 624–632. [CrossRef]
16. Zhang, R.; Castagna, J. Seismic sparse-layer reflectivity inversion using basis pursuit decomposition. *Geophysics* **2011**, *76*, R147–R158. [CrossRef]
17. Zhang, R.; Sen, M.K.; Srinivasan, S. A prestack basis pursuit seismic inversion. *Geophysics* **2013**, *78*, R1–R11. [CrossRef]
18. Wang, L.; Zhou, H.; Liu, W.; Yu, B.; He, H.; Chen, H.; Wang, N. Data-driven multichannel poststack seismic impedance inversion via patch-ordering regularization. *Geophysics* **2021**, *86*, R197–R210. [CrossRef]
19. Wang, G.; Chen, S. Pre-stack seismic inversion with L1-2-norm regularization via a proximal dc algorithm and adaptive strategy. *Surv. Geophys.* **2022**, *43*, 1817–1843. [CrossRef]
20. Ma, M.; Zhang, R.; Yuan, S.Y. Multichannel impedance inversion for nonstationary seismic data based on the modified alternating direction method of multipliers. *Geophysics* **2019**, *84*, A1–A6. [CrossRef]
21. Zhang, F.C.; Liu, J.; Yin, X.Y.; Yang, P.J. Modified Cauchy-constrained seismic blind deconvolution. *Oil Geophys. Prospect.* **2008**, *43*, 391–396.
22. Dai, R.; Zhang, F.; Liu, H. Seismic inversion based on proximal objective function optimization algorithm. *Geophysics* **2016**, *81*, R237–R246. [CrossRef]
23. Sun, J.; Li, Y. Adaptive Lp inversion for simultaneous recovery of both blocky and smooth features in a geophysical model. *Geophys. J. Int.* **2014**, *197*, 882–899. [CrossRef]
24. Hamid, H.; Pidlisecky, A. Multitrace impedance inversion with lateral constraints. *Geophysics* **2015**, *80*, M101–M111. [CrossRef]
25. Yuan, S.Y.; Wang, S.X.; Luo, C.M.; He, Y.X. Simultaneous multitrace impedance inversion with transform-domain sparsity promotion. *Geophysics* **2015**, *80*, R71–R80. [CrossRef]
26. Hamid, H.; Pidlisecky, A. Structurally constrained impedance inversion. *Interpretation* **2016**, *4*, T577–T589. [CrossRef]
27. Dai, R.; Zhang, F.; Yin, C.; Hu, Y. Multi trace post stack seismic data sparse inversion with nuclear norm constraint. *Acta Geophys.* **2021**, *69*, 53–64. [CrossRef]
28. Yang, Y.; Xia, X.; Yin, X.; Li, K.; Wang, J.; Liu, H. Data-driven fast prestack structurally constrained inversion. *Geophysics* **2022**, *87*, N31–N43. [CrossRef]
29. Karimi, P. Structure constrained relative acoustic impedance using stratigraphic coordinates. *Geophysics* **2015**, *80*, A63–A67. [CrossRef]
30. Yin, X.Y.; Yang, Y.M.; Li, K.; Man, J.; Li, H.; Feng, Y. Multitrace inversion driven by cross-correlation of seismic data. *Chin. J. Geophys.* **2020**, *63*, 3827–3837.
31. Zhang, Y.; Wu, W.; Zhang, M.; Liang, M.; Feng, B. Multitrace Impedance Inversion Based on Structure-Oriented Regularization. *IEEE Geosci. Remote Sens. Lett.* **2021**, *19*, 1–5. [CrossRef]
32. Donoho, D.L. Compressed sensing. *IEEE Trans. Inf. Theory* **2006**, *52*, 1289–1306. [CrossRef]
33. Baraniuk, R.G. Compressive sensing. *IEEE Signal Process. Mag.* **2007**, *24*, 118–121. [CrossRef]
34. Lan, N.Y.; Zhang, F.C.; Li, C.H. Robust high-dimensional seismic data interpolation based on elastic half norm regularization and tensor dictionary learning. *Geophysics* **2021**, *86*, V431–V444. [CrossRef]
35. Lions, P.L.; Mercier, B. Splitting algorithms for the sum of two nonlinear operators. *SIAM J. Numer. Anal.* **1979**, *16*, 964–979. [CrossRef]
36. Lan, N.Y.; Zhang, F.C. Seismic data recovery using deep targeted denoising priors in an alternating optimization framework. *Geophysics* **2022**, *87*, V279–V291. [CrossRef]
37. Berteussen, K.; Ursin, B. Approximate computation of the acoustic impedance from seismic data. *Geophysics* **1983**, *48*, 1351–1358. [CrossRef]
38. Barrett, R.; Berry, M.; Chan, T.F.; Demmel, J.; Donato, J.M.; Dongarra, J.; Eijkhout, V.; Pozo, R.; Romine, C.; Van der Vorst, H. *Templates for the Solution of Linear Systems: Building Blocks for Iterative Methods*; Society for Industrial and Applied Mathematics: Philadelphia, PA, USA, 1994; pp. 64–68.
39. Candes, E.; Romberg, J.; Tao, T. Robust uncertainty principles: Exact signal reconstruction from highly incomplete frequency information. *IEEE Trans. Inf. Theory* **2006**, *52*, 489–509. [CrossRef]

40. Huang, H.; Misra, S.; Tang, W.; Barani, H.; Al-Azzawi, H. Applications of compressed sensing in communications networks. *arXiv* **2014**, arXiv:1305.3002.
41. Xu, Z.B.; Chang, X.Y.; Xu, F.; Zhang, H. $L_{1/2}$ regularization: A thresholding representation theory and a fast solver. *IEEE Trans. Neural Netw. Learn. Syst.* **2012**, *23*, 1013–1027.
42. Sorensen, H.V.; Jones, D.L.; Burrus, C.S.; Heideman, M.T. On computing the discrete Hartley transform. *IEEE Trans. Acoust. Speech Signal Process.* **1985**, *33*, 1231–1238. [CrossRef]

Article

Research on the Tectonic Characteristics and Hydrocarbon Prospects in the Northern Area of the South Yellow Sea Based on Gravity and Magnetic Data

Wenqiang Xu [1], Changli Yao [1,*], Bingqiang Yuan [2], Shaole An [3], Xianzhe Yin [1] and Xiaoyu Yuan [1]

[1] School of Geophysics and Information Technology, China University of Geosciences, Beijing 100083, China; xuwq@email.cugb.edu.cn (W.X.); yinxz@email.cugb.edu.cn (X.Y.); xiaoyuyuan1991@gmail.com (X.Y.)

[2] School of Earth Sciences and Engineering, Xi'an Shiyou University, Xi'an 710065, China; yuanbingqiang@sohu.com

[3] Xinjiang Research Centre for Mineral Resources, Xinjiang Institute of Ecology and Geography, Chinese Academy of Sciences, Ürümqi 830011, China; slan16@ms.xjb.ac.cn

* Correspondence: clyao@cugb.edu.cn

Abstract: To further explore the geological structure and the Mesozoic–Paleozoic hydrocarbon prospects in the northern area of the South Yellow Sea (SYS), multiple geological and geophysical data were systematically gathered and compiled, including gravity and magnetic data, seismic surveys, drilling data, and previous research results. The characteristics and genesis of the gravity and magnetic anomalies are examined. This study employs residual gravity anomalies and multiple edge detection methods to identify fault lineament structures and assess the tectonic framework. Moreover, the study utilizes 2.5D gravity-seismic joint modellings and regression analysis to estimate the basement depth. Additionally, the study examines the basement characteristics and discusses the thickness of the Mesozoic–Paleozoic strata. Finally, the study further identifies prospects for hydrocarbons in the Mesozoic–Paleozoic. Our findings show that the faults are incredibly abundant and that the intensity of fault activity weakens gradually from NW to SE. Specifically, NE (NEE) trending faults are interlaced and cut off by NW (NWW), near-EW, and near-SN trending secondary faults, which form an en-echelon composite faults system with a dominant NE (NEE) orientation. Thick Mesozoic–Paleozoic strata are preserved, but we observe distinct variations in basement characteristics and the pre-Cenozoic structural deformation along the N-S direction. Therefore, the Northern Basin of SYS (NBSYS) and the Middle Uplift of SYS (MUSYS) are characterized by alternating sags and bulges in the S-N direction and in the E-W direction, respectively, forming a chessboard tectonic framework. Considering the oil and gas accumulation model, we identify three target hydrocarbon prospects in the NBSYS and two favorable hydrocarbon prospects in the MUSYS.

Keywords: gravity modelling; gravity and magnetic anomalies; tectonic characteristic; hydrocarbon prospect; the northern area of the South Yellow Sea

Citation: Xu, W.; Yao, C.; Yuan, B.; An, S.; Yin, X.; Yuan, X. Research on the Tectonic Characteristics and Hydrocarbon Prospects in the Northern Area of the South Yellow Sea Based on Gravity and Magnetic Data. *Minerals* **2023**, *13*, 893. https://doi.org/10.3390/min13070893

Academic Editors: Luan Thanh Pham, Saulo Pomponet Oliveira and Le Van Anh Cuong

Received: 13 May 2023
Revised: 23 June 2023
Accepted: 27 June 2023
Published: 30 June 2023

1. Introduction

The northern area of the South Yellow Sea (SYS) is mainly located in the Lower Yangtze Plate (LYP) east of the Tancheng–Lujiang fault zone (TLFZ), the north of which is close to the Shandong Peninsula (SP). This belongs to the Sino–Korean Plate (SKP), and the south of the SKP is tightly connected to the South China Plate (SCP) (Figure 1). Large numbers of geological and geophysical investigations for hydrocarbon exploration have been conducted in the SYS since the 1960s. Over the years, considerable geological and geophysical data have been collected. Exploration practices and research results indicate that the SYS is rich in hydrocarbons and other potential resources [1–13]. However, since the pre-Sinian period, the Mesozoic-Paleozoic marine sedimentary basin of the SYS has experienced the forceful action of multi-period tectonic movements, including the Caledonian, Hercynian, Indosinian, Yanshan, and Himalayan movements [4–7,14,15]. Moreover, the strata have experienced multiple tectonic events, including extrusion uplift and denudation, extensional rift sedimentation, and igneous rock intrusion across different ages, resulting in large structural fluctuations, fault development, significant lateral variations, and substantial heterogeneity. The pre-Cenozoic residual basins developed through these processes, creating a complex sedimentary system, fault system, and distinctive tectonic evolution characteristics [5,6,15–18]. As a result, these processes have made it significantly more challenging to understand the geological structure and to prospect for hydrocarbon resources in the pre-Cenozoic residual basins.

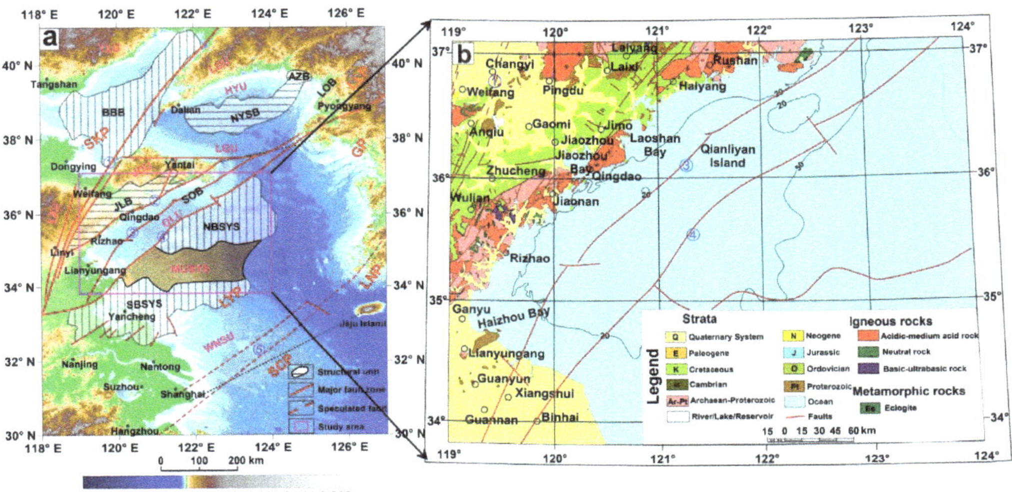

Figure 1. Map of geotectonic locations in the SYS and adjacent areas: (**a**) geological sketch map of the study area modified from [18]; (**b**) modified from [19]; topography and bathymetry data were derived from SRTM30_PLUS V11, which is sourced from the Scripps Institution of Oceanography, the University of California San Diego [20]. ① Tancheng–Lujiang Fault zone (TLFZ); ② Wulian–Qingdao–Rongcheng Fault zone (WQRFZ); ③ Qianliyan Fault zone (QLFZ); ④ Jiashan–Xiangshui Fault zone (JXFZ); ⑤ Jiangshan–Shaoxing Fault zone (JSFZ). SKP—Sino-Korean Plate; LYP—Lower Yangtze Plate; SCP—South China Plate; LNP—Lingnan Plate; GP—Gyeonggi Plate; LLP–Langlin Plate; SYS—the South Yellow Sea. YLU—Yanliao Uplift; BBB—Baohai Bay Basin; LDU—Liaodong Uplift; HYU—Haiyangdao Uplift; AZB—Anzhou Basin; NYSB—the North Yellow Sea Basin; LGU—Liugongdao Uplift; JDU—Jiaodong Uplift; LXU—Luxi Uplift; JLB—Jiaolai Basin; SOB—Sulu Orogenic Belt; LOB—Linjinjiang Orogenic Belt; QLU—Qianliyan Uplift; NBSYS—the Northern Basin of the South Yellow Sea; MUSYS—the Middle Uplift of the South Yellow Sea; SBSYS—Subei-the Southern Basin of the South Yellow Sea; WNSU—Wunansha Uplift.

In recent years, several studies have utilized seismic and drilling data to investigate geological structures, sedimentary layers, and oil and gas potential [2–7,10,11,16,17,21–28]. However, the Paleozoic can suffer from low reflectivity in deep seismic imaging due to the strong shielding effect of the carbonate's powerful reflection interface in a seismic profile, the weak reflection amplitude of the target strata, the low signal-to-noise ratio, and the poor lateral resolution. Although the results of two wide-angle seismic profiles have also been applied to the research of the SYS [29–32], they have only revealed the distribution of sediments, structural characteristics, and the splicing relationship between the plates in the local area. Furthermore, the geological features of the Mesozoic–Paleozoic are challenging to reveal due to the scarcity of drillings, which only comprise 26 wells with shallow depths.

Thus, all these factors have hindered the effective determination of geological features, such as fault structures, tectonic evolution, the Mesozoic–Paleozoic distribution, and basement properties.

On the other hand, gravity and magnetic data play significant factors in regional geological structures and hydrocarbon exploration research. Refs. [33–35] effectively detected oil and gas potential areas by employing the normalized full gradient (*NFG*) method, which has been commonly used in structural studies. For example, Ref. [36] used the improved gravity *NFG* method to detect hydrocarbon reservoirs in the Shengli oil field in eastern China. Ref. [37] conducted a comprehensive investigation of the geological structure and oil and gas properties of the Hasankale Horasan petroleum exploration province, employing the *NFG* method in conjunction with the Filon method. Ref. [38] identified the hydrocarbon exploration targets in the Tabas basin in Yazd province, eastern Iran, utilizing the gravity *NFG* method.

Furthermore, significant progress has been achieved in studying geological structures and other aspects by applying gravity and magnetic data. For instance, Ref. [39] applied gravity and magnetic data to study the structural characteristics and oil and gas prospects in the Tobago Basin. Ref. [40] analyzed the geological structure of the Red Sea region through the use of gravity and magnetic data. Ref. [41] used gravity data and multiple field boundary detectors to map the detailed structure of the Wadi Umm Ghalqa area in the southeastern desert of Egypt. Ref. [42] described the lineament of the Thua Thien Hue area using Bouguer Gravity anomaly data. Ref. [43] used gravity and magnetic data to determine the structural characteristics of the Qattara Compression area in Egypt. Ref. [44] utilized Bouguer gravity data and 15 field boundary enhancement techniques to better understand the structural features of the southern Red Sea. Ref. [45] interpreted the gravity data of the study area to investigate the sedimentary cover and structural configuration in the eastern portion of the Suez area, Sinai, and Egypt. Ref. [46] employed advanced gravity edge detection methods to map the structural configuration of the western part of the Gulf of Guinea. Likewise, significant progress has been made in researching geological structures, hydrocarbon potential, and other aspects in the SYS using gravity and magnetic data [14,47–57]. Nevertheless, considerable controversies still exist in understanding the fault characteristics, basement structure, and stratigraphic distribution due to the long-time span, inconsistent quality, data precision, poor short-wave signal, and single or limited gravity and magnetic data processing methods.

Exploration for industrial hydrocarbon resources in the SYS remains challenging, despite substantial geophysical and geological research on faults, basement structure, stratigraphic distribution, and oil and gas prospects. Several fundamental reasons underlie this stagnation: there is relatively insufficient geophysical work; the target layers for oil and gas prospecting are not yet thoroughly determined; and little attention is paid to uplift areas or bulge belts. Furthermore, a unified understanding of geological problems such as fault systems, basement structure, and stratigraphic distribution is lacking. It is imperative to carry out comprehensive geological-geophysical research to study these geological features, including structural characteristics, basement structure, and the distribution of the Mesozoic–Paleozoic, and reduce the ambiguity of geophysical methods. Therefore, we integrate recent gravity and magnetic data with existing aero-gravity data (1:250,000) [54] for corresponding geophysical and geological interpretation. This study redetermines and constructs the faults system using high-precision geopotential field separation methods and edge detection techniques. By combining this information with known seismic constraint data, the regression analysis method is used to calculate the basement depth, from which the Mesozoic–Paleozoic thickness is obtained. In addition, we analyze the stratigraphic development characteristics and structural contact relationship. Then, based on these analyses, the oil and gas prospecting areas are determined. Hence, our findings provide the primary geophysical basis for exploring the structure and new strata resources of the Mesozoic–Paleozoic in the SYS.

2. Geological and Geophysical Setting

2.1. Geological Setting

The study area is predominantly situated in LYP in the east of the Sulu Orogenic Belt [58]. It has undergone the consolidated diagenesis and reactivation of the Late Sibao Movement in the Mesozoic–Proterozoic and the Jinning Movement in the Neo-Proterozoic, forming a two-layer crystalline basement of Meso-Neo-Proterozoic shallow metamorphic rock and Archean–Paleo-Proterozoic deep metamorphic rock [59,60]. Under the influence of the multi-period tectonic movement reformations and the action of various tectonic stresses, the study area has successively experienced various types of tectonic activity, such as stable sedimentation, gradual collage, oscillation migration, collision orogenesis, extrusion folds, compressive stress and tensile stress conversion, extensional fault depressions, and strike-slip subsidence [17]. Overall, it represents a marine-continental multi-cycle superimposed petroliferous basin formed based on a pre-Sinian metamorphic basement [5,6,14,17,21,25]. Simultaneously, frequent tectonic-magmatic activities and sustained magmatism have led to the development of multi-period igneous rocks. These strata and rocks are extensively distributed in the surrounding areas in the northern region of the SYS, of which Archean and Proterozoic magmatic rocks dominate basement igneous rocks, and igneous rock masses in the sedimentary layers are mainly rock stocks and rock branches. Although rocks of various properties are present, acidic and medium-acidic rocks remain dominant [17]. In addition, the complex fault development and tectonic patterns determine the distribution of various geological features, such as uplift and depression structures (secondary sag-bulge structures), strata, igneous rocks, and hydrocarbon resources.

Until now, according to drilling and seismic data, three sets of tectonic sequences have been revealed in the SYS: marine strata such as the Sinian, Paleozoic, and Middle–Lower Triassic; continental faulted formations such as the Upper Triassic, Jurassic, Cretaceous, Paleogene; and the Neogene depression sedimentary layer and Quaternary marine sedimentary structural layer. Ref. [61] studied the geothermal field characteristics and determined that the average heat flow value was 67.0 mW/m^2, which is slightly higher than the global average heat flow value of 65.0 mW/m^2. The study also revealed the thermal background characteristics of "cold crust and hot mantle". The heat flow value formed by the combination of various sedimentary layers was 5.85 mW/m^2, which comprised 8.9% of the surface heat flow. These chiefly influences the maturity of organic matter in the sedimentary layers, which is conducive to forming hydrocarbon resources. Extensive research has shown that the Mesozoic–Paleozoic provides excellent material foundation and geological structure conditions for hydrocarbon accumulation. Hence, it holds considerable potential for hydrocarbon exploration [5–12,17,62–67]. Moreover, this paper also summarizes the features of the sources, reservoirs, and cap layers in the Mesozoic-Paleozoic. As a whole, the tectonic activity over multiple periods contributed to the formation of five sets of source rocks, three sets of reservoirs, and five sets of cap layers in the Mesozoic–Paleozoic. These formations formed three sets of complete source-reservoir-cap assemblages, which exhibit characteristics of multi-source and multi-period oil and gas accumulations.

2.2. Physical Properties

The physical property data of the northern area of the SYS are principally sourced from the neighbouring SP and northern Jiangsu region, while some physical properties are acquired from the drillings and the conversion of seismic wave velocity and density in the sea (Figure 2) [17,50–52,54,55,68,69]. Based on previous research, we propose that the density of the strata (rocks) in the study area gradually decreases with sedimentary age, from old to new, and with depth, from deep to shallow (Figure 2a). However, there is a reverse phenomenon in the Mesozoic and Paleozoic. Igneous rock densities increase gradually from acidic to ultrabasic, ranging from 2.63 to 2.82 × 10^{-3} kg/m^3 (abbreviated as 2.63~2.82, the same below) (Figure 2b). Metamorphic rock densities increase with the degree of metamorphism, and the density values of shallow and deep metamorphic rocks range from 2.67 to 2.86 (Figure 2b). However, strata (rocks) densities in different structural units are different in horizontal and vertical directions according to the comprehensive statistical analysis of this study, which is prone to cause a strong gravity effect. While the macro density models exhibit general similarities across the northern region of the SYS and its surrounding areas, the simple density model can better represent the actual situation of the major density interfaces causing gravity anomalies. The strata can be, therefore, divided into four density layers and three density interfaces (Figure 2a).

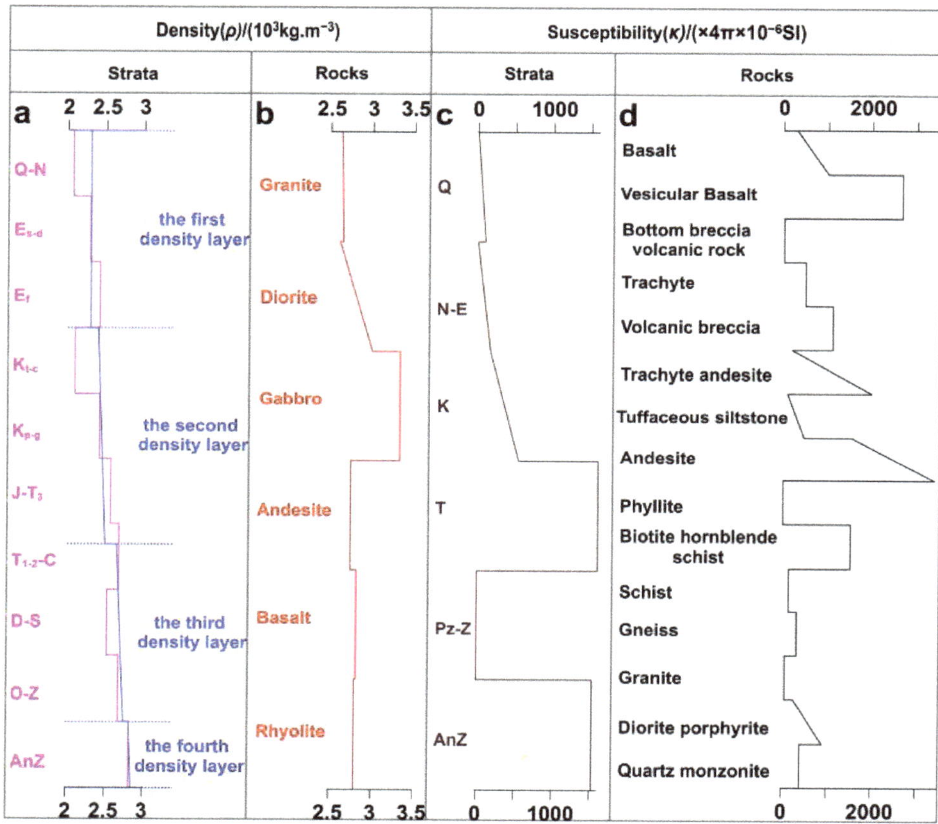

Figure 2. Histogram of strata (rocks) density and susceptibility in the northern area of the SYS: (**a**) histogram of sediment density; (**b**) histogram of igneous rocks density; (**c**) histogram of sediment susceptibility; and (**d**) histogram of igneous rocks susceptibility.

The Cenozoic, Mesozoic, and Paleozoic normally have either weak or no magnetization, and only a few strata have moderate to strong magnetization (Figure 2c). For example, the susceptibility of the Upper Jurassic andesite is as high as $3410 \times 10^{-6} \times 4\pi$ SI (abbreviated as 3410, the same below), which is classified as a relatively strong magnetization. In contrast, the pre-Sinian crystalline basement composed of the Mesozoic–Paleozoic–Proterozoic and Archean has medium to strong magnetization. In addition, medium and shallow metamorphic rocks ordinarily present either negligible or weak magnetization, whereas deep metamorphic rocks generally display strong magnetization, with a susceptibility value exceeding 3000. In addition, the susceptibilities of igneous rocks are usually strong and vary greatly (Figure 2d). For instance, ultrabasic, basic, neutral intrusive rocks, and extruded rocks possess strong susceptibilities (especially basic and ultrabasic rocks have the strongest susceptibility), ranging primarily between 1000 and 3000. However, the susceptibility of acidic igneous rocks is commonly between the range of 500 and 1000, while granite porphyry reveals very weak magnetization, with a susceptibility of 5~500, and the magnetism of medium-acid igneous rocks is between that of the neutral and acid rocks. In addition, the Cenozoic and Mesozoic have weak or no magnetization in the eastern Shandong area on the land, while the Mesozoic Qingshan Group is composed of eruptive volcanic rocks with uneven susceptibility and significant variations. In igneous rocks, the ancient basic and ultrabasic complexes manifest strong susceptibility, and the Mesozoic igneous rocks present the characteristics of uneven susceptibility; however, the susceptibility dispersion of basic, medium and acid igneous rocks vary widely. The susceptibility of various intrusive rocks in the Yanshanian Period is uneven and varies widely, while the susceptibility of intrusive rocks in the Indosinian and Sinian Periods is weak and uniform.

3. Data and Methodology

3.1. Data

The gravity and magnetic data obtained from the National Geological Archives of the China Geological Survey are grid longitude and latitude. When integrating data, researchers perform various correction processes to remove deviations from between types of data. Gravity data consists of aero-gravity, land gravity, and marine shipborne gravity data. Among these, aero-gravity data are the major type (blue dashed range, Figure 3), and land gravity and marine shipborne gravity data supplement at the periphery. The magnetic data consist of aeromagnetic and marine magnetic survey data. In addition, WGS84 coordinate conversion is conducted on the gravity and magnetic data, in which the Gauss-Krüger projection is used to project to a plane coordinates system. Using the minimum curvature method, gravity and magnetic data are interpolated into a 2 km × 2 km grid Bouguer gravity anomaly (Figure 3a) and magnetic anomaly, and the magnetic anomaly data are reduced to the pole (RTP parameters: geomagnetic declination: −6.09°, geomagnetic inclination: 52.14°) (Figure 4a). Finally, the gravity and magnetic data are evaluated. The resolution of the gravity and magnetic data is equivalent to 1:500,000, which is higher than previously (1:1,000,000).

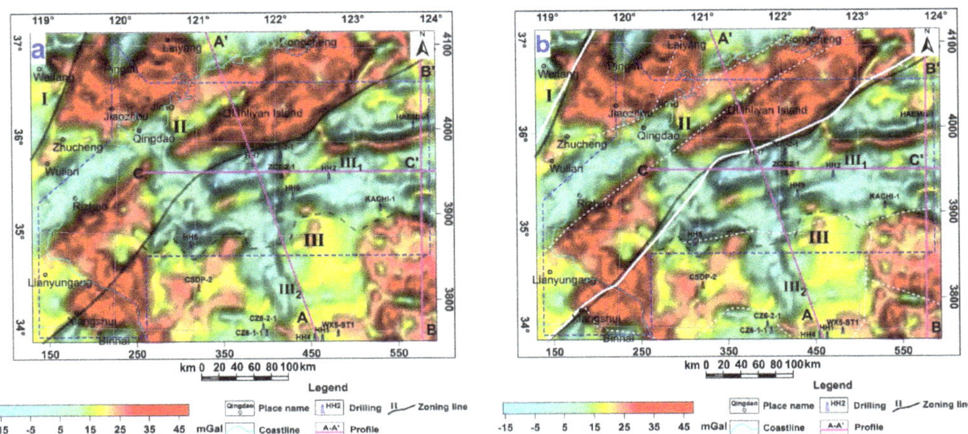

Figure 3. Map of Bouguer gravity anomaly: (**a**) original zoning map; (**b**) zoning comparison map. The blue dashed line represents the aero-gravity range; the purple line represents the gravity modelling profile; the grey line represents the latitude and longitude; the white line represents the zoning result of [52]; and the black line represents the result of this zoning.

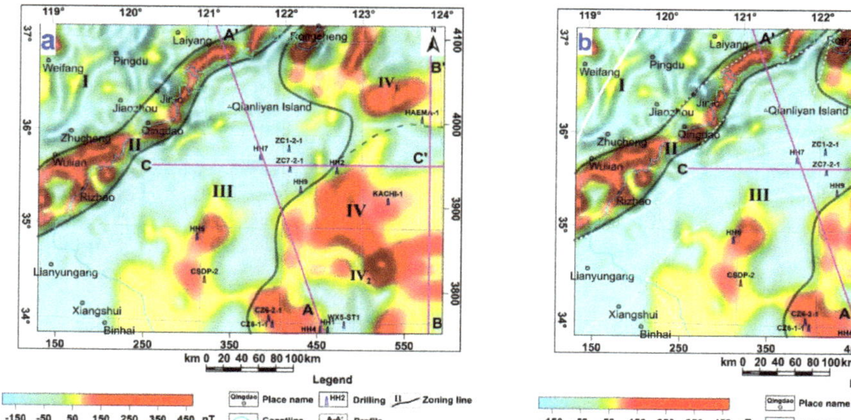

Figure 4. Map of RTP magnetic anomaly: (**a**) original zoning map; (**b**) zoning comparison map. The purple line represents the gravity modelling section; the grey line represents the latitude and longitude; the white line represents the zoning result of [52]; and the black line represents the result of this zoning.

3.2. Methodology

The methods applied in this paper include regularization filtering (*RF*), vertical second derivative (*VSDR*), normalized vertical derivative of total horizontal derivative (*NVDR-THDR*), normalized standard deviation (*NSTD*), enhanced gradient amplitude (*EHGA*), improved logistic filter (*IL*), the constrained inversion of the *2.5D* gravity profile, and regression analysis (*RA*).

3.2.1. Regularization Filtering (*RF*)

RF is a geopotential field separation method proposed by Ref. [70] based on the regularization theory [71]. The stable regularization factor can be expressed as follows:

$$f_\alpha^{mn} = 1 / \left\{ 1 + \alpha e^{\beta(f-f_0)\lambda_x} \right\} \tag{1}$$

where $\beta \geq 2$; f_0 refers to the minimum wavenumber of the high-frequency interference to be eliminated, which is equal to the reciprocal of its largest horizontal size λ_0^{-1}. $f = \sqrt{m^2 + n^2 \cdot (\lambda_x/\lambda_y)} / \lambda_x$, λ_x and λ_y are the fundamental wavelengths; the wavenumbers are $u = \frac{m}{\lambda_x}$, $v = \frac{n}{\lambda_y}$, $f = \sqrt{u^2 + v^2}$, ($m = 0, 1, \ldots, M - 1$; $n = 0, 1, \ldots, N - 1$). In addition, the regularization parameter α is determined through theoretical model tests and practical data applications [39,71], and in the 2D stable regularizing filter, α usually can be set as 2–3 ($2 \leq \alpha \leq 3$). In this paper, α is set as 2.0, and $\beta \geq \alpha$.

This method uses regularization stable filtering factors to perform low-pass filtering on the geopotential field data, which can obtain anomalies with different filter scales (the size of the local field to be eliminated) from the gravity and magnetic geopotential field. Then high-frequency and low-frequency signals can be separated. The filter has an ideal low-pass filtering feature that can be used to filter the high frequency for all physical quantity evaluations when using the discrete Fourier transform. On the other hand, the filter can obtain a stable approximation, and the actual data processing effect is relatively good [67,69]. Therefore, it is a method with ideal low-pass filtering characteristics and strong adaptability.

3.2.2. Vertical Second Derivative (*VSDR*)

The calculation of vertical derivative is widely used in the geological interpretation of gravity and magnetic fields. Theoretical models prove that the calculation results of *VSDR* can more accurately describe the edge position of the geological body than the calculation results of the first vertical derivative. *VSDR* can suppress the anomaly influences caused by deep regional geological factors, thus highlighting the anomaly characteristics of small and shallow structures and distinguishing the superimposed anomalies caused by different sizes and depths of anomaly bodies. The position of the zero value can be used to detect the edge position of anomaly bodies [72,73].

There are many different numerical formulas for calculating *VSDR*. The derivation principles of these formulas are all approximate solutions using the technical approximation method, and only the processing methods are different, which causes the effect of the calculation results to be different. In particular, the Rosenbach formula retains the R^4 term in the derivation and directly solves *VSDR*, which has the advantage of high calculation accuracy. Therefore, the Rosenbach formula [74] is applied in this study:

$$g_{zz} = \frac{1}{24R^2}\left[96g(0) - 72g(R) - 32g\left(\sqrt{2}R\right) + 8g\left(\sqrt{5}R\right)\right] \tag{2}$$

where R refers to the number-picking radius, and g refers to the gravity and magnetic geopotential field.

3.2.3. Normalized Vertical Derivative of Total Horizontal Derivative (*NVDR-THDR*)

The *NVDR-THDR* edge detection method is proposed by Ref. [75]. This method integrates the characteristics of the total horizontal derivative, n-order vertical derivative, and total horizontal derivative peak methods. Firstly, the n-order vertical derivative is obtained by calculating the total horizontal derivative, and a threshold value greater than zero is applied to calculate the peak value of the total horizontal derivative. Secondly, the ratio of the peak value of the normalized total horizontal derivative to the total horizontal derivative is used to calculate the total horizontal derivative. The detailed calculation process is as follows:

(1) Total horizontal derivative (*THDR*) of gravity and magnetic data is calculated:

$$THDR(x,y) = \sqrt{\frac{\partial f(x,y)}{\partial x}^2 + \left(\frac{\partial f(x,y)}{\partial y}\right)^2} \tag{3}$$

where $f(x,y)$ refers to the gravity and magnetic field.

(2) n-order vertical derivative (VDR_n) of total horizontal derivative (*THDR*) is calculated:

$$VDR_n(x,y) = \frac{\partial^n THDR(x,y)}{\partial z^n} \tag{4}$$

n refers to the order of vertical derivative, n = 1, 2, 3, ... Moreover, the larger the order, the higher the lateral resolution. Theoretically, it is more appropriate for gravity and magnetic anomalies when n is 2.

(3) Threshold values greater than 0 are used to calculate the peak value of total horizontal derivative (*PTHDR*):

$$PTHDR(x,y) = \begin{cases} 0, VDR_n(x,y) < 0 \\ VDR_n(x,y) \geq 0 \end{cases} \tag{5}$$

(4) The ratio of total horizontal derivative peak (*PTHDR*) to total horizontal derivative (*THDR*) is calculated:

$$VDR - THDR(x,y) = \begin{cases} 0, PTHDR(x,y) < 0 \\ \frac{PTHDR(x,y)}{THDR(x,y)}, PTHDR(x,y) \geq 0 \end{cases} \tag{6}$$

(5) The maximum value of the vertical derivative of the total horizontal derivative $(VDR - THDR_{max})$ is calculated, and the maximum value is used to normalize the vertical derivative of a total horizontal derivative. Eventually, *NVDR-THDR* is acquired.

$$NVDR - THDR(x,y) = \frac{VDR - THDR(x,y)}{VDR - THDR_{max}} \tag{7}$$

The *NVDR-THDR* method identifies the fault structures with the position of the maximum values and improves the lateral resolution of the total horizontal derivative. Threshold technology is used to eliminate the information outside the edge of the geological body, making the lineament structure anomaly simple, straightforward, and easy to determine.

3.2.4. Normalized Standard Deviation (*NSTD*)

The *NSTD* method is a new method proposed by Ref. [76] from the perspective of mathematical statistics. The standard deviation is slight when the data are relatively smooth; the anomaly will be immense when data varies greatly (such as in the presence of boundaries). Therefore, this method is also essential for enhancing the edge of gravity and magnetic anomalies and has received extensive attention. The formula can be expressed as:

$$NSTD = \frac{\sigma\left(\frac{\partial \Delta f}{\partial z}\right)}{\sigma\left(\frac{\partial \Delta f}{\partial x}\right) + \sigma\left(\frac{\partial \Delta f}{\partial y}\right) + \sigma\left(\frac{\partial \Delta f}{\partial z}\right)} \tag{8}$$

where Δf refers to the gravity and magnetic field, and $\sigma(\ )$ refers to the standard deviation of the correlation quantity in a moving window of size $m \times n$. The calculation result of the centre point in the current filter window is further determined by taking the ratio of the standard deviation of the vertical first derivative to the sum of the standard deviations of the first derivative in the x, y, and z directions. Its essence is to detect and enhance the field source boundary by utilizing the characteristics of the boundary position with the drastic change in gravity and magnetic anomalies and the significant standard deviation. The sliding window is traversed on the whole grid data to obtain the calculation results of all grid points. Finally, the maximum value positions of the results are obtained, which are the edge positions of the identified geological body.

3.2.5. Enhanced Gradient Amplitude (*EHGA*)

Ref. [77] proposed *EHGA* (Enhanced Gradient Amplitude) to accurately determine the horizontal boundary of the field source. This method is based on calculating the derivatives of the gradient amplitude of the gravity data. The maxima in the *EHGA* map is used to delineate the horizontal boundary of the source. The formula for the edge detector is as follows:

$$EHGA = Re\left(asin\left(p\left(\frac{\frac{\partial HG}{\partial z}}{\sqrt{\left(\frac{\partial HG}{\partial x}\right)^2 + \left(\frac{\partial HG}{\partial y}\right)^2 + \left(\frac{\partial HG}{\partial z}\right)^2}} - 1\right) + 1\right)\right) \tag{9}$$

with

$$HG = \sqrt{\left(\frac{\partial F}{\partial x}\right)^2 + \left(\frac{\partial F}{\partial y}\right)^2} \tag{10}$$

where Re refers to the real part, and p is a positive dimensionless scalar number defined by the interpreter and is a constant greater than or equal to 2 [77]. This study used $p = 3$ to yield the best results [41,46,78].

3.2.6. Improved Logistic Filter (*IL*)

Another method introduced by Ref. [79] is based on the logic function of analyzing signals [80] to balance differential signals, using maxima to delineate the horizontal source boundaries. This is called the improved logic function (*IL*) and is represented by the following formula:

$$IL = \frac{1}{1 + exp[-p(R_{HG} - 1) + 1]} \tag{11}$$

where R_{HG} is the ratio of the vertical gradient to the horizontal gradient amplitude of *HG*, given by the equation formula:

$$R_{HG} = \frac{\frac{\partial HG}{\partial z}}{\sqrt{\left(\frac{\partial HG}{\partial x}\right)^2 + \left(\frac{\partial HG}{\partial y}\right)^2}} \tag{12}$$

and p is a normal number defined by the interpreter. Model validation shows that an p-value between 2 and 5 will yield the optimal results [79].

3.2.7. Constrained Modelling of 2.5D Gravity Profile

The modelling technology of the 2.5D gravity anomaly profile is an important method for identifying structural features, stratigraphic densities, and rock mass distributions. This method employs a group of 2.5D prisms to simulate underground geological body structures with different densities. The technique combines model parameters modified through human-machine interactions and automatic computer iterative calculation to invert and forward the gravity anomalies of the profile. Moreover, the methodology involves forward calculating gravity field modelling through human-computer interaction and nonlinear optimization inversion solutions. A nonlinear optimization method is based on linearizing the inversion residual objective function, which applies the generalized inverse theory and solves the linear equations using singular value decomposition. Thus, this process yields physical properties and morphological corrections of the model, which are then updated to fulfill the inversion purpose. In practice, we first construct the initial geometry density model using preliminary geometric modelling obtained from seismic data and initial physical property information. The geometric and physical parameters of the model are then iteratively modified by using the residual between the measured gravity and the forward gravity. When the residual between the measured gravity and the forward gravity of the initial model is gradually reduced to the set threshold, the modification is stopped, and the geometric shape and physical properties of the model are obtained. Finally, the technique forms the gravitational geological interpretation profile in combination with the existing geological data.

3.2.8. Regression Analysis (*RA*)

The Parker-Oldenburg interface inversion method [81,82] is widely used to estimate the continuous density interface of horizontally uniform density strata; however, the stability of this method significantly decreases in the presence of complex tectonic structures. Since the Cenozoic sediments remain stable in the study area, the fluctuation and physical properties of strata change insignificantly. Therefore, the depth of the Cenozoic interface is calculated by the Parker-Oldenburg inversion method. Nevertheless, the Mesozoic–Paleozoic structures are characterized by complex features with numerous faults, and the basement fluctuation and physical properties across the north-south direction are significantly different; thus, the Parker-Oldenburg inversion method is not applicable. *RA* is a simple and effective method for calculating strata interface inversion, which is ideal for use in areas with complex structures and a lack of constraints. Refs. [39,83] used *RA* to calculate the basement depth of the Tobago basin and Pannonian basin, respectively, achieving good results. We found that the Bouguer gravity anomalies exhibited a significant correlation with the basement depth in the study area. Consequently, we utilized *RA* to estimate the basement interface of the study area and further acquire the Mesozoic–Paleozoic thickness.

4. Results and Analysis

4.1. Characteristics and Geological Significance of Gravity and Magnetic Anomalies

4.1.1. Characteristics and Geological Significance of Bouguer Gravity Anomaly

The Bouguer gravity anomaly (Figure 3) indicates that the study area exhibits anomaly zoning characteristics; the northern part has high gravity anomaly features, while the southern part has lower values. Notably, the edge regions of the study area display large-area mass high anomaly values, while the basin anomalies show low values. Moreover, the high and low anomalies exhibit significant fluctuation, with an amplitude contrast of 64 mGal. Furthermore, different sizes of horizontal gradient zones form apparent zoning characteristics, which mainly trend to the NE direction, as well as locally NW, near-EW, and near-SN directions. Based on the shape of the anomaly and the gradient zone range, it can be divided into three gravity anomaly zones: the western complex anomaly zone (I), the central high anomaly zone (II), and the eastern high–low alternating anomaly zone (III). These are generally similar to the zoning results of [52].

The western complex anomaly zone (I) is in the eastern part of the southwestern Shandong Uplift Fold Belt. It is characterized by low regional anomalies, accompanied by alternating high and low anomalies locally. In addition, the primary anomaly trend is towards the NE direction, with NW trending anomalies locally present. According to Ref. [84], the NE trending anomalies stem from the large-scale strike-slip activity of the TLFZ, while the NW trending anomalies are due to the tectonic transition from extrusion to tension. Hence, by analyzing the configuration and orientation of extensive gravity anomalies on land, the evolution of the regional tensile stress field, and the alteration of the strike-slip direction within the TLFZ, we infer that the NE trending structures likely formed at an earlier stage, whereas the NW trending tectonic deformations emerged at a relatively later stage.

The central high anomaly zone (II) is located in the holding area of TLFZ and JXFZ, and the overall anomaly trend is in the NE direction, with a regional high–low interphase distribution. A high anomalies area predominates this anomalies zone, and the anomaly shapes manifest wide and gentle features and large amplitudes. Low anomalies along Rizhao–Qingdao–Qianliyan–Rongcheng are mainly composed of mass low anomaly traps and NE trending beaded low anomalies, which roughly correspond to the Sulu ultra-high pressure (UHP) metamorphic belt. Large igneous rocks are in this anomaly area, and eclogites are exposed in many places on the surface (Figure 1b). Therefore, the consensus is that the positive anomaly zone may be the evidence of collision and splicing between the SKP and LYP, combined with the joint geological interpretation results of the wide-angle seismic profile velocity model and gravity [32].

The eastern high-low anomaly interphase zone (III) is located southeast of JXFZ, which is the principal research goal. The anomalies in this area are characterized by large-area regional low anomalies and the alternate distribution of high and low anomalies in part. In light of the anomaly characteristics, it can be further divided into the Northern Basin of SYS (NBSYS) anomaly area (III_1) and the Middle Uplift of SYS (MUSYS) anomaly area (III_2) (Figure 3). In the NBSYS (III_1), the gravity anomalies mainly present as low, and the anomaly trends mostly show NE–NEE and near-EW directions. The gradient zones are arranged in parallel, resembling a string of beads, and are alternatingly distributed. The massive low anomaly traps are apparent. The anomalies converge towards the west and radiate towards the northeast, producing a 'trumpet-shaped' figure that opens to the east. However, the overall anomalies of the MUSYS (III_2) manifest as relatively stable. They are mainly characterized by the gentle block high anomalies, and the anomaly gradient zone is rarely developed or undeveloped. In addition, the anomaly trends of the MUSYS gradually turn from NNE (NE) to SN directions, and are separated by a near-SN trending low anomaly in the middle. The anomaly trends in the west are primarily in the NE direction, and the anomalies in the east turn into NW and near-EW directions. In general, it is considered that the structural deformation of the MUSYS is weaker than that of the NBSYS. Therefore, combined with the seismic profiles [17,85], it is revealed that the tectonic stress decreases along the SN direction and gradually turns to EW-trending tectonic action in this area.

4.1.2. Characteristics and Geological Significance of the Magnetic Anomaly

The magnetic RTP anomaly of the study area (Figure 4) reveals belt-shaped anomalies mainly trending toward the NE direction in the northwest part. Moreover, the local anomalies in the northwest part manifest NW directions. Conversely, blocky-shaped anomalies characterize the southeast region, and the trend of the anomaly is complex and variable. In addition, the anomalies in the depression of the sea area present low-negative magnetic anomalies surrounded by high-positive anomalies. Overall, high-positive and low-negative magnetic anomalies change visibly and display apparent zoning. Similarly, the magnetic anomaly can be divided into four anomaly zones: a low-negative anomaly zone in East Shandong (I), a high-positive anomaly zone in the Sulu UHP (II), a central low-negative anomaly zone (III), and an eastern high-positive anomaly zone (IV) (Figure 4a). However, notable disparities exist between the zoning outcomes presented by [41] (Figure 4b) and our findings, particularly in the eastern region of the SYS. Fundamentally, the division in this study is identified by using the characteristics of the magnetic anomaly gradient zones. The results show that the magnetic anomaly zoning exhibits an east-west division in the sea area, while the zoning results reported by [52] show north-south zoning along the NEE direction.

The low-negative anomaly area in eastern Shandong (I) is located northwest of the Wulian–Jimo line. The magnetic anomalies in this area primarily manifest as varying sizes of positive anomaly traps on the low-negative background field, and the anomaly ridges show NE and NW directions. On the strength of the susceptibility (Figure 2c,d), the basement of this area is composed of a set of weak or no-magnetization pre-Cambrian and Archean–Neoproterozoic shallow metamorphic rock series. The magnetization of the overlying Cenozoic, Mesozoic Laiyang Group, and Wangshi Group is either weak or negligible. Thus, these strata (rocks) form the background of negative magnetic anomalies. In contrast, the magnetization of volcaniclastic rocks of the Cretaceous Qingshan Group is strong. It is inferred that the small-scale high-positive magnetic anomalies are caused by the volcaniclastic rocks of the Qingshan Group, basic and ultrabasic complexes, including amphibolite, and amphibolite gneiss with strong magnetization.

The high-positive anomaly zone in the Sulu UHP (II) is a conspicuous NE trending belt-shaped high-positive magnetic anomaly zone along the line of Rizhao-Qingdao-Rongcheng, which comprises multiple local positive anomaly traps. Under the influence of the Indosinian-Yanshan Movement, the mantle thermal material upwelled, and the basement continuously uplifted, resulting in the Paleoproterozoic–Archean deep metamorphic–shallow metamorphic basement being exposed to the surface (Figure 1). The Paleozoic strata above the basement were eroded and denuded, and the Mesozoic–Cenozoic was incompletely developed. Drilling results revealed that continental clastic sediments and pyroclastic sediments of the Early Cretaceous Qingshan Group and the Late Jurassic Laiyang Group were formed in this area. Furthermore, multi-period igneous rocks have developed in this area since the Yanshanian period. Consequently, it is speculated that the metamorphic basement and multi-period igneous rocks cause this high-positive magnetic value zone. However, large areas of high magnetic and low gravity anomalies have an excellent negative correlation in the southwest of this high-positive anomaly zone. We deduce that the low gravity anomalies are caused by the colossal low-density granite-like rocks overlying the high-density basement.

The central low-negative anomaly zone (III) is a broad and gentle low-negative magnetic anomaly area extending from Lianyungang–Xiangshui to the northeast sea of the SYS, trending in a NE direction. It is deduced that the anomaly zone is comprehensively caused by the combination of the weak susceptibility of the Sinian Yuntai Group, the non or weak susceptibility of strata in the NBSYS and surrounding areas, and the strong susceptibility of the crystalline basement composed of the underlying pre-Sinian metamorphic and igneous rocks. In contrast, small-scale low-positive anomalies sporadically distributed in part of the areas are caused by various intrusive rocks, igneous rocks, and pyroclastic materials formed in the basement or sedimentary layers by magmatic activities. In addition, the low magnetic anomaly on the east side of Qianliyan Island indicates that there are few volcanic intrusions and that the basement is not visibly damaged.

The eastern high-positive anomaly zone (IV) is located east of the study area, which is divided into two sub-anomaly zones in the north and south, with the line between well HH-2 and well HAEMA-1. Numerous NEE trending positive magnetic anomaly traps exist in the north (IV_1), principally caused by the medium to strong magnetism of some sandstones formed by thermal fluid alteration. Cretaceous volcanic rocks and Cenozoic basalts were revealed in well HAEMA-1, which indicated that the sedimentary layers might be destroyed in this area. In addition, there are large-scale positive anomalies with unobvious trends and gentle changes in the south (IV_2), but trap centres with negative anomalies are developed locally. According to drilling and seismic data, a large set of Cretaceous basalt has been discovered in the eastern part of the SYS. Especially in the east well KACHI-1, a layer of rhyolitic tuff was drilled in the Lower Cretaceous sandstone in the 2362–2368 m segment, and cryptocrystalline mid-basic volcanic rocks were found in the lower part. Furthermore, there are also light grey and light green grey granites on the south side of the NBSYS and the east side of the MUSYS. Therefore, we speculate that the massive positive anomaly in the southeast of the study area may be caused by the thinning of the lithosphere, materials with a high basic composition formed by the upwelling of thermal mantle materials, and igneous rocks intruding into the shallow layer during the subduction of the Pacific Plate.

4.2. Fault Structure

To effectively extract the linear anomaly information of the fault structures, the gravity and magnetic data are processed by derivative calculation and numerical statistics edge detection methods, such as *VSDR* (Figure 5b), *NVDR-THDR* (Figure 5c), *NSTD* (Figure 5d), *EGHA* (Figure 5e) and *IL* (Figure 5f). The linear structure information of various anomalies is extracted, and a wealth of fault structure information is obtained. After comprehensive analysis and comparison, *NVDR-THDR*, *NSTD*, *EGHA*, and *IL* anomalies exhibit higher resolution, sharper anomaly peaks, better linear structure continuity, and some weak linear anomaly information was strengthened, which is superior to *VSDR*. The fault trends reflected by this linear anomaly information are relatively clear and effective. The research results (Figure 5) show that the linear gravity anomaly information in the NE (NEE) directions and near-EW directions shows wide anomaly gradients, large peak anomaly zones, and preferable anomaly continuity; the NW (NWW) and near-SN trending linear gravity anomaly information shows narrow anomaly gradients, blurred traces of anomaly peaks, poor anomaly continuity and scattered distribution. Consequently, the fault structures system of the study area is redetermined (Figure 5), in combination with the results of geopotential field separation (Figure 5a), regional geological structure and tectonic evolution characteristics, as well as seismic, geology, and other previous research results [17,22–24,49,52,57].

Comparing the faults identified in this study with those previously delineated by researchers [17,49] and seismic profiles [17,21–24,30–32,85], the results of our fault division exhibit greater specificity and demonstrate excellent consistency with the seismic profiles and established structures. Furthermore, these findings align with the multi-period structural evolution and tectonic deformation characteristics of the study area. Thus, the results of this study could be more reliable and superior to previous research results.

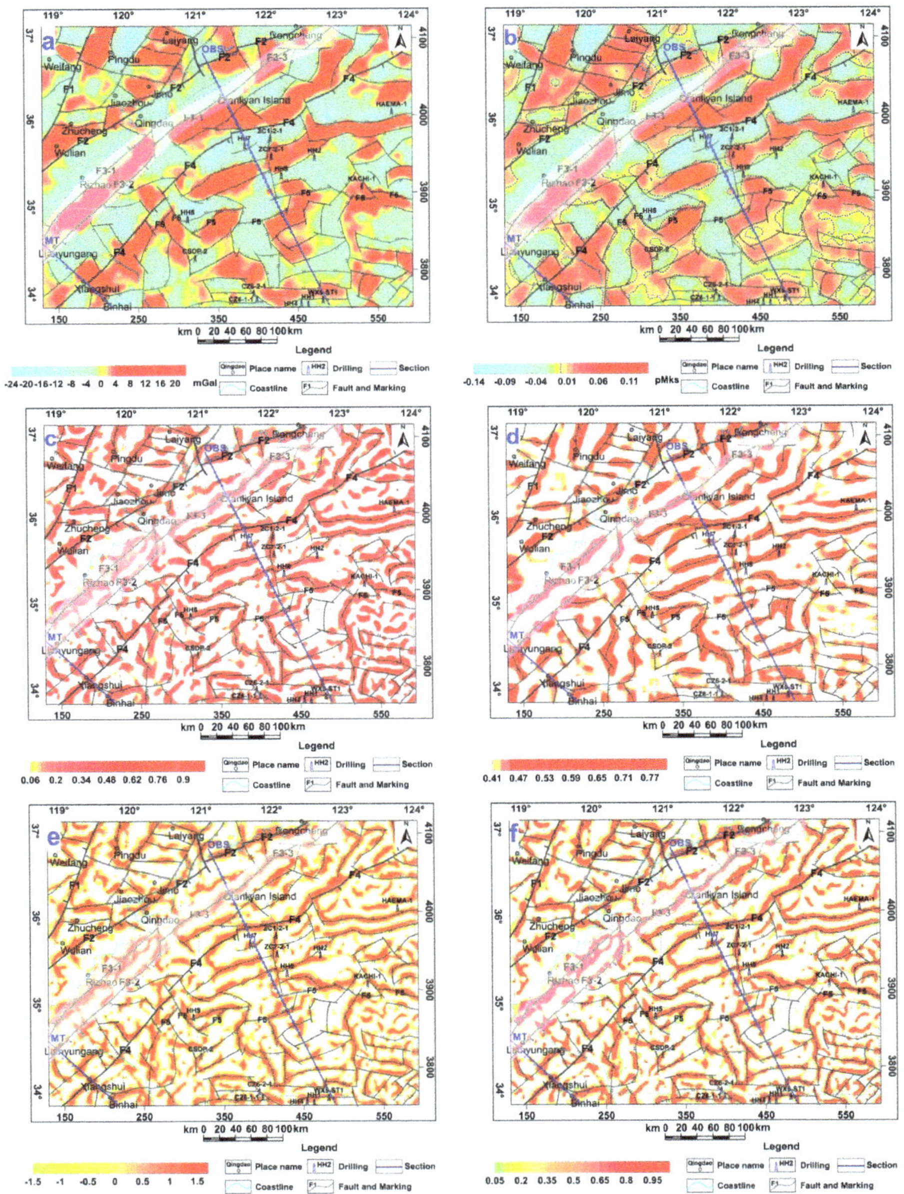

Figure 5. Map of faults: (**a**) residual gravity anomalies of *RF* (filter factor 50 km); (**b**) *VSDR* of Bouguer gravity anomaly (*R* = 28 km); (**c**) *NVDR-THDR* of Bouguer gravity anomaly, (**d**) *NSTD* of Bouguer gravity anomaly; (**e**) *EGHA* of Bouguer gravity anomaly and (**f**) *IL* of Bouguer gravity anomaly. F1: Tancheng–Lujiang fault; F2: Wulian–Qingdao–Rongcheng fault; F3: Lianyungang–Qianliyan fault; F4: Jiashan–Xiangshui fault; F5: fault in the southern margin of NBSYS. The black lines represent faults; the blue lines represent wide-angle seismic line and MT line; the blue dots represent the OBS points and crosses represent MT points.

Using the F3 Lianyungang–Qianliyan fault zone (LQFZ) (F3-1, F3-2, F3-3, the shaded area in Figure 5) as an example, the fault zone shows a clear gravity anomaly gradient zone in the residual gravity anomaly (Figure 5a), with NE (NEE) directions. The north and south sides in the southwest segment of F3 are shown as parallel anomaly gradient belts, and the middle segment anomalies show negative gravity anomalies with NW directions and apparent dislocation traces. Additionally, F3 shows zero lines of *VSDR*; the zero lines on the north side of F3 exhibit reasonable continuity, but the continuity on the south side of the southwest segment is poor, and the zero lines in the middle segment are staggered. On the northern side of F3, the lineament anomaly peak belts of *NVDR-THDR*, *NSTD*, *EGHA*, and *IL* are also wide, conspicuous, and greatly continuous on the north side of F3. In contrast, the anomaly trends on the south side of the southwest segment of F3 are variable, with linear anomaly discontinuity. The anomaly trends are disconnected on the middle segment of F3, presenting NWW trending anomaly peak belts. Therefore, according to the analysis results of the various anomalies, the trend of F3 is in a NE (NEE) direction and is divided into three faults, and the southwest segment is divided into two branches. F3 is shown in both the interpretation results of MT [86] and the wide-angle seismic profile [32] (Figure 6); it is considered that the division result of F3 is reasonable and credible. Thus, the reliability of these division results is better than that of the previous fault results (Figure 7).

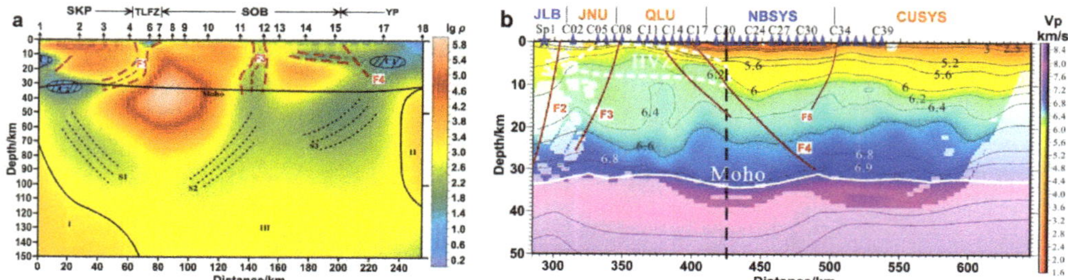

Figure 6. The characteristics of F3 on MT and wide-angle seismic profile. The profile position is shown in Figure 5: (**a**) Pingyi–Binhai MT profile (modified from [86]); (**b**) Vp velocity inversion profile of OBS2013-SYS Line (modified from [32]). F1: Tancheng–Lujiang fault; F2: Wulian–Qingdao–Rongcheng fault; F3: Lianyungang–Qianliyan fault; F4: Jiashan–Xiangshui fault; F5: fault in the southern margin of NBSYS. SKP—Sino-Korea Plate; TLFZ—Tancheng–Lujiang Fault Zone; SOB—Sulu Orogenic Belt; YP—Yangtze Plate; JLB—Jiaolai Basin; JNU—Jiaonan Uplift; QLU—Qianliyan Uplift; NBSYS—Northern Basin of the South Yellow Sea; MUSYS—Middle Uplift of the South Yellow Sea.

The MT interpretation results (Figure 6a) indicate apparent massive high-resistance bodies on both sides of the southwest segment of F3. The volume of the high-resistance body on the north side is more significant than that on the south side, and there is a low-resistance zone of nearly 20 km between the two high-resistance bodies. The interpretation results of the wide-angle seismic profile (Figure 6b) demonstrate that the Vp wave is twisted northwest of F3, and there is a low-speed body in the southeast of F3. The crust thickness on the northwest side of F3 is greater than that on the southeast side. Based on the comprehensive analysis results, we believe that the two branches on the north side of F3 are the major faults, cutting off the Moho interface and controlling the stratigraphic distribution and structural boundary, which is the boundary fault between the Jiaonan Uplift (JNU) and Qianliyan Uplift (QLU) (Figure 6). In comparison, F3-2 on the south side of F3 is a secondary fault that controls the rocks and structures in the JNU.

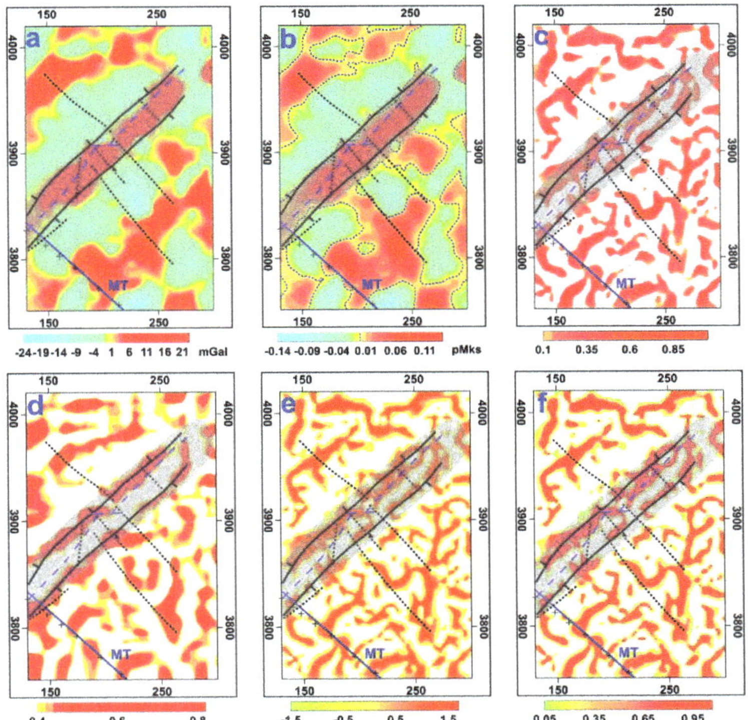

Figure 7. Comparison map between F3 fault zone on the south side and previous fault division results: (**a**) residual gravity anomaly of *RF* (filter factor 50 km); (**b**) *VSDR* of Bouguer gravity anomaly (R = 28 km); (**c**) *NVDR-THDR* of Bouguer gravity anomaly; (**d**) *NSTD* of Bouguer gravity anomaly; (**e**) *EGHA* of Bouguer gravity anomaly and (**f**) *IL* of Bouguer gravity anomaly. The blue line represents MT; the crosses represent MT point positions; the blue dashed lines represent the faults of Cai et al. (2005) [18], the black dotted lines represent the faults of Zhang et al. (2007) [49], the black lines represent the faults of this study.

Based on the aforementioned analyses, we assert that the faults can be effectively identified by combining multiple lineament structural anomaly detection methods. This application is particularly effective in complex structural areas, providing superior application and powerful adaptability. Moreover, this approach enables accurate determination of the plane extension characteristics and controlling the range of faults while facilitating an understanding of the movement patterns and tectonic frameworks.

The understanding of the newly determined faults system is as follows:

(1) The fault structures are numerous, and their development varies in scale. In addition, faults with varying directions interlace and intersect, creating a composite fault tectonic system that exhibits prominent hierarchical and regional features and multi-period activities. These characteristics are consistent with previous studies in the area.

(2) The statistical analysis of faults with different trends (Figure 8) indicates that NE (NNE) trending faults are the dominant faults in terms of long extension distances and large scales. However, NW (NWW), near-EW, and near-SN trending faults exhibit more minor scales but are more numerous and widely distributed, with shorter extension distances. Furthermore, the major faults in the NE (NEE) direction are typically cut off and staggered by the NW (NWW), near-EW, and near-SN trending faults, creating a "chessboard-type" structural system.

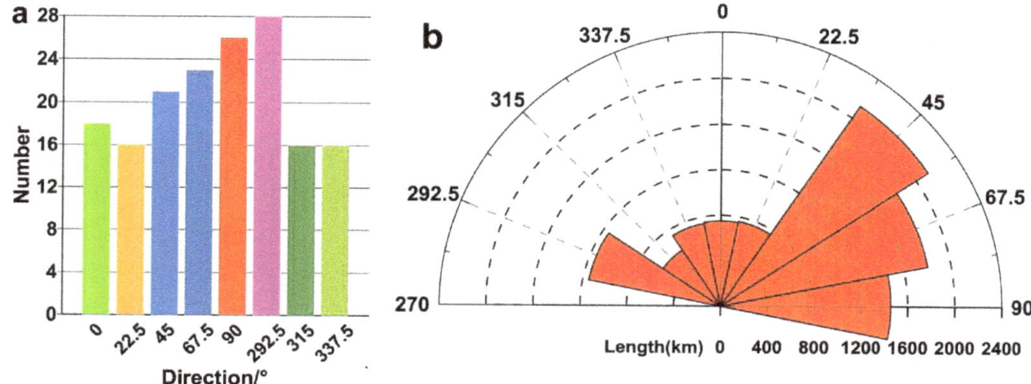

Figure 8. Statistical map of fault characteristics in the northern area of the SYS: (**a**) statistical diagram of the fault number and directions; and (**b**) rose diagram of fault directions.

(3) Several NE (NEE) trending major faults are consistent with the structural trend of the study area. Comparing the data from the wide-angle seismic interpretation profile [32] and MT interpretation profile [86], these faults exhibit early formation times, multiple tectonic periods, considerable fault distances, and deep vertical extensions. They govern the main structural framework of the study area, controlling the stratigraphic distribution, structural trends, and major boundaries of each secondary structural unit. In contrast, the NW (NWW), near-EW, and near-SN trending faults exhibit substantially shallower vertical extensions. These faults are typically interpreted as the basement or sedimentary layer faults based on seismic profiles [17,85] and form control boundaries of secondary structures in different sizes. Additionally, there are numerous near-EW trending faults in the MUSYS, which form fault-block structures. This results in the secondary structural units alternating between sags and bulges along the near-EW direction. The study area is thus divided into zones along NE (NNE) direction and blocks in the EW direction.

(4) The fault structures in the region exhibit differences and imbalances influenced by multi-period tectonic activities and various dynamics. NE(NEE) trending faults are distributed throughout the area, while NW (NWW) trending faults are mainly concentrated in the northwest corner of the study area, the NBSYS, and the east and west sides of the MUSYS. Near-EW trending faults are primarily distributed in the western regions of the NBSYS and the MUSYS. The triangular structure of the area consists of NE(NEE), near-EW, and NW (NWW) trending faults. It is speculated that the regional stress presents a north-south hedging pattern in the SN direction, with the EW direction of the MUSYS central to this pattern. This results from the collision of the SKP-LYP and the left-strike slip activity of TLFZ, alongside the strong almost N-S compression of the Indosinian–Yanshan Movement during the pre-rift period. Consequently, from south to north, tectonic deformation weakens, and the strata become shallower, thus forming the triangular structure pattern caused by these specific regional tectonic stresses.

(5) A new understanding of the fault division results has been achieved. The southwest segment of F3 (LQFZ) is divided into two branches: the west branch is the major fault and serves as the boundary between the JNU and QLU, whereas the east branch is the secondary fault that acts as a stress adjustment and buffering fault in QLU. In addition, the fault results in the southern margin of the NBSYS differ from the results of the previous divisions and show northerly results in this analysis.

(6) Analysis of the surface geological map (Figure 1b) and seismic interpretation of igneous rocks from [87,88] revealed that the distribution of igneous rocks is controlled by NE (NEE) trending large regional faults. Moreover, the development of igneous or intrusive rocks in sedimentary layers is regulated by the junction of NW, near-EW, and NE (NEE)

trending. Consequently, we speculate that the deep magmatic source area is connected by NE (NEE) trending deep faults. Furthermore, the magmatic activity around these faults is affected by multi-period tectonic movements.

4.3. 2.5D Gravity Modelling

To further ascertain the characteristics of stratigraphic development and the relationship between structural units, the geometric constraints of the two seismic interpretation profiles XQ7-9 and HB09-5 are used (Figures 3 and 4) in the gravity modelling. Also, the forward and inverse fitting of gravity anomaly is carried out (Figures 9 and 10). (See Section 3.2.7 for the modelling process.) The calculated gravity anomaly is gratifyingly consistent with the measured anomaly. In addition, based on the known geological interpretation results, such as drillings and seismic profiles, an EW trending geological geometric constraint model (Figures 3 and 4) is constructed, and gravity modelling is conducted (Figure 11).

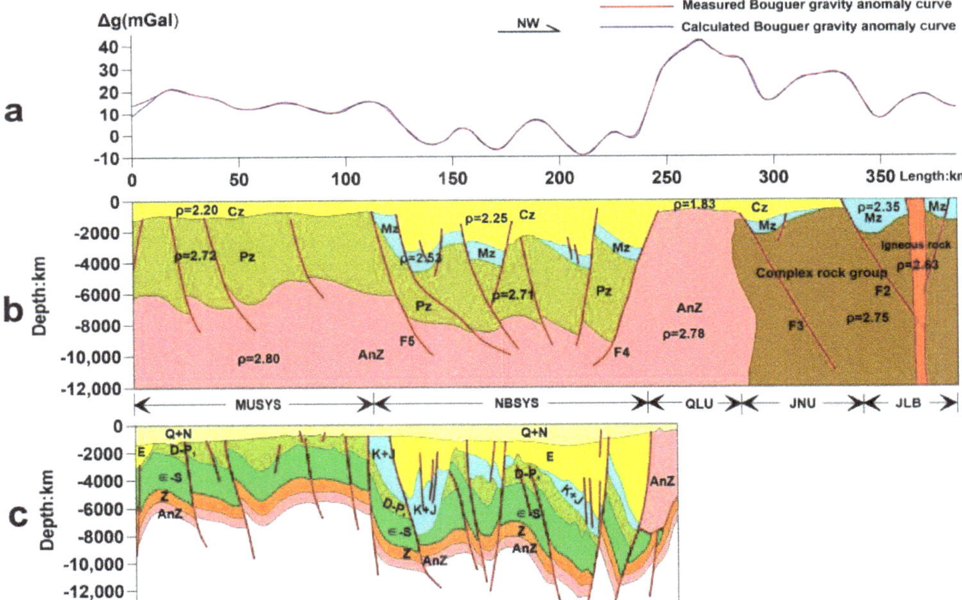

Figure 9. Map of A–A′ profile gravity modelling (the profile position is shown in Figures 3 and 4): (**a**) gravity anomaly curve; (**b**) gravity modelling results; (**c**) seismic geological results. F2: Wulian–Qingdao–Rongcheng fault; F3: Lianyungang–Qianliyan fault; F4: Jiashan–Xiangshui fault; F5: fault in the southern margin of NBSYS. JLB—Jiaolai Basin; JNU—Jiaonan Uplift; QLU—Qianliyan Uplift; NBSYS—the Northern Basin of the South Yellow Sea; MUSYS—the Middle Uplift of the South Yellow Sea.

The geological interpretation results of three gravity models are comprehensively analyzed (Figures 9–11), and the structural features are as follows:

(1) High-positive gravities are primarily caused by the basement and the Paleozoic uplifts or bulges, the Mesozoic erosion, and the thin growth of the Cenozoic. In contrast, Low gravities are caused by the sedimentary layers with weak sedimentary diagenesis and low stratigraphic density in the Mesozoic–Cenozoic fault depressions, formed by the basement fault depression or faults.

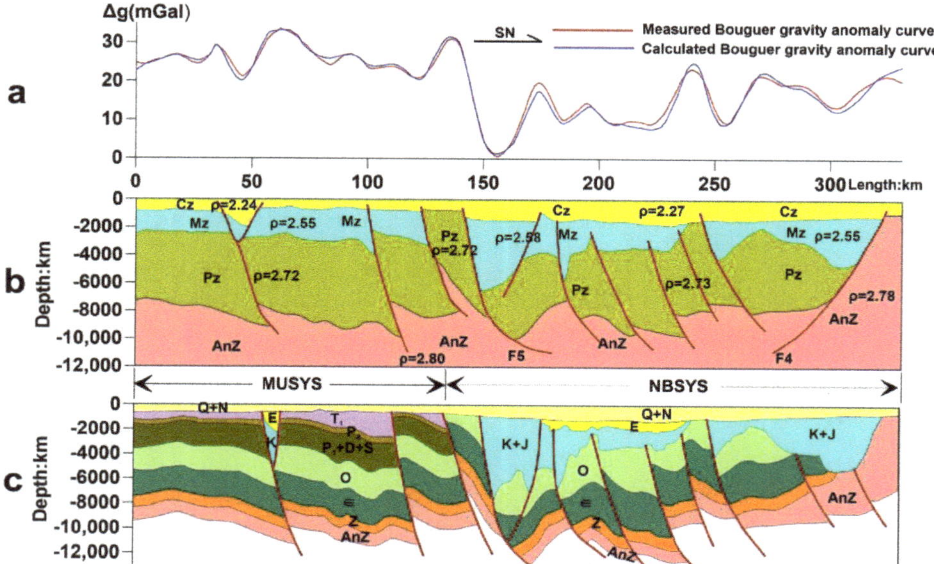

Figure 10. Map of B–B′ profile gravity modelling (the profile position is shown in Figures 3 and 4): (**a**) gravity anomaly curve; (**b**) gravity modelling results; (**c**) seismic geological results. F4: Jiashan–Xiangshui fault; F5: fault in the southern margin of NBSYS. NBSYS—the Northern Basin of the South Yellow Sea; MUSYS—the Middle Uplift of the South Yellow Sea.

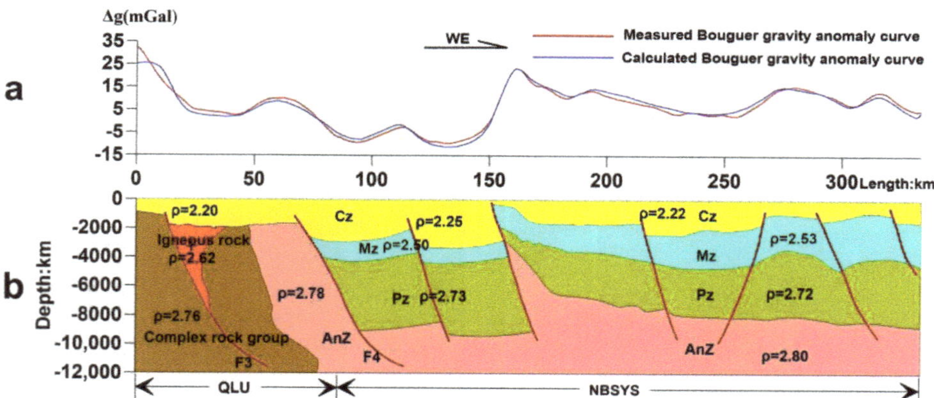

Figure 11. Map of C–C′ profile gravity modelling (the profile position is shown in Figures 3 and 4): (**a**) gravity anomaly curve; (**b**) gravity modelling results. F3: Lianyungang–Qianliyan fault; F4: Jiashan–Xiangshui fault. QLU—Qianliyan Uplift; NBSYS—the Northern Basin of the South Yellow Sea.

(2) The QLU and JNU are pre-Sinian basement uplifts, where many large regional faults are developed and accompanied by magmatic rocks. Moreover, the Mesozoic is partly developed in local areas of the JNU, the partly faulted depression deposits of the QLU are thick, and the Cenozoic shows unconformity in the basement. Refs. [32,89] inferred that the QLU was a collision composite splicing zone between the SKP and YP based on seismic profiles. According to the characteristics of gravity and magnetic anomalies and

the outcomes of gravity modelling, we conclude that the QLU and JNU belong to a part of the Sulu UHP belt, which represents the composite amalgamation zone between the SKP and YP collisional orogenic belts.

(3) The fault bulges or fault sags of the NBSYS are controlled by faults in the graben and half-graben types. Particularly, JXFZ is the primary controlling fault for the formation and development of the NBSYS, which controls the development of the northern fault sag. Furthermore, close to JXFZ, the sedimentary layers in the NBSYS are relatively thick, and it is the most developed in the north fault sag, which is a dustpan-shaped fault sag.

(4) The Cenozoic, Mesozoic, and Paleozoic are all widely developed in the SYS. In the NBSYS, the Paleozoic is widely distributed and thick but with significant formation fluctuations; the Mesozoic is widely distributed, but its continuity is generally poor, and there are substantial variations in strata thickness. These variations follow a pattern of gradual thickening from west to east while thinning or even absence can be observed in localized uplifted (bulge) areas. Especially with the central bulge zone as the boundary, the strata in the northwest part of the bulge zone are developed as a fault on the northwest segment and overlap on the southeast segment. However, the strata in the southeast part of the bulge zone are formed as northwest trending fault steps. The Cenozoic is well-developed in the whole basin, with good continuity and stable deposition, which however, fluctuates mainly in the form of a dustpan. In contrast, the Cenozoic–Mesozoic in the MUSYS is relatively thin, and the Mesozoic is absent or thin in some uplift (bulge) areas; however, the Paleozoic is thickly developed and stably subsided, and minor faults are relatively extensive in the Paleozoic.

(5) From the undulation characteristics of the basement and Paleozoic and the absence of the Mesozoic, we believe that the strata distribute in a complex, and the basement fluctuates significantly in the study area. In the pre-Cenozoic period, S-N trending tectonic stress weakened and gradually turned into E-W trending tectonic action in the study area. As a result, the Northern Basin of SYS (NBSYS) experienced more intense tectonic deformation than the Middle Uplift of SYS (MUSYS). In addition, the intensity of the N-S trending tectonic movement is more significant than that in the W-E direction, and the eastern basement is more stable than the western basement.

4.4. Basement Interface Inversion and Hydrocarbon Prospect Area

4.4.1. Interface Inversion

This study employs the Parker-Oldenburg inversion method [81,82] to calculate the Cenozoic depth. Instead of gravity anomalies, previous researchers usually utilized magnetic anomalies to compute the pre-Sinian basement depth [14,49,90,91]. In this work, however, we separated the Bouguer gravity anomaly by employing *RF* and calculated the residual gravity anomaly by the regularization scale factor of 50 km. An *RA* zoning gravity inversion method is used to estimate the basement depth of the NBSYS and the MUSYS for the first time (Figure 12a,b,d). Based on the gravity anomaly characteristics and the profile positions of the seismic interpretation results, the basement depth in the study area is further divided into the following two parts for inversion: the north-south trending zone of the NBSYS, and the west-east trending zone of the MUSYS (Figure 12). The correlation coefficient R between the basement and the residual gravity anomaly in the NBSYS is 0.84, and the quadratic polynomial regression formula for the basement depth of the NBSYS is $H = 0.0023\Delta g^2 + 0.148\Delta g - 7.856$ (Figure 12a). Moreover, the correlation coefficient R between the basement and the residual gravity anomaly in the MUSYS is 0.85, and the cubic polynomial regression formula for the basement depth of the MUSYS is $H = -0.00186\Delta g^3 + 0.00406\Delta g^2 + 0.390\Delta g - 7.194$ (Figure 10b) (Δg: mGal, basement depth: km).

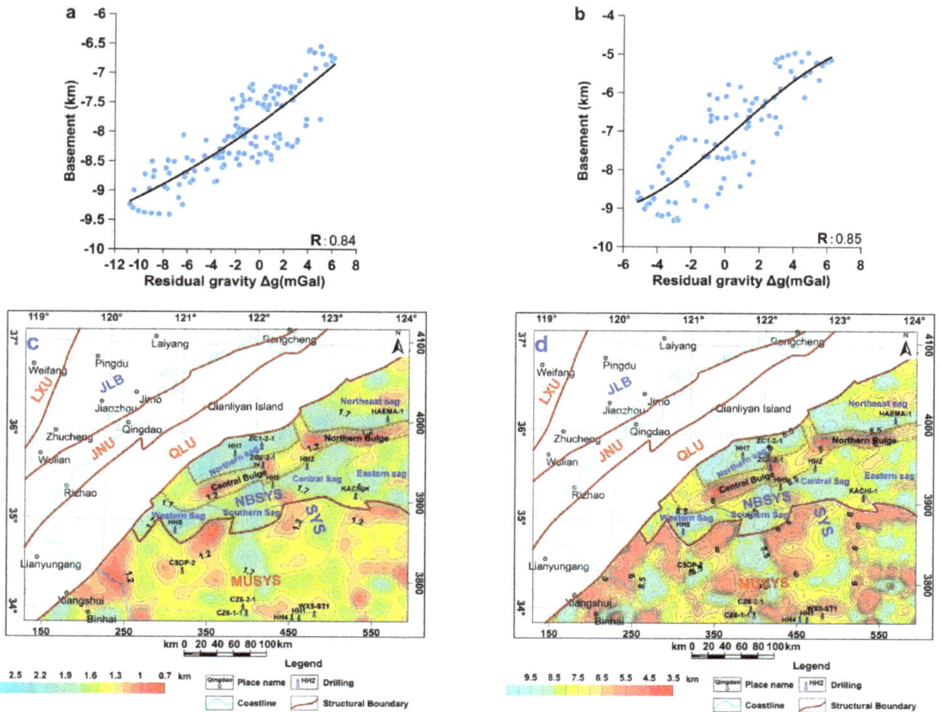

Figure 12. Map of the interface depth and tectonic unit division: (**a**) *RA* curve of gravity anomaly and pre-Sinian basement depth in the north; (**b**) *RA* curve of gravity anomaly and pre-Sinian basement depth in the south; (**c**) the map of Cenozoic depth; (**d**) the map of pre-Sinian basement depth. SYS—the South Yellow Sea; LXU—Luxi Uplift; JLB—Jiaolai Basin; JNU—Jiaonan Uplift; QLU—Qianliyan Uplift; NBSYS—the Northern Basin of the South Yellow Sea; MUSYS—the Middle Uplift of the South Yellow Sea. The red lines represent structural boundary; the red dashed lines represent secondary structural boundary.

The pre-Sinian basement depth calculated by *RA* is shown in Figure 12d. To verify the reliability of the basement depths, we extracted 24 basement depth points of the seismic geological interpretation profiles [17,21]. Then we compared them with the basement depth obtained by *RA* at the exact location. The comparison results indicate that the maximum and minimum errors between them are 899 m and 34 m, respectively, with a mean square error of ±532.11 m (Table 1). Therefore, we conclude that the depths calculated by *RA* are relatively reliable and reflect the undulation features of the basement. However, the difference in basement depths is great in the depressions (sags) and the structural boundaries, and slight in the uplifts (bulges) region, which indicates that the areas with intense changes in strata depth can affect the results of *RA* when calculating the basement.

4.4.2. Basement Features

The basement depths (Figure 12d) show that the JXFZ bounds the north of the NBSYS, and the area to the south of that adjacent to the MUSYS bounded by faults is a "trumpet-shaped" basin opening eastward. Moreover, the basement depth of the basin fluctuates greatly, and the basement is cut into secondary structures of different sizes by faults. Macroscopically, the NBSYS presents a pattern of alternating sag and bulge from north to south (Figure 12d). Along the north-south direction, therefore, the NBSYS can be further divided into eight sub-units, such as northeast sag, northern bulge, eastern sag, central sag, northern sag, central bulge, western sag, and southern sag.

Table 1. Comparison of the basement depth between seismic interpretation and the results of *RA*.

Number	Basement Depth by Seismic Interpretation (m)	Basement Depth Calculated by the Regression Formula (m)	Difference (m)	Difference %
1	7599.25	7564.64	34.61	0.46
2	7818.26	8468.2	−649.94	7.68
3	7781.76	8515.16	−733.4	8.61
4	6832.74	7538.48	−705.74	9.36
5	7307.25	7441.83	−134.58	1.81
6	7562.75	8093.74	−530.99	6.56
7	6650.24	7549.32	−899.08	11.81
8	6467.74	6664.87	−197.13	2.96
9	8475.27	7889.26	586.01	7.42
10	8438.77	8773.33	−334.56	3.81
11	9533.78	8841.07	692.71	7.84
12	8949.77	8781.61	168.16	1.91
13	8781.06	9376.66	−595.6	6.35
14	7622.42	7460.35	162.07	2.17
15	6463.78	6502.93	−39.15	0.60
16	7359.09	6506.29	852.8	13.11
17	8201.74	9091.95	−890.21	9.79
18	7464.42	8252.78	−788.36	9.55
19	9149.72	8941.66	208.06	2.33

$\sigma = \sqrt{\frac{\sum_{i=1}^{N}(\sigma_i - \overline{\sigma})^2}{n}} = \pm 532.11$, σ: mean square error.

By contrast, the MUSYS and NBSYS are connected by faults and overlaps, and also the basement presents a pattern of alternating sag and bulge from west to east. However, the fluctuation difference in the regional basement is slight (Figure 12d). Furthermore, the two bulge areas exhibit a significant spatial extent, with relatively small variations in basement depth. These regions are characterized by a relatively stable structure and the presence of multiple small-scale internal faults within strata. From the basement characteristics, therefore, it is implied that the tectonic conditions of the NBSYS are relatively complex, and the later structural reformation is intense; hence, it is a region with intense tectonic activities. In contrast, the tectonic conditions of the MUSYS are relatively stable, and the later reformation is weak, indicating that it represents a region of reasonable stability on the LYP.

5. Discussion

5.1. Relationship between Tectonic Movement and Hydrocarbons

At the end of the Early Paleozoic, the Caledonian Movement induced significant vertical differential movement in the SYSB, resulting in the underlying Early Paleozoic strata being eroded and flattened to a certain extent. Therefore, the Upper Silurian and Middle and Lower Devonian sediments are lacking in the SYSB [5,21]. The SYSB experienced long-term subsidence during the Late Paleozoic to Early Triassic as it primarily accumulated marine carbonate rocks and marine continental interactive transitional clastic rock. According to geophysical data and drilling revelation, the SYS has extensively preserved vast thickness and relatively gentle strata of the Paleozoic and Middle and Lower Triassic [1,5,28], which are well-preserved. Additionally, it has multiple sets of source rocks and hydrocarbon reservoir-caprock combinations. The Indosinian Movement ended the long-term stable marine sedimentation in the SYSB. The collision of the LYP with the SKP caused the entire basin to uplift and undergo intense compression and deformation. The folding and uplift of the marine Mesozoic and Paleozoic led to these strata being eroded and transformed, forming the sedimentary basement of later continental Mesozoic and Cenozoic basins [14]. The marine strata experienced multi-period reformation and alteration, with most of the Mesozoic–Paleozoic faulting occurring during the Indosinian Stage [21].

Meanwhile, this tectonic movement in the SYSB led to the widespread development of sedimentary cover, which developed the Indosinian tectonic system [12,14]. The Indosinian Movement resulted in varying degrees of damage and modification of the Mesozoic–Paleozoic hydrocarbon reservoirs, causing the already accumulated oil and gas to be redistributed and re-migrated again. Caprocks play a critical factor in preserving oil and gas reservoirs in this region.

Following the Mesozoic, the NBSYS experienced intense tectonic activity and multi-period stretching and rifting under the influences of the Yanshan and Himalayan movements, creating fault-depression regions where the Paleogene–Neogene and Cretaceous strata were deposited. These dustpan-shaped fault-depression systems with thicker sediments protected previously damaged oil and gas systems and prevented the diffusion of oil and gas, which is conducive to preserving hydrocarbon reservoirs. The basement faults formed by tectonic movements not only facilitate the sealing and preservation of oil and gas but also provide pathways for oil and gas migration, thereby further changing the pattern of oil and gas preservation and distribution. The marine petroleum system could be rebuilt. In contrast, after the Indosinian Movement, the MUSYS underwent long-term uplift resulting in the absence of many strata. However, according to the results of gravity modelling and seismic data interpretation, the tectonic deformation of the Mesozoic–Paleozoic is relatively weak, the stratigraphic sedimentation is stable, and the influence of faults and magmatic activity on the MUSYS is relatively small. MUSYS has well-developed multiple sets of inner cover layers and is distributed with stable upper cover layers [1,5–8]. In addition, the Mesozoic–Paleozoic burial depth is shallow, the source rock is of late maturity, and the hydrocarbon generation is late, while the sealing of the Mesozoic–Cenozoic caprocks is good. These conditions are conducive to the enrichment and preservation of oil and gas in marine structural layers [12,21].

5.2. Hydrocarbon Prospect Area

The basement depth (Figure 12d) and the Mesozoic-Paleozoic thickness (Figure 13) show that the basement of the NBSYS is deep, and the Mesozoic–Paleozoic is widely distributed and thick. Although the basement and the Mesozoic–Paleozoic thickness fluctuate significantly, and the fault activities present are intense, the magmatic activity is weak, and there are few igneous rock intrusions. Therefore, the internal destruction of the Mesozoic–Paleozoic was not severe, and the reservoir conditions of oil and gas are relatively good. The basement depth (Figure 12d) in the sags of the NBSYS is relatively deep, with a depth of 3.5~10.0 km, and the Mesozoic–Paleozoic thickness is relatively thick (Figure 13), with a thickness of 3.0~8.0 km. Additionally, based on the oil and gas geological characteristics of the SYS, it is apparent that the oil and gas system in the NBSYS is well developed and has superior caprocks. Although fault activities have altered the distribution of oil and gas, their redistribution through fault structures has resulted in multiple fault traps. Consequently, the secondary structures in the NBSYS provide favorable geological conditions for oil and gas accumulation.

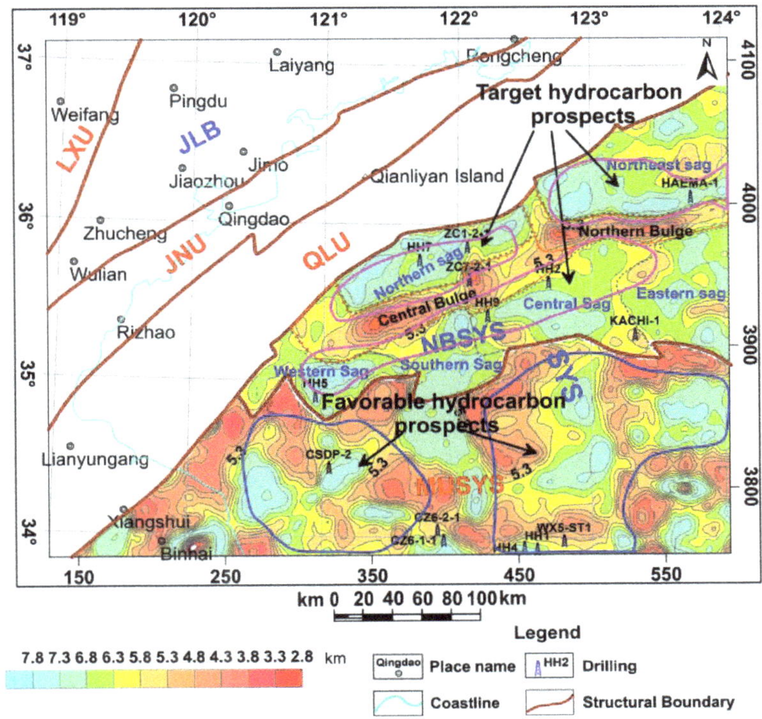

Figure 13. Map of the hydrocarbon prospects. The base map is the Mesozoic-Paleozoic thickness; the purple ranges represent the target prospect areas; the blue ranges represent the favorable prospect areas. SYS—the South Yellow Sea; LXU—Luxi Uplift; JLB—Jiaolai Basin; JNU—Jiaonan Uplift; QLU—Qianliyan Uplift; NBSYS—the Northern Basin of the South Yellow Sea; MUSYS—the Middle Uplift of the South Yellow Sea. The red lines represent structural boundary; the red dashed lines represent secondary structural boundary; the purple lines represent target hydrocarbon prospects; the blue lines represent favorable hydrocarbon prospects.

The depth of the basement in the central bulge and the northern bulge, adjacent to the north and south sides of the sag areas, is relatively shallow. According to the oil and gas accumulation model in the NBSYS [17,64], combined with the fault structures and stratigraphic evolution, we consider that the oil and gas in the sags can be migrated vertically or horizontally to the structural highs of the northern bulge and the central bulge through the faults. Under the influence of fault structures, oil and gas storage structures such as fault anticlines or fault block traps can develop. Based on the above analysis, therefore, three target prospect areas for oil and gas exploration of the Mesozoic-Paleozoic are determined in the NBSYS. These include the transition zone between the northern sag and central bulge, the transition zone between the northeastern sag and northern bulge, and the transition areas between the western sag, southern sag, central sag and eastern sag, as well as between the northern bulge and central bulge.

The basement depth of the MUSYS (Figure 12d) is relatively shallow; however, the Mesozoic-Paleozoic strata are thick, widely distributed, and well preserved (Figure 13), with thicknesses between 2.5 and 7.8 km. In addition, the Mesozoic–Paleozoic in the MUSYS is characterized by gentle fluctuations, stable structural conditions, weak faults, and poor magmatic activity; only the southeast corner of the MUSYS has igneous intrusions, which may destroy the basement and the Paleozoic. Moreover, the Mesozoic–Paleozoic is alternately distributed along the east-west direction with highs and lows, and the strata are imbricated under the influence of the north-south trending compression. In addition, the Mesozoic–Paleozoic has well-developed internal sedimentary faults, and two large-scale fault bulges exist in the MUSYS (Figure 13) that are characterized by extensive areas, mild deformation, stable structures, and numerous source rock formations. In this area, there are not only several sets of inner caprocks with favorable conditions, but also the overlying caprocks with superior conditions are widely distributed. In the later period, the oil and gas preservation conditions are superior, with multiple types of oil and gas resource reservoir-caprock assemblages.

Combined with the fault division results (Figure 5) and gravity modelling geological results, we consider that the sediments and structures of the MUSYS are relatively stable, the hydrocarbon storage conditions are favorable, and that there is a favorable combination of source, reservoir, and caprocks (Figures 9–11). Additionally, according to the oil and gas accumulation pattern of the MUSYS [11], the faults can provide good passage and motive power for oil and gas migration and accumulation, as well as considerable structural traps for oil and gas accumulation. Furthermore, both the east and west sides of the two bulges are close to the sag areas, and the oil and gas in the sags can be migrated to the structural highs. Hence, the two bulges in the MUSYS should also be favorable prospective areas for further oil and gas exploration of the Mesozoic–Paleozoic.

It should be noted that the northern South Yellow Sea area has a complex geological setting, with significant variations in the strata thickness both vertically and horizontally. Therefore, calculating density interfaces and the thickness of the Mesozoic and Paleozoic strata using gravity data is based on approximation and inference. The results may differ from actual geological conditions. However, the macroscopic interface depth and strata thickness characteristics are relatively reliable. In addition, due to the low resolution of gravity data and the non-uniqueness of geophysical inversion methods, although we apply multiple geophysical data for joint interpretation, the determined fault structures, density interfaces, and predicted oil and gas favorable areas in this study may be differences due to these limitations. Nevertheless, our results still reflect the structural characteristics and strata features of the study area.

To improve the reliability of interpretation of the results and deepen the understanding of geological structures, it is necessary to obtain further high-resolution seismic profiles, drilling and geological data, and higher precision gravity and magnetic data in future work. Furthermore, the joint inversion method of gravity and seismic data should be enhanced to minimize the non-uniqueness inherent in geophysical inversion.

6. Conclusions

We analyzed the characteristics and genesis of gravity and magnetic anomalies. Based on higher precision gravity data, we refined the results of previous fault structures by implementing reasonable geopotential field separation and edge detection processing methods. In addition, we utilized the constraint information from the seismic profiles to estimate the basement depth and the Mesozoic–Paleozoic thickness by gravity inversion. The findings can be summarized as follows:

(1) In our study area, the structures exhibited considerable complexity, and regional tectonism varies. Specifically, the intensity of tectonic activities weakened gradually from NW to SE. During the pre-Cenozoic, the N-S trending tectonic movement was more powerful than the E-W trending movement, resulting in greater structural deformation in the NBSYS than that in the MUSYS. Additionally, the eastern basement was more stable than the western basement in the study area. Consequently, these factors led to the development of a chessboard tectonic pattern characterized by N-S trending zonings in the NBSYS and E-W trending blockings in the MUSYS.

(2) The rationality and reliability of our fault division findings have been enhanced through our study. The area features abundant faults, and the dominant faults in NE (NEE) directions were staggered and cut off by secondary faults in NW (NWW), near-EW, and

near-SN directions. This geometric pattern gave rise to a complex en echelon-typed fault system that extends along the NE(NEE) directions. Notably, we successfully discovered a new fault, F3-2, trending in the NE direction.

(3) This study represents the first instance of utilizing *RA* and gravity data to estimate the pre-Sinian basement depth and the Mesozoic–Paleozoic thickness. The basement and strata features differed significantly along the N-S direction. The NBSYS displayed significant fluctuations in the basement and Mesozoic–Paleozoic thicknesses, coupled with intense fault activity. Conversely, the Mesozoic–Paleozoic in the MUSYS was characterized by mild fluctuations and relatively stable structural conditions.

(4) This paper assessed the potential for hydrocarbon exploration in the Mesozoic–Paleozoic strata. The investigation utilized geological information, including fault structures, stratigraphic distribution, and hydrocarbon accumulation models to identify three transitional areas between the bulges and the sags connected by faults and overlaps in the NBSYS, leading to the identification of prospective target areas for hydrocarbon exploration. Additionally, the study also identified two stable bulges in the MUSYS which are favorable prospective areas for further oil and gas exploration.

(5) The study found that the fault structures, density interfaces, and predicted favorable oil and gas areas inferred from the analysis of gravity and magnetic data are relatively reliable. However, the complex structures in the SYS, in conjunction with the low accuracy of the gravity and magnetic data and the inherent non-uniqueness of geophysical inversion methods, may lead to potential discrepancies from actual geological formations. Therefore, future investigations should supplement our findings with high-resolution seismic profiles, drillings, geological data, and high-precision gravity and magnetic data to enhance the reliability their interpretation. Furthermore, additional quantitative analyses of the gravity and magnetic data with geological structures are necessary. Additionally, it is crucial to improve the joint inversion method of gravity, magnetic, and seismic data to minimize the non-uniqueness of geophysical inversion methods.

Author Contributions: W.X.: Data collection and compiling, data processing and analyses, methodology, interpretation, and editing. C.Y.: Leading the study, data analyses, methodology, and review. B.Y.: Data collection, interpretation, and review. S.A., X.Y. (Xianzhe Yin) and X.Y. (Xiaoyu Yuan): Data collection, methodology and interpretation. All authors have read and agreed to the published version of the manuscript.

Funding: National Key R&D Program of China (2021YFB3900200).

Data Availability Statement: The data is available by request from the corresponding author.

Acknowledgments: This study has been partly supported by the Aero Geophysical Prospecting and Application for Marine Geology Project (Grant No. GZH2009005002) of the China Aero Geo-physical Survey and Remote Sensing Center for Land and Resources. We thank the National Geological Archives of China Geological Survey for permission to apply and release the data.

Conflicts of Interest: The authors declare that the research was conducted in the absence of any commercial or financial relationships that could be construed as a potential conflict of interest.

References

1. Zhang, J.Q. Exploration prospects of Mesozoic and Paleozoic oil and gas in the South Yellow Sea. *Mar. Geol. Lett.* **2002**, *18*, 25–27. (In Chinese)
2. Feng, Z.Q.; Yao, Y.J.; Zeng, X.H.; Wang, Q.; Wang, L.L.; Chen, Q.; Yi, H.; Jin, H.F. New understanding of Mesozoic-Paleozoic tectonics and hydrocarbon potential in Yellow Sea. *China Offshore Oil Gas (Geol.)* **2002**, *16*, 367–373. (In Chinese)
3. Wu, S.G.; Ni, X.L.; Cai, F. Petroleum geological framework and hydrocarbon potential in the Yellow Sea. *Chin. J. Oceanol. Limnol.* **2008**, *26*, 23–34. [CrossRef]
4. Yao, Y.J.; Chen, C.F.; Feng, Z.Q.; Zhang, S.Y.; Hao, T.Y.; Wan, R.S. Tectonic evolution and hydrocarbon potential in northern area of the South Yellow Sea. *J. Earth Sci.* **2010**, *21*, 71–82. [CrossRef]
5. Liang, J.; Zhang, P.H.; Chen, J.W.; Gong, J.M.; Yuan, Y. Hydrocarbon preservation conditions in Mesozoic-Paleozoic marine strata in the South Yellow Sea Basin. *Nat. Gas Ind. B* **2017**, *4*, 432–441. [CrossRef]

6. Chen, J.W.; Xua, M.; Lei, B.H.; Liang, J.; Zhang, Y.G.; Wu, S.Y.; Shi, J.; Yuan, Y.; Wang, J.Q.; Zhang, Y.X.; et al. Prospective prediction and exploration situation of marine Mesozoic-Paleozoic oil and gas in the South Yellow Sea. *China Geol. (Engl. Ed.)* **2019**, *1*, 67–84. [CrossRef]

7. Lei, B.H.; Chen, J.W.; Liang, J.; Zhang, Y.G. Structural Characteristics and Favorable Zone for Marine Hydrocarbon Exploration of the Laoshan Uplift in the South Yellow Sea Basin. *Acta Geol. Sin. (Engl. Ed.)* **2019**, *93*, 102–103.

8. Yuan, Y.; Chen, J.W.; Liang, J.; Xu, M.; Lei, B.H.; Zhang, Y.X.; Cheng, Q.S.; Wang, J.Q. Hydrocarbon geological conditions and exploration potential of Mesozoic–Paleozoic marine strata in the South Yellow Sea Basin. *J. Ocean Univ. China (Ocean. Coast. Sea Res.)* **2019**, *18*, 1329–1343. [CrossRef]

9. Pang, Y.M.; Guo, X.W.; Han, Z.Z.; Zhang, X.H.; Zhu, X.Q.; Hou, F.H.; Han, C.; Song, Z.G.; Xiao, G.L. Mesozoic–Cenozoic denudation and thermal history in the Central Uplift of the South Yellow Sea basin and the implications for hydrocarbon systems: Constraints from the CSDP-2 borehole. *Mar. Pet. Geol.* **2019**, *99*, 355–369. [CrossRef]

10. Cai, L.X.; Xiao, G.L.; Guo, X.W.; Wang, J.; Wu, Z.Q.; Li, B.G. Assessment of Mesozoic and Upper Paleozoic source rocks in the South Yellow Sea Basin based on the continuous borehole CSDP-2. *Mar. Pet. Geol.* **2019**, *101*, 30–42. [CrossRef]

11. Cai, L.X.; Xiao, G.L.; Zeng, Z.G.; Zhang, X.H.; Guo, X.W.; Wang, S.P. New insights into marine hydrocarbon geological conditions in the South Yellow Sea Basin: Evidence from borehole CSDP-2. *J. Oceanol. Limnol.* **2020**, *38*, 1169–1187. [CrossRef]

12. Wang, J.Q.; Sun, J.; Xiao, G.L.; Wu, Z.Q.; Li, S.L. Tectonic characteristics of the South Yellow Sea basin and implications for petroleum geology. *Mar. Geol. Front.* **2014**, *30*, 34–39. (In Chinese)

13. Zhang, X.H.; Guo, X.W.; Wu, Z.Q.; Xiao, G.L.; Zhang, X.H.; Zhu, X.Q. Preliminary results and geological significance of Well CSDP-2 in the Central Uplift of South Yellow Sea Basin. *Chin. J. Geophys.* **2019**, *62*, 197–218. (In Chinese)

14. Zhang, X.H.; Yang, J.Y.; Li, G.; Yang, Y.Q. Basement structure and distribution of Mesozoic-Paleozoic marine strata in the South Yellow Sea basin. *Chin. J. Geophys.* **2014**, *57*, 4041–4051. (In Chinese)

15. Lei, B.H.; Xu, M.; Chen, J.W.; Liang, J.; Zhang, Y.G. Structural characteristics and evolution of the South Yellow Sea Basin since Indosinian. *China Geol.* **2018**, *4*, 466–476. [CrossRef]

16. Feng, Z.Q.; Chen, C.F.; Yao, Y.J.; Zhang, S.Y.; Hao, T.Y.; Wan, R.S. Tectonic evolution and exploration target of the northern foreland basin of the South Yellow Sea. *Earth Sci. Front.* **2008**, *15*, 219–231. (In Chinese)

17. Zhang, X.H.; Xiao, G.L.; Wu, Z.Q.; Li, S.L. *Recognition and Discussion of Several Geological Problems in Oil&Gas Exploration of the South Yellow Sea—New Progress and Challenges in the Mesozoic-Paleozoic Marine Oil&Gas Exploration of the South Yellow Sea*; Science Press: Beijing, China, 2017. (In Chinese)

18. Cai, Q.Z. *Oil & Gas Geology in China Seas*; China Ocean Press: Beijing, China, 2005. (In Chinese)

19. Ma, L.F. *Geological Atlas of China (Shandong Geological Map & Jiangsu Geological Map)*; Geological Publishing House: Beijing, China, 2002. (In Chinese)

20. Becker, J.J.; Sandwell, T.D.; Smith, W.H.F.; Braud, J.; Binder, B.; Depner, J.; Fabre, D.; Factor, J.; Ingalls, S.; Kim, S.-H.; et al. Global bathymetry and elevation data at 30 arc seconds resolution: SRTM30_PLUS. *Mar. Geod.* **2009**, *32*, 355–371. [CrossRef]

21. Yao, Y.J.; Feng, Z.Q.; Hao, T.Y.; Xu, X.; Li, X.J.; Wan, R.S. A new understanding of structural layers in the South Yellow Sea Basin and their hydrocarbon-bearing characteristics. *Earth Sci. Front.* **2008**, *15*, 232–240. (In Chinese)

22. Zhang, H.Q.; Chen, J.W.; Li, G.; Li, G.; Wu, Z.Q.; Zhang, Y.G. Discovery from seismic survey in Laoshan Uplift of the South Yellow Sea and the significance. *Mar. Geol. Quat. Geol.* **2009**, *29*, 107–113. (In Chinese)

23. Wang, L.Y. Research on Pre-Tertiary Seismic Stratigraphy and Structural Characteristics of the South Yellow Sea. Master's Thesis, Ocean University of China, Qingdao, China, 2011. (In Chinese)

24. Li, N.; Li, W.R.; Long, H.Y. Tectonic evolution of the north depression of the South Yellow Sea Basin since Late Cretaceous. *J. Ocean Univ. China (Ocean. Coast. Sea Res.)* **2016**, *15*, 967–976. [CrossRef]

25. Xu, H.; Zhang, H.Y.; Zhang, B.L.; Yan, G.J.; Shi, J.; Yang, Y.Q.; Sun, H.Q.; Li, J.W.; Dong, G.; Lu, S.C. Characteristics of the 26 wells from the South Yellow Sea Basin. *Mar. Geol. Front.* **2015**, *31*, 1–6. (In Chinese)

26. Xu, X.; Sun, L.P.; Xiao, M.C.; Li, X.; Zhu, X.Y.; Gao, S.L. Structural characteristics and activity rules of faults in the Cenozoic faulted-basins in the Lower Yangtze Region. *Geol. J. China Univ.* **2018**, *24*, 723–733. (In Chinese)

27. Yuan, Y.; Chen, J.W.; Zhang, Y.X.; Liang, J.; Zhang, Y.G.; Zhang, P.H. Tectonic evolution and geological characteristics of hydrocarbon reservoirs in marine Mesozoic-Paleozoic strata in the South Yellow Sea Basin. *J. Ocean Univ. China (Ocean. Coast. Sea Res.)* **2018**, *17*, 1075–1090. [CrossRef]

28. Xu, M.; Chen, J.W.; Liang, J.; Zhang, Y.G.; Lei, B.H.; Shi, J.; Wang, J.Q. Basement structures underneath the northern South Yellow Sea basin (East China): Implications for the collision between the North China and South China blocks. *J. Asian Earth Sci.* **2019**, *186*, 104040. [CrossRef]

29. Qi, J.H. Research on the Crustal Velocity Structure in the Southern Yellow Sea. Ph.D. Thesis, Ocean University of China, Qingdao, China, 2015. (In Chinese).

30. Hou, F.H.; Guo, X.W.; Wu, Z.Q.; Zhu, X.Q.; Zhang, X.H.; Qi, J.H.; Wang, B.J.; Wen, Z.H.; Cai, L.X.; Pang, Y.M. Research progress and discussion on formation and tectonics of South Yellow Sea. *J. Jinlin Univ. (Earth Sci. Ed.)* **2019**, *49*, 96–105. (In Chinese)

31. Zhao, W.N.; Wang, H.G.; Shi, H.C.; Xie, H.; Wu, Z.Q.; Chen, S.S.; Zhang, X.H.; Hao, T.Y.; Zheng, P.H. Crustal structure from onshore-offshore wide-angle seismic data: Application to Northern Sulu Orogen and its adjacent area. *Tectonophysics* **2019**, *770*, 1–12. [CrossRef]

32. Liu, L.H.; Hao, T.Y.; Lü, C.C.; Wu, Z.Q.; Zheng, Y.P.; Wang, F.Y.; Xu, Y.; Kim, K.; Kim, H. Crustal deformation and detachment in the Sulu Orogenic Belt: New constraints from Onshore-Offshore Wide-Angle seismic data. *Geophys. Res. Lett.* **2021**, *48*, 1–10. [CrossRef]

33. Berezkin, V.M. Method of analytical continuation of total vertical gradient of gravity field for the study of density masses distribution in the earth crust. *Geol. I Razved. (Geol. Explor.)* **1968**, *12*, 104–110. (In Russian)

34. Berezkin, W.M. Application of the full vertical gravity gradient to determination to sources causing gravity anomalies. *Expl. Geopys.* **1967**, *18*, 69–76. (In Russian)

35. Berezkin, W.M. *Application of Gravity Exploration to Reconnaissance of Oil and Gas Reservoir*; Nedra Publishing House: Moscow, Russia, 1973. (In Russian)

36. Zeng, H.L.; Meng, X.H.; Yao, C.L.; Li, X.M.; Lou, H.; Guan, Z.N.; Li, Z.P. Detection of reservoirs from normalized full gradient of gravity anomalies and its application to Shengli oil field, east China. *Geophysics* **2002**, *67*, 1138–1147. [CrossRef]

37. Aydin, A. Interpretation of gravity anomalies with the normalized full gradient (NFG) method and an example. *Pure Appl. Geophys.* **2007**, *164*, 2329–2344. [CrossRef]

38. Aghajani, H.; Moradzadeh, A.; Zeng, H.L. Detection of high-potential oil and gas fields using normalized full gradient of gravity anomalies: A case study in the Tabas Basin, Eastern Iran. *Pure Appl. Geophys.* **2011**, *168*, 1851–1863. [CrossRef]

39. Yuan, B.Q.; Song, L.J.; Han, L.; An, S.L.; Zhang, C.G. Gravity and magnetic field characteristics and hydrocarbon prospects of the Tobago Basin. *Geophys. Prospect.* **2018**, *66*, 1586–1601. [CrossRef]

40. Saada, S.A.; Mickus, K.; Eldosouky, A.M.; Ibrahim, A. Insights on the tectonic styles of the Red Sea rift using gravity and magnetic data. *Mar. Pet. Geol.* **2021**, *133*, 105253. [CrossRef]

41. Eldosoukya, A.M.; Pham, L.T.; Duong, V.-H.; Eitel, F.; Ghomsid, K.; Henaish, A. Structural interpretation of potential field data using the enhancement techniques: A case study. *Geocarto Int.* **2022**, *37*, 16900–16925. [CrossRef]

42. Pham, L.T.; Nguyen, D.A.; Eldosouky, A.M.; Abdelrahman, K.; Vu, T.V.; Al-Otaibi, N.; Ibrahim, E.; Kharbish, S. Subsurface structural mapping from high-resolution gravity data using advanced processing methods. *J. King Saud Univ.-Sci.* **2021**, *33*, 101488. [CrossRef]

43. Saada, S.A.; Eldosouky, A.M.; Kamel, M.; Khadragy, A.E.; Abdelrahman, K.; Fnais, M.S.; Mickus, K. Understanding the structural framework controlling the sedimentary basins from the integration of gravity and magnetic data: A case study from the east of the Qattara Depression area, Egypt. *J. King Saud Univ.-Sci.* **2022**, *34*, 101808. [CrossRef]

44. Eldosouky, A.M.; Pham, L.T.; Henaish, A. High precision structural mapping using edge filters of potential field and remote sensing data: A case study from Wadi Umm Ghalqa area, South Eastern Desert, Egypt. *Egypt. J. Remote Sens. Space Sci.* **2022**, *25*, 501–513. [CrossRef]

45. El-Sehamy, M.; Gawad, A.M.A.; Aggour, T.A.; Orabi, O.H.; Abdella, H.F.; Eldosouky, A.M. Sedimentary cover determination and structural architecture from gravity data: East of Suez Area, Sinai, Egypt. *Arab. J. Geosci.* **2022**, *15*, 109. [CrossRef]

46. Pham, L.T.; Ghoms, F.E.K.; Vu, T.V.; Oksum, E.; Steffen, R.; Tenzer, R. Mapping the structural configuration of the western Gulf of Guinea using advanced gravity interpretation methods. *Phys. Chem. Earth* **2023**, *129*, 103341. [CrossRef]

47. Liang, R.C.; Pei, Y.L.; Zheng, Y.P.; Wei, J.W.; Liu, Y.G. Gravity and magnetic field and tectonic structure character in the southern Yellow Sea. *Chin. Sci. Bull.* **2003**, *48*, 64–73. [CrossRef]

48. Wang, W.Y.; Liu, J.L.; Qiu, Z.Y.; Huang, Y.J.; Cai, D.S. A research on Mesozoic thickness using satellite gravity anomaly in the southern Yellow Sea. *China Offshore Oil Gas* **2004**, *16*, 151–156,169. (In Chinese)

49. Zhang, M.H.; Xu, D.S.; Chen, J.W. Geological structure of the yellow sea area from regional gravity and magnetic interpretation. *Appl. Geophys.* **2007**, *4*, 75–83. [CrossRef]

50. Yao, Y.J. Tectonic of the South Yellow Sea: Application of Joint Gravity-Magnetic-Seismic Inversion. Ph.D. Thesis, Guangzhou Institute of Geochemistry, Chinese Academy of Sciences, Guangzhou, China, 2008. (In Chinese)

51. Hao, T.Y.; Huang, S.; Xu, Y.; Li, Z.W.; Zhang, L.L.; Wang, J.L.; Suh, M.; Kim, K. Geophysical understandings on deep structure in Yellow Sea. *Chin. J. Geophys.* **2010**, *53*, 1315–1326. (In Chinese)

52. Yang, J.Y. Research on the Tectonic Relation between the South Yellow Sea Basin and Its Adjacent Area and Distribution Characteristic and Tectonic Evolution of the Mesozoic-Paleozoic Marine Strata. Ph.D. Thesis, Zhejiang University, Hangzhou, China, 2009. (In Chinese).

53. Choi, S.C.; Ryu, I.-C.; Gotze, H.-J. Depth distribution of the sedimentary basin and Moho undulation in the Yellow Sea, NE Asia interpreted by using satellite-derived gravity field. *Geophys. J. Int.* **2015**, *202*, 41–53. [CrossRef]

54. Zhang, X.J.; Zhang, W.; Fan, Z.L.; Zhu, W.P.; Tong, J.; Yao, G.T. Characteristics of airborne gravity field and the main geological discovery in the northern South Yellow Sea. *Geol. Surv. China* **2017**, *4*, 50–56. (In Chinese) [CrossRef]

55. Tong, J.; Zhang, X.J.; Zhang, W.; Xiong, S.Q. Marine strata morphology of the South Yellow Sea based on high-resolution aeromagnetic and airborne gravity data. *Mar. Pet. Geol.* **2018**, *96*, 429–440. [CrossRef]

56. Zhang, R.Y.; Yang, F.L.; Zhao, C.J.; Zhang, J.W.; Qiu, E.K. Coupling relationship between sedimentary basin and Moho morphology in the South Yellow Sea, East China. *Geol. J.* **2020**, *55*, 6544–6561. [CrossRef]

57. Xu, W.Q.; Yuan, B.Q.; Liu, B.L.; Yao, C.L. Multiple gravity and magnetic potential field edge detection methods and their application to the boundary of fault structures in northern South Yellow Sea. *Geophys. Geochem. Explor.* **2020**, *44*, 962–974. (In Chinese)

58. Jin, X.L.; Yu, P.Z. *Geological Structure of the Yellow Sea and East China Sea*; Science Press: Beijing, China, 1982. (In Chinese)

59. Qin, Y.S.; Zhao, Y.Y. *The Yellow Sea Geology*; China Ocean Press: Beijing, China, 1989. (In Chinese)
60. Chen, H.S.; Zhang, Y.H. *The Lithospheric Textural and Structural Features as Well as Oil and Gas Evaluation in the Lower Yangtze Area and Its Adjacent Region*; Geological Publishing House: Beijing, China, 1999. (In Chinese)
61. Sun, X.D.; Guo, X.W.; Zhang, X.H.; Li, Z.Y.; Liu, H.S.; Zhang, S.S. Radiogentic heat production of formation and thermal structure of lithosphere in the South Yellow Sea Basin. *Earth Sci.* **2023**, *48*, 1040–1057. (In Chinese)
62. Xiao, G.L.; Zhang, Y.G.; Wu, Z.Q.; Lei, B.H.; Sun, J.; Wang, J.Q. Comparative assessment of hydrocarbon resource potential in the South Yellow Sea Basin. *Mar. Geol. Front.* **2014**, *30*, 25–33. (In Chinese)
63. Xu, Z.Q.; Liang, J.; Chen, J.W.; Zhang, P.H.; Zhang, Y.G.; Wang, J.Q.; Lei, B.H.; Yang, C.S. Evaluation of hydrocarbon preservation on the Laoshan uplift. *Mar. Geol. Quat. Geol.* **2018**, *38*, 125–133. (In Chinese)
64. Zhang, M.Q.; Gao, S.L.; Tan, S.Z. Geological characteristics of the Meso-Paleozoic in South Yellow Sea Basin and future exploration. *Mar. Geol. Quat. Geol.* **2018**, *38*, 24–34. (In Chinese)
65. Zhang, X.H.; Zhang, X.H.; Wu, Z.Q.; Guo, X.W.; Xiao, G.L. New understanding of Mesozoic-Paleozoic strata in the central uplift of the South Yellow Sea basin from the drilling of well CSDP-02 OF THE "Continental Shelf Drilling Program". *Chin. J. Geophys.* **2018**, *61*, 2369–2379. (In Chinese)
66. Tan, S.Z.; Chen, C.F.; Xu, Z.Z.; Hou, K.W.; Wang, J. Geochemical characteristics and hydrocarbon generation potentials of Paleozoic source rocks in the Southern Yellow Sea basin. *Mar. Geol. Quat. Geol.* **2018**, *38*, 116–124. (In Chinese)
67. Fu, Y.X.; Tan, S.Z.; Hou, K.W. Formation conditions and resource potentiality of Taizhou formation in north sag of South Yellow Sea Basin. *J. Jilin Univ. (Earth Sci. Ed.)* **2019**, *49*, 230–238. (In Chinese)
68. Chen, Q.H.; Song, R.W.; Dai, J.S.; Lu, K.Z.; Xiong, J.H. Jiaolai basin gravity and magnetic data interpretation and structure analysis. *Prog. Geophys.* **1994**, *9*, 70–79. (In Chinese)
69. Huang, T.L. Characteristic of regional geophysical field in Jiaolai Basin and its tectonic unit division. *Shandong Geol.* **2000**, *16*, 41–47. (In Chinese)
70. An, Y.L.; Guan, Z.N. The regularized stable factors of removing high frequency disturbances. *Comput. Tech. Geophys. Geochem. Explor.* **1985**, *7*, 13–23. (In Chinese)
71. Morozov, V.A. *Methods for Solving Incorrectly Posed Problems*; Springer: New York, NY, USA, 1984; pp. 1–31.
72. Boschetti, F. Improved edge detection and noise removal in gravity maps via the use of gravity gradients. *J. Appl. Geophys.* **2005**, *57*, 213–225. [CrossRef]
73. Zeng, H.L. *Gravity Field and Gravity Exploration*; In Chinese with English Contents; Geological Publishing House: Beijing, China, 2005. (In Chinese)
74. Rosenbach, O. A Contribution to the computation of the "Second Derivative" from Gravity Data. *Geophysics* **1953**, *18*, 3113–3123. [CrossRef]
75. Wang, W.Y. The Research on the Edge Recognition Methods and Techniques for Potential Field. Ph.D. Thesis, Chang'an University, Xi'an, China, 2009. (In Chinese).
76. Cooper GR, J.; Cowan, D.R. Edge enhancement of potential-field data using normalized statistics. *Geophysics* **2008**, *73*, 1–4. [CrossRef]
77. Pham, L.T.; Eldosouky, A.M.; Oksum, E.; Saada, S.A. A new high resolution filter for source edge detection of potential field data. *Geocarto Int.* **2020**, *37*, 3051–3068. [CrossRef]
78. Ekwok, S.E.; Eldosouky, A.M.; Achadu, O.I.M.; Akpan, A.E.; Pham, L.T.; Abdelrahman, K.; Gómez-Ortiz, D.; Ben, U.C.; Fnais, M.S. Application of the enhanced horizontal gradient amplitude (EHGA) filter in mapping of geological structures involving magnetic data in southeast Nigeria. *J. King Saud Univ. Sci.* **2022**, *34*, 102288. [CrossRef]
79. Pham, L.T.; Vu, T.V.; Thi, S.L.; Trinh, P.T. Enhancement of potential field source boundaries using an Improved Logistic Filter. *Pure Appl. Geophys.* **2020**, *177*, 5237–5249. [CrossRef]
80. Pham, L.T.; Oksum, E.; Do, T.D.; Le-Huy, M. New method for edges detection of magnetic sources using logistic function. *Geofiz. Zhurnal.* **2018**, *40*, 127–135. [CrossRef]
81. Parker, R.L. The rapid calculation of potential anomalies. *Geophys. J. Int.* **1973**, *31*, 447–455. [CrossRef]
82. Oldenburg, D.W. The inversion and interpretation of gravity anomalies. *Geophysics* **1974**, *39*, 526–536. [CrossRef]
83. Florio, G. Mapping the depth to basement by Iterative Rescaling of Gravity or Magnetic Data. *J. Geophys. Res. Solid Earth* **2018**, *123*, 9101–9120. [CrossRef]
84. Liu, Y.M.; Wu, Z.P. Diachronous characteristics and genetic mechanism of the Paleogene tectonic transition in Bohai Bay basin: A case study of the southwest Bohai Sea and Jiyang Depression. *J. Cent. South Univ. (Sci. Technol.)* **2022**, *53*, 1095–1110.
85. Qi, J.H. Mesozoic-Paleozoic Tectonic Evolution in the South Yellow Sea Basin and the Comparative Analysis with Sichuan Basin. Master's Thesis, China University of Geosciences, Beijing, China, 2012. (In Chinese)
86. Xiao, Q.B.; Zhao, G.Z.; Wang, J.J.; Zhan, Y.; Chen, X.B.; Tang, J.; Cai, J.T.; Wan, Z.S.; Wang, L.F.; Ma, W.; et al. Deep electrical structure of the Sulu Orogen and neighboring areas. *Sci. China Ser. D Earth Sci.* **2009**, *52*, 420–430. [CrossRef]
87. Pang, Y.M.; Zhang, X.H.; Xiao, G.L.; Guo, X.W.; Wen, Z.H.; Wu, Z.Q.; Zhu, X.Q. Characteristics of Meso-Cenozoic igneous complexes in the South Yellow Sea Basin, Lower Yangtze Craton of Eastern China and the tectonic setting. *Acta Geol. Sin. (Engl. Ed.)* **2017**, *91*, 971–987. [CrossRef]

88. Pang, Y.M.; Zhang, X.H.; Xiao, G.L.; Shang, L.N.; Guo, X.W.; Wen, Z.H. The Mesozoic-Cenozoic igneous intrusions and related sediment-dominated hydrothermal activities in the South Yellow Sea Basin, the Western Pacific continental margin. *J. Mar. Syst.* **2018**, *180*, 152–161.
89. Chen, J.W.; Xu, M.; Lei, B.H.; Shi, J.; Liu, J. Collision of North China and Yangtze Plates: Evidence from the South Yellow Sea. *Mar. Geol. Quat. Geol.* **2020**, *40*, 1–12. (In Chinese)
90. Xing, T.; Zhang, X.H.; Zhang, X.Y. Magnetic basement and structure of the Southern Yellow Sea. *Oceanol. Limnol. Sin.* **2014**, *45*, 946–953. (In Chinese)
91. Li, W.Y.; Liu, Y.X.; Xu, J.C. Onshore–offshore structure and hydrocarbon potential of the South Yellow Sea. *J. Asian Earth Sci.* **2014**, *90*, 127–136. [CrossRef]

MDPI

St. Alban-Anlage 66

4052 Basel

Switzerland

www.mdpi.com

Minerals Editorial Office

E-mail: minerals@mdpi.com

www.mdpi.com/journal/minerals